EASY SCHEDULING

A Construction Scheduling Resource Handbook

BNi Building News

By: Dan Ramsey and Stephen Matzen

BNi® Building News

Editor-In-Chief

William D. Mahoney, P.E.

Technical Services

Andrew Atkinson

Sara Gustafson

Design

Robert O. Wright

BNi PUBLICATIONS, INC.

990 Park Center Drive, Ste E

Vista, CA 92081

LOS ANGELES

10801 National Blvd., Ste. 100

Los Angeles, CA 90064

ANAHEIM

1612 S. Clementine St.

Anaheim, CA 92802

1-800-873-6397

ISBN 978-155701-615-7

Table of Contents

Table of Contents

Table of Contents

Table of Contents

Introduction

Why You Need This Book

You have enough to do, managing a small to medium sized construction or remodeling company. During a typical day you talk with clients and subs, meet with your leader or crew, discuss projects with lenders and inspectors, follow up on material orders, interview prospects, and keep checking the weather forecast to make sure Thursday's foundation pour will be on schedule. You may also be the lead framer or painter with projects to complete before the sun goes down. So why in the world would you ever want to spend your valuable time and money learning about scheduling?

Because you want to make your business more profitable and easier to run. That's the bottom line.

And that's why you need this book. You probably already have some type of scheduling system in place: a calendar, project files, the back of a piece of scrap paper. But you know that things are falling through the cracks and that you would have more time and higher profits if you could just keep ahead of the scheduling. The problem is: How do you schedule time to learn about better scheduling?

Read this book!

Your business requires accurate scheduling to increase profits and avoid waste. By investing as little as half an hour per day in learning and applying proven project scheduling skills, you can expand your day's productivity by an hour or more. That's a 100% return on your investment!

To do so, you need a practical get-it-done system that's easy to learn. If you are comfortable using computers, consider the many options available to contractors for computerized scheduling. If you don't want to depend on computers, there are many proven choices for manual scheduling that will make your job easier and more profitable. If you're not into computers but want to consider them for your business tool belt, you'll discover your options in this book.

The problem, again, is that you may not have the time to even consider learning a new task. Fortunately, the fact that you bought this book means you realize the need to learn scheduling. (If not, Chapter 1 includes many specific bottom-line benefits.) So this book is written in a conversational style that will make the pill easier to swallow. You'll discover how real-world contractors just like yourself have used proven scheduling tools to make their efforts more profitable. You'll get easy-to-apply specifics that you can take to work with you tomorrow.

Using Scheduling Tools

Computers have dramatically changed the way that business is transacted today. By using a computer you can manage your banking, pay bills, correspond with lenders and suppliers, and even pay your taxes. Of course, you also can use computers to schedule your construction and remodeling projects more efficiently. You can even use integrated contractor software that handles bidding, estimating, contracts, ordering, and scheduling from a single program.

Do you need a computer to manage your business? No, not really. Many successful contracting firms over the years have made their profits without using computers. But they did use the best tools available to them.

Introduction

The most popular scheduling software is Microsoft Project. This book focuses on this common business tool and how to put it to work for you. In addition, you'll learn about scheduling templates that you can put to work immediately to save many hours of input. You'll also learn about alternatives to MS Project that are more cost-effective and may be better suited to your contracting business. And you'll see how to schedule smaller jobs more efficiently using manual systems.

Chapter 1 offers an introduction to the benefits and requirements of scheduling systems including, how to schedule for profits. Chapter 2 introduces you to the most popular scheduling program, Microsoft Project, and Chapter 3 offers other computer scheduling options. Chapter 4 shows you how to set up your own Project templates and how to use the ConstructionScheduling.com templates.

Chapters 5-9 take you *step-by-step* through pre-construction scheduling, organizing, planning, controlling, and consolidating schedules using Microsoft Project. Chapter 10 puts you ahead of your competition with advanced scheduling techniques.

There are eight information-filled appendices that you can reference as you develop your scheduling system. Four of them offer the four levels of the Construction Specification Institute (CSI) coding, the standard for construction planning, specifications, and scheduling. The next three appendices cover three levels of residential scheduling templates: master, general, and summary. For those who prefer manual scheduling, reprintable scheduling and related forms are provided in Appendix H. Finally, Appendix I documents the programs, templates, demos, and forms.

What You Can Expect From This Book

This book doesn't promote one type of scheduling over another. It gives you the basic information you need to decide for yourself which tools work best for your small to medium-sized construction or remodeling business, then shows you how to use them effectively.

In addition, this book offers real-world models that are specific to the small contractor trade. You'll get examples that you can use, whether you are a small builder, a remodeler, a trade sub, or all of the above.

Flip through the pages of this book and you will see that it is filled with illustrations. There are numerous clear examples of forms and schedules used by small contractors every day. In fact, there are numerous appendices with scheduling forms that you can use right now.

The purpose of this book is to make your life easier, by helping you develop new skills that can give you an advantage over your competition, add more profit to your bottom line, and even give you some extra time to enjoy. Your investment will be well rewarded.

Introduction

About the Authors

Dan Ramsey is a licensed general building and home improvement contractor as well as the author of 20 books for contractors and consumers. Dan's titles include ***Builder's Guide to Foundations*** and ***Floor Framing and Builder's Guide to Barriers*** (McGraw-Hill) as well as ***The Complete Idiot's Guide to Building Your Own Home*** (Alpha Books, two editions) and Electrical Contractor: ***Start and Run a Money-Making Business*** (McGraw-Hill). Dan is also a former president of the National Association of Home and Workshop Writers.

The Microsoft Project scheduling methods and templates in this book were developed by Stephen Matzen, co-author of this book. All procedures, formatting and scheduling techniques will work with the scheduling templates offered on Stephen's website www.ConstructionScheduling.com. Portions of this book were first published in Quick Schedule Guide by Stephen C. Matzen.

Chapter One

Scheduling for Profits

Scheduling for Profits

Today's contracting business presents a real challenge. Customers expect speed, quality, and reduced costs -- and "the customer is always right." The challenge becomes getting the job done on time, with high quality work and within budget. The success of your building or remodeling business depends on it.

So smart contractors must change the way they manage their projects, using the latest tools and methods available. To achieve a competitive edge, they must focus on how things are done and find ways to get work completed more quickly without compromising cost or quality.

The answer to controlling the speed, quality, and costs of the construction process is in scheduling. Without good planning and communication, the inevitable happens –the cost, time, and frustration start to run the project resulting in the contractor's reaction to emotions rather than targeting a proactive resolution. This book will show you how to quickly create and manage time through an effective and easy-to-use scheduling system using proven tools. It illustrates construction scheduling using Microsoft Project 2000 or higher versions integrating the coding of the Construction Specification Institute. You may use the popular Microsoft Project, another software program, or even a manual scheduling system. Whatever the tool, the results will be greater planning and greater profitability for your building or remodeling business.

Of course, a schedule is much more than a spreadsheet that has a bunch of dates on it. Properly laid out and communicated, a schedule will synchronize everyone's efforts into an organized system that will quickly become the backbone of all of your communication and planning. A schedule is not a replacement for communication; it is a major tool of communication. A well 'laid-out' schedule that is communicated and updated regularly will become the basis of all planning and communication for all projects -- and increase your profits. This book was written to create a comprehensive and useful scheduling system, easy to understand with tools that guide anyone to organizing, planning, and managing time using Microsoft Project -- eliminating months and years of learning. This book will teach you the elements of smart scheduling within a system that enables you to put it to work quickly-- and to communicate the schedule effectively.

Benefits of Scheduling

The benefits of effective scheduling are well known:

- Increased time
- Increased profits
- Improved cash flow projections
- An atmosphere of organization
- Better resource management
- Ability to forecast workload
- Adjusted profit margins for future work
- Respect from your customers and team members
- Ability to schedule subcontractors, material deliveries, and manpower effectively
- Reduced stress
- Improved communication between contractors, employees, subs, suppliers, and customers
- Reduced lead-time problems
- Decreased costs of purchasing materials and subcontract work
- Increased accountability
- Decreased or eliminated penalties

Scheduling helps you manage your business's primary resource: time. Scheduling provides an opportunity to organize of activities and communicate project goals. Aside from creating organization, you will be able to create and establish business planning tools that will help view and forecast cash flow and responsibilities to maintain accountability. Scheduling is the most important part in organizing and maintaining control in a project. Most people do not plan. They don't realize the importance of scheduling and consequently deal with delays, increased costs, stress, etc. Effective scheduling forces you to look ahead, plan, and communicate with other team members. Without effective scheduling, you are destined to react instead of act.

Schedules that don't promote action are not effective. Schedules must create an atmosphere where everyone maintains their work schedules in order to uphold project goals. If all team members know that everyone else is maintaining the schedule it will force them to be accountable and preserve their personal schedules to work with your schedule. Unfortunately, many contractors don't know how to schedule effectively, or they have their own way of doing things that do not mesh well with scheduling systems used by others in the process. Perhaps the schedule isn't clearly communicated to all participants. Or schedule changes don't get passed along. There are many reasons why schedules -- and the projects they attempt to control -- fail.

In addition to helping you control your individual projects, smart scheduling will give you insight across all projects to control your workload, adjust profit margins, keep people accountable, and have access to the reports that are necessary to look ahead and adjust your actions accordingly. This insight is unavailable without effective scheduling.

Thinking Like a Scheduler

Let's take a quick look at the steps to good scheduling:

1. Establishing the project goals
2. Determining the objectives
3. Establishing checkpoints, activities, relationships, and time estimates
4. Creating the schedule
5. Empowering people individually and as a team
6. Communicating the schedule to keep everyone focused
7. Using agreements and commitments
8. Empowering yourself and others
9. Approaching problems creatively and objectively
10. Acting instead of reacting

Scheduling for Profits

The scheduling mindset is an important part of creating and maintaining an effective schedule. Managing time is one of the most challenging tasks that one undertakes, not only in a schedule but also in life.

The first step in managing time is to establish your goal and the intermediate steps that are required to reach that goal. Task management is directly affected by the time constraints that are placed on a project and the effort that it takes to communicate the expectations of the project. It is this linear picture (schedule) that keeps everyone focused. Without a realistic and well-communicated schedule, projects are destined for failure.

The creation of a realistic schedule as well as the method and attitude in which these goals are conveyed to the participants of the project, sets the tone and expectations of the project. These expectations set up an accountability that measures performance and forces people to act. Without accountability or established expectations, things happen in their own time when it is convenient to the individual schedules of the participants of a project.

In addition, the attitude in which a schedule is conveyed can dictate the success of a project. A well-prepared schedule with all of the tasks and milestones in place that isn't presented with an attitude of expectation has a greater chance of failing. It takes firm communication from the start of the project to let everyone know that they are being held accountable to the deadline and success of the project. If this attitude is not maintained by the leaders of the project, the project will run them instead of the other way around.

In most cases, the final version of the schedule should be an attachment to all contracts along with a procedure for communicating and documenting any changes as the project progresses. If the schedule is linked financially to performance, contractors will see increased attention and accountability. Projects can be controlled with great success if all team members know that their performance will be measured.

This book will give you the tools you need not only to schedule for greater profitability but also to communicate the schedule with authority and manage changes as needed for team success. Your efforts at understanding and applying this proven process will empower you to be a better and more profitable contractor.

Managing Projects

Leaders make things happen. Effective project managers are the leaders who take the time to plan their projects with the team and to constantly work with and revise the plan to reach the project goals. No project or plan will produce successful results without a skillful leader.

The success of a project manager depends largely on their ability to motivate team members to cooperate on the project. No system or effective plan will work with ineffective people and poor communication. You can have all of the tools and systems in the world, but people respond best to leadership. That's you!

Project management requires constant communication and the ability to communicate clearly regarding the direction of the project. Involving the customer in the process, for example, is essential. The schedule is the single most important tool to communicate with clarity to project team members. In essence, project management is the defining, planning, scheduling, and controlling of all the tasks that must be completed to reach the project goals, and the allocation and management of the resources to perform those tasks.

- Defining and planning are necessary so you know what you and others will do.
- Scheduling is important so you know when you will do it.
- Management is important because things never work out exactly as planned.

Project management is the key to a successful project; however, the project management process starts the moment a decision is made to proceed with building the project, not the moment that actual construction begins. Management must start at this early stage to set up the organization, planning, and control for the project.

Far too often project management does not start until the project goes into actual physical construction. A project at this point may already be destined for major problems, and no one knows it because no one has planned ahead and coordinated all of the activities required to complete the project on time. The early phase of project management is referred to as Project Development because this process is actually the development phase of a project and is separate from the typical Project Management phase that begins with the actual construction work.

Developing an effective schedule is an important part of the pre-construction process. The pre-construction activities must be organized and communicated in the same way that tasks are scheduled in construction. Each task, phase, and milestone must be identified and input into the schedule so that everyone knows the project expectations at this phase of development. All of the scheduling processes and templates in this book emphasize this pre-construction planning and development.

Chapter 5 includes typical Pre-Construction Checklists. The checklists offered are just a guideline of typical pre-construction tasks to get you started. You should add, edit, or delete items as necessary to suit the particular needs of your project.

At the pre-construction phase, the emphasis is on getting the project ready for construction and handling the critical planning tasks rather than on the details of how to get the project completed after it goes to construction. The pre-construction tasks and phases are very similar on most projects but require a lot of attention to make sure that team members know their responsibilities and that they are being held accountable.

Remember to think of the time spent on scheduling as an investment that will reduce costly problems.

Communicating Project Requirements

Communication is essential to a successful project. Effective communication requires insight into what the priorities are that need to be dealt with at any given moment and how these priorities are managed and communicated.

The project developer or manager needs to organize a system of communication so that all team members are sharing project information. The flow of this information is critical and will determine whether the team members are informed and are receiving the necessary feedback required for evaluation and action. Every project should have a flow chart of communication so that there is no question of who is responsible to whom and where each responsibility lies. This will also help reduce liability and eliminate a lot of wasted time and energy in answering and responding to items that should be directed to someone else.

To reach the project goals, it is important to be on top of changes, tracking, rescheduling, and communication as the project progresses.

The three keys to a successful project are:

1. Quality communication: making sure participants are aware of the schedule and any subsequent changes

2. Priority management: knowing what to do and when

3. Total responsibility, while never assuming liability

Once in place, a well-managed project schedule serves as:

- Communication to all construction or remodeling team members to cooperate toward maximum production with minimum effort.

- A guide to show the customer how the project or service will meet defined requirements.

- An indicator of the need for additional personnel, management of resource work loads, and alternate ways of keeping the project on track.

- A baseline against which the project's progress can be analyzed.

- A record of what happened on the project, to be compared to the original plan and used as the basis of future schedules for similar projects.

As an added benefit, proper communication will minimize or eliminate the risk of being held accountable for project delays. Managing and communicating a dynamic schedule is the key to profitability.

Using Scheduling Tools

There are many powerful scheduling tools used by businesses. One of the most popular is Microsoft Project. The power and abilities of Microsoft Project enable you to use the software to any level desired. It is a shell product that has a lot of power and many features that most people do not understand. This book will introduce you to the various features, capabilities, and power of Microsoft Project in construction. As you continue to create and work with schedules you will find new ways to enhance and add to your business planning. It is this book's goal for you to establish a scheduling system that will get you started in the shortest time possible so that you can get up and running and learn by doing. If you've never used Microsoft Project before, Chapter 2 will give you a crash course. You can also visit www.ConstructionScheduling.com for additional Microsoft Project templates, education, services and support.

In addition, there are other computer software programs written especially for construction and remodeling contractors. Many have a scheduling component that works with estimating and other components to help you manage your business successfully. This book will help you use the scheduling components of these programs to their best advantage. If you just don't want to learn computers and software, you can use this book to learn the basics of scheduling. It includes numerous reprintable scheduling forms that you can photocopy and put to work on your next job.

Using CSI Codes

CSI stands for the Construction Specification Institute. This organization has organized the various tasks and items common to the construction industry and has coded them each individually. Each code represents a specific method, material, or task. This coding system is the basis for most specification writing. It is an outline that is used for organizational purposes throughout the construction industry.

This book uses coding as presented by CSI as the basis for all coding within the schedules. With coding, you can filter and easily identify tasks or groups of tasks that coordinate with the specifications for a particular project or across all projects.

Whether your contracting business is general, subcontract, residential, or another related construction trade, you will find great value in using this coding structure. Even if you do not use or follow a CSI specification coding system, it is still a useful tool for organizing information in a format that the industry uses and understands. The tasks as identified in the templates provided with this book are based on the CSI coding structure as follows:

Master Format -- Level 1 (Appendix A)

These are the top level tasks as defined be CSI. These are the summary headings of levels 2 - 4.

CSI Master Format -- Level 2 (Appendix B)

These are the level 2 tasks as defined by CSI with a more detailed breakdown, although they still follow the same top level coding and structure.

CSI Master Format -- Level 3 (Appendix C)

These are the level 3 tasks as defined by CSI with an even more detailed breakdown and still follow the same coding and structure of Level 1 and Level 2.

CSI Master Format -- Level 4 (Appendix D)

These are the most expanded tasks within the coding structure. This is the greatest detail of coding breakdown by CSI and still follows the coding and structure of Level 1, 2, and 3.

For additional information on the Construction Specifications Institute code system, visit their web site at http://www.csinet.org.

Using Custom Coding

The Construction Specification Institute coding structure is used in this book as the basis for all templates and scheduling task items to create uniformity between schedules. In many cases, terms better suited for the construction industry have been added, and summary items are broken down into definable terms that need to be reported individually.

All construction and remodeling contractors face the challenge of creating schedules that have sufficient detail but are not overwhelming. The structure of the schedule needs to be laid out in a way that all team members can view, understand, and relate to. Your schedules need to be designed to help communicate to all team members and trades in order to facilitate getting the job done. Too many times contractors create schedules that are actually their personal to-do lists. These schedules are not understood and are meaningless to the other team members and are consequently ineffective.

The following are the Master lists derived from the CSI coding that serve as the basis of the scheduling templates included with this book:

Residential Master (Appendix E)

This is the most detailed coding structure and includes most of the common CSI tasks needed to report within a schedule. Some CSI items have been left out because they do not need to be included within a schedule and many items have been renamed as familiar construction terms.

Residential General (Appendix F)

This template is a 'filtered-down' version of the Residential Master and provides general item descriptions. This template can be used for most scheduling purposes.

Residential Summary (Appendix G)

This is a summary listing of tasks and coding. This template provides a summary breakdown for quick schedules when you do not have the time, desire, or information to create detailed schedules.

Using Templates

While CSI provides the general coding structure, it does not break down the scheduling tasks within each section. The templates are developed by co-author Stephen Matzen and provided by www.ConstructionScheduling.com.

These template schedules can be used to create your own schedules or as a basis for creating your own templates. Each company or organization has different tasks and responsibilities that need to be scheduled. You can modify these templates as required to fit your basic needs. Doing so will save a lot of time on future scheduling.

For best results, follow the coding as it has been presented. If you add items, you should use the same number as a task or item that is similar to the one you are adding. This will create a grouping of similar tasks and will be useful when you filter for code numbers. In addition, if you share your project files with other team members who use the same scheduling and coding system, they will be able to use the schedule and communicate it effectively. This is especially beneficial if you have several people scheduling in an organization locally or globally. Schedules with the same coding structure will also be valuable when you share them with other people or organizations that use and are familiar with the coding structure.

Our goal is to promote and enhance communication. If you have several project managers in various locations using the same scheduling system and coding structure, you can have a centralized scheduler who can combine all associated projects into a Master Consolidated Project. This is extremely valuable in ensuring that management can get reports and view information across all projects, i.e. Project Status, Cash flow, Workload, Resource Information, etc. Using the same coding structure and methods of scheduling with others you are working with will save time and money. This uniformity will help create a consistency within your schedules.

Understanding Projects

A project is an undertaking requiring concerted effort. It is a plan toward completion. A project can be anything that requires coordination to reach a common or consistent goal. Most people think of a project as a plan with a definite end. However, a project can be an ongoing process of tasks designed to reach a common goal such as streamlining a process, coordinating the efforts of a group for an ongoing process, or the fabrication and delivery of a product.

A successful project requires the coordination and use of a system that leads an organized effort toward a specific objective or toward an ongoing process. For example, a project can be an ongoing manufacturing and delivery process or the coordination of a group of individuals working on different projects, such as a drafting group.

The simplest method of visualizing a schedule is with a matrix of tasks (vertical) and time (horizontal). This method was developed a century ago by Henry Gantt and is called the Gantt chart (Figure 1-1). Project scheduling systems and software typically use the Gantt chart view with the calendar format in coordination with tables, filters, views, and resources. Whether it is a specific project, group of projects, or a system, you can use the built-in features of the software to reach any desired outcome.

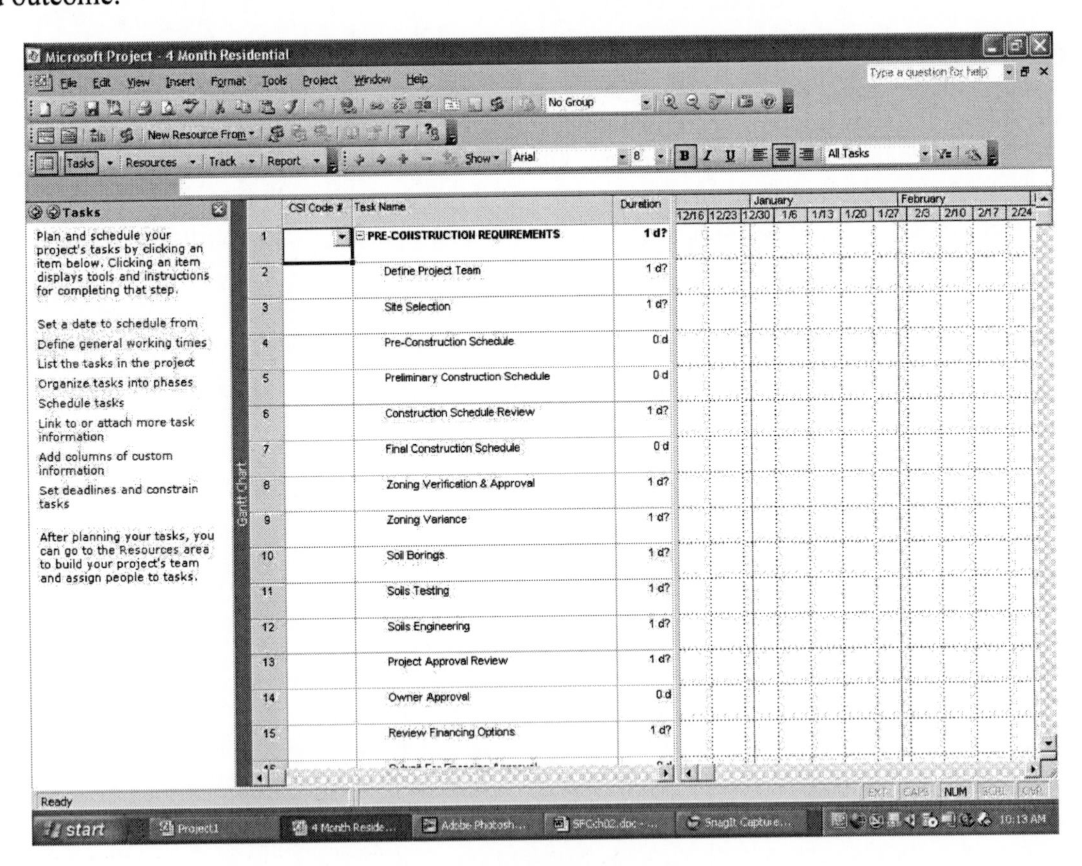

1-1: Gantt Chart

Construction and remodeling projects have many subprojects that are required to complete the main goal. Managing the various steps and planning within the subprojects to implement the master project takes a lot of coordination and communication to maximize profits and minimize cost. You will learn how Microsoft Project can be used to manage any process.

The coordination of all team members and subcontractors is essential to reaching the project goals. A project therefore is the coordination of all of these separate projects and activities. Efforts must be combined to reach individual goals as well as the employee, subcontractor, and customer goals.

Understanding Microsoft Project

Microsoft Project (Figure 1-2) is the leading scheduling software for business and industry. It uses the Gantt chart system.

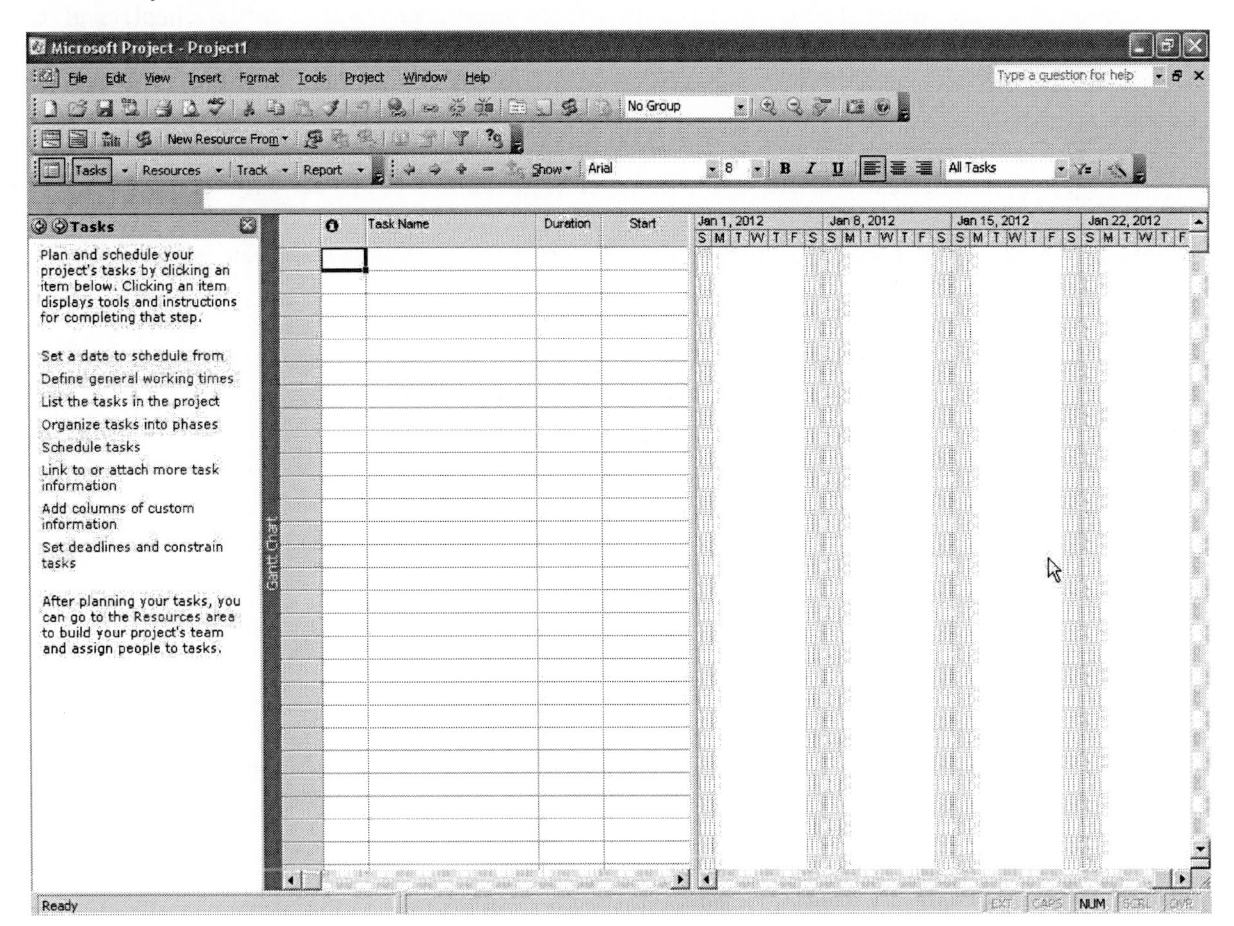

1-2: Microsoft Project

A schedule is made for all team members. Microsoft Project provides a platform to work with many other office programs and has a lot of resources available. Microsoft Project is similar to many other Windows programs that computer users are familiar with and it lends itself to easy customization and communication. There are many features that in conjunction with e-mail, the Internet, and other Microsoft Office programs, can maximize efficiency and communication.

Many contractors, employees, and customers have difficulty visualizing schedules. Effective schedules must be laid out in a format that is easy to understand by everyone who is a part of the project, regardless of their experience and level of understanding. If the schedule cannot be read or understood by the people working on the project it forces the sole responsibility of communication on the scheduler. Microsoft Project provides an easy interface and has many features that allow contractors to filter and view any information required to share with other team members.

Chapter 2 will introduce you to Microsoft Project. If you do not have Microsoft Project, you can order or download a free '60-day' trial version, as well as check out other Microsoft Project deals at www.ConstructionScheduling.com.

Putting This Book to Work

As mentioned in the introduction, this book is divided into chapters. The initial chapters give an outline and description of the importance of scheduling throughout the various phases of the construction process. Chapter 2 introduces you to Microsoft Project and Chapter 3 offers other scheduling programs.

Chapter 4 on Scheduling Basics offers instructions on using Microsoft Project with templates provided to produce job schedules. Each step outlines the process, working with Microsoft Project and the templates provided. The interface of the menus, etc. may vary slightly between Microsoft Project versions.

Chapter 5 covers Pre-Construction Scheduling, considered the most important part of structuring a successful project. Too little attention is given to this early phase; however, this process directly affects the outcome of a project. The organization and communication at the pre-construction phase is critical to the project's success. This phase needs to be scheduled and communicated with the same level of organization as the actual construction project.

Chapter 6 describes Organizing Schedules. The organization process is the foundation of your scheduling structure, methods of communication and reporting. Each project is different in its nature and requirements. This chapter will guide you in identifying the level of detail required for your project and in organizing the structure of your schedules.

Chapter 7 covers Planning Schedules. This process involves estimating task durations and developing a working copy of your schedule. The Planning phase requires that you communicate with the project team members to ensure that you have accurate information. Communication is the key to preparing a realistic and successful project schedule.

Chapter 8 introduces you to Controlling Schedules. The control phase is the process of maintaining a schedule, constantly updating and communicating project information, and distributing the schedule via e-mail, internet, or direct contact in a manner that is understood by other team members. With good control, you will have the opportunity to foresee conflicts and problems so that you can act before they happen instead of reacting after they happen.

Chapter 9 helps you in Scheduling Multiple Projects. Consolidating your active projects into a master project gives you unlimited flexibility to communicate multi-project information, view workloads, create reports, and make intelligent business planning decisions across all projects. This chapter will provide you with the tools needed to create and use the necessary views and filters.

Finally, the customizations in this book and in the scheduling templates are outlined within the appendices at the end of the book. You can constantly refer back to these sections for details on how each scheduling phase is structured so that you can use, modify, or recreate the original settings.

Use this book as a reference while you are working on your project schedules. It will answer any questions and guide you through the scheduling process. The goal of this book is to provide a standard scheduling and reporting system that contractors, subcontractors, architects, and owners can use for efficient planning and communication. The book will make scheduling easier and more efficient, saving you time and money.

Getting Additional Help

Whether you are experienced or are new to scheduling, you may need additional help to understand and apply what you are learning. The appendices in this book include additional reference material and resources that will help you learn and apply smart scheduling.

Microsoft Project software includes an extensive Help system (use the F1 function key). Other software discussed in this book also includes Help systems and tutorials. In addition to this book and the various other scheduling books available as reference, you can receive support through the Members section at www.ConstructionScheduling.com.

The Members section also offers technical support and training for Construction Scheduling templates and products.

Better scheduling can make your job as a project manager easier and your contracting business more profitable.

Chapter Two

Microsoft Project for Scheduling

Microsoft Project for Scheduling

As introduced in Chapter 1, Microsoft Project is a scheduling program that uses the Gantt chart system for tracking tasks, resources, and time. It is widely used in business and is especially useful for scheduling projects such as building and remodeling, as it can easily be modified to reflect ongoing changes to a job.

A potential problem with Microsoft Project is that it includes more scheduling features than small contractors typically need, so users can get lost in the details; it requires training to become proficient. Fortunately, you can learn the basics quickly and add features and skills as you grow in experience.

This chapter offers an introduction to using Microsoft Project 2000 as well as newer versions. It also serves as a refresher course for contractors who have used this or other scheduling software but need a quick overview to get back into the habit. The learning curve for contractors new to computers will be greater than for those who have computer experience and have used other Microsoft Office tools such as Word (word processor), Excel (spreadsheet), and Access (database). Those who have some Project or Office experience can skim this chapter to pick up enough to put Microsoft Project to work.

Microsoft Project (Figure 2-1) is available from Microsoft Corp. (One Microsoft Way, Redmond WA 98052, 800-MICROSOFT; www.microsoft.com) $599 (Standard), $999 (Professional) or bundled with other Microsoft Office products at a package price. Substantial discounts are available when purchased with a new computer as part of the Microsoft Office Suite or ordered through www.ConstructionScheduling.com.

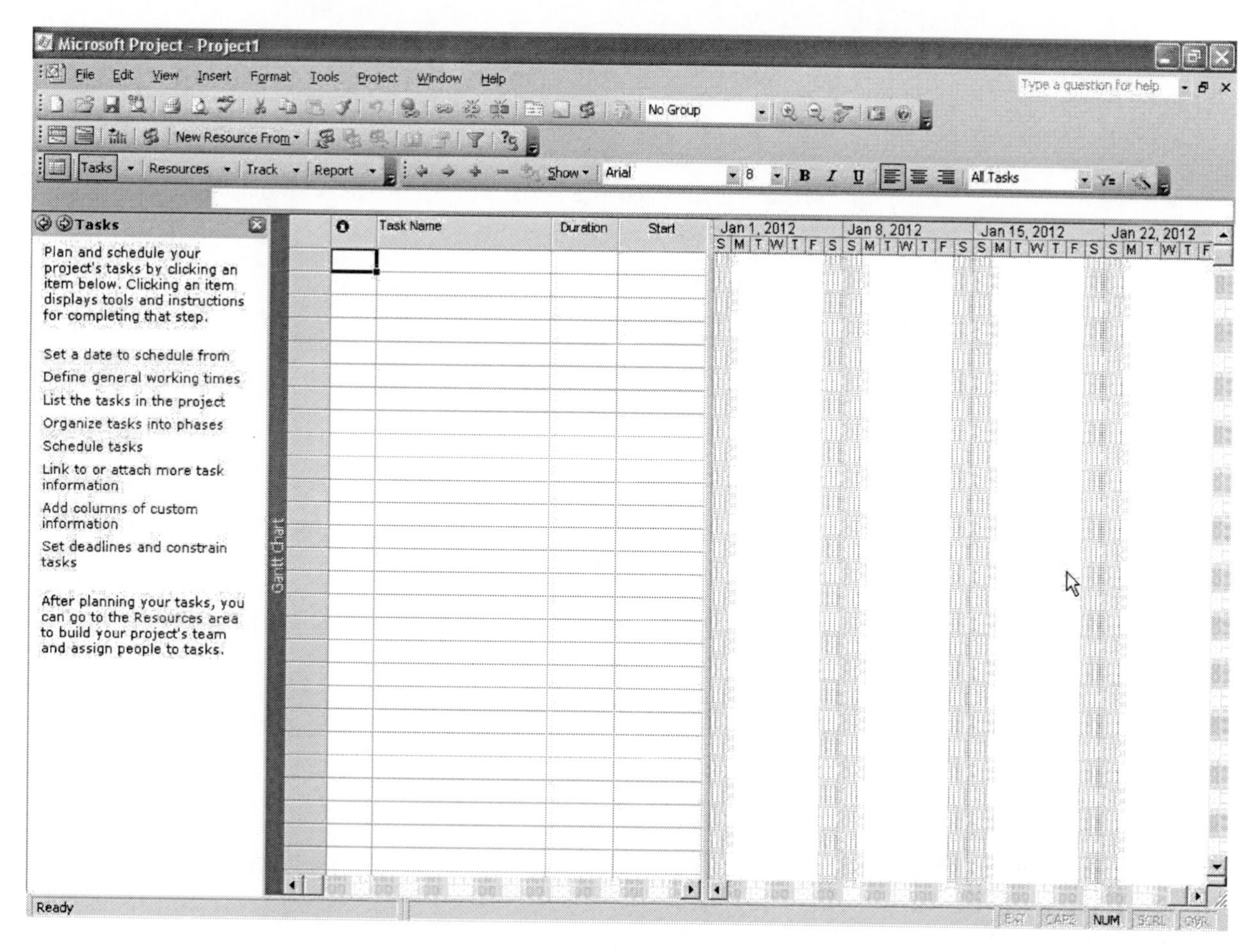

2-1: Microsoft Project

First, some basics. Projects such as specific construction or remodeling jobs can be stored in process-specific files called templates. These templates include all the data you've input regarding a particular type of project, such as a bathroom remodel or a spec house you often build. These templates will save you time and effort, as you will then only have to enter the specific criteria of an individual job, such as the Smith's bathroom remodel or the Brooktrails spec house.

In addition, you can purchase pre-built templates from sources such as ConstructionScheduling.com. A residential template package, for example, will include time-specific construction schedules (4-, 5-, 6-month) as well as general construction and pre-construction schedules. Add the specifics of the job and you have a full-blown schedule without all the effort. You can then modify or customize the template as needed and save it for future jobs.

Interfacing with a Project

Opening up a Microsoft Project file can be scary. There are all those menus, sidebars, and other components. Fortunately, soon you will know how it all works together to efficiently schedule jobs with the least effort.

Files can be identified by their three-character extension, such as .MPP or .MPT (Figure 2-2). An MPP file is a Microsoft Project Plan, the standard file format for this program. An .MPT file is a Microsoft Project Template, a boilerplate file that includes a variety of information that can be used as a starting point for MPP files. If you need to transfer an .MPP file to another scheduling, spreadsheet, or database program, you will first need to export or convert it to the .MPX or Microsoft Project Exchange format. These are the most common file extensions and formats for Microsoft Project, but you'll discover others as you become a Project guru.

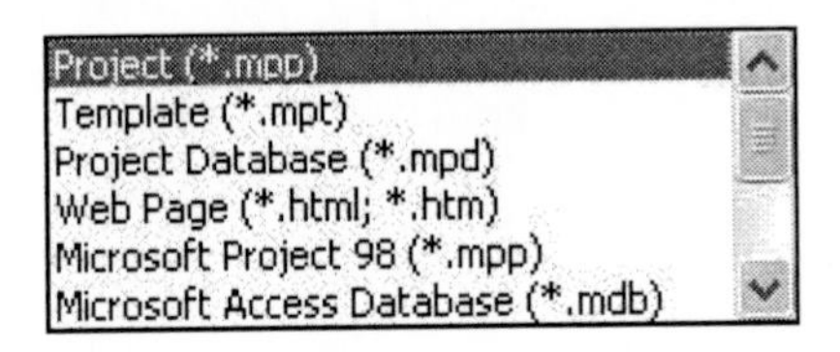

2-2: Microsoft Project file extensions

Note: Descriptions and illustrations refer specifically to Microsoft Project 2003, but many of the features are the same on previous and subsequent versions.

Menu Bar: Similar in function to other Microsoft Office products, the menu bar (Figure 2-3) allows you to open, edit, view, insert, format, and perform other functions. Simply point the mouse arrow to one of these menu items and click the right mouse button for pull-down menus that display additional options. You won't hurt anything by simply opening up each of these menu items and selecting various options.

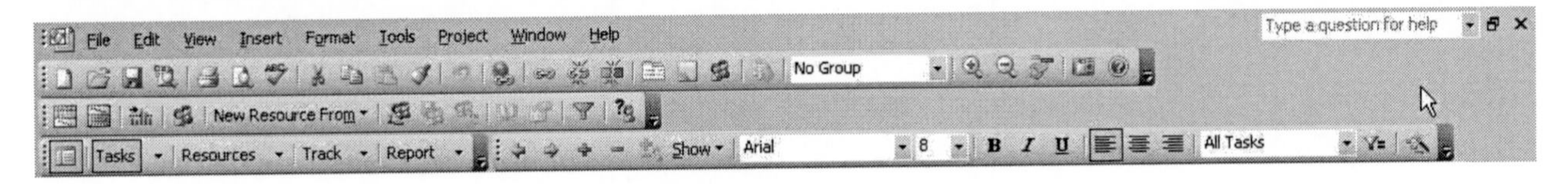

2-3: Menu Bar

Tip: Start by selecting the Help menu to see what assistance is available: Help, Office Assistance, Online Help, and What's This?

Tool bars: Toolbars are quick shortcuts to the most commonly used commands. Once you're more familiar with Project, you can customize the Toolbars (View, Toolbars) for greater efficiency.

Table: A table (Figure 2-4), also known as a sheet, shows you most of the data related to the tasks and resources. Tables can be customized.

Columns: Columns are fields that allow you to select and control what data you show on the table.

Chart: The Gantt chart displays the timeline and tasks for the project.

Time: You can change and customize the time frame for the project, starting it on a specific day, excluding weekends or other days, and even setting the work hours to make the schedule more accurate.

View bar: The view sidebar is a handy location for selecting the most common views of your schedule.

Status bar: The status bar indicates that the chart is "Ready" or is being "Saved".

2-4: Table

File (Figure 2-5): Includes commands for creating, opening, saving, printing, and exporting projects.

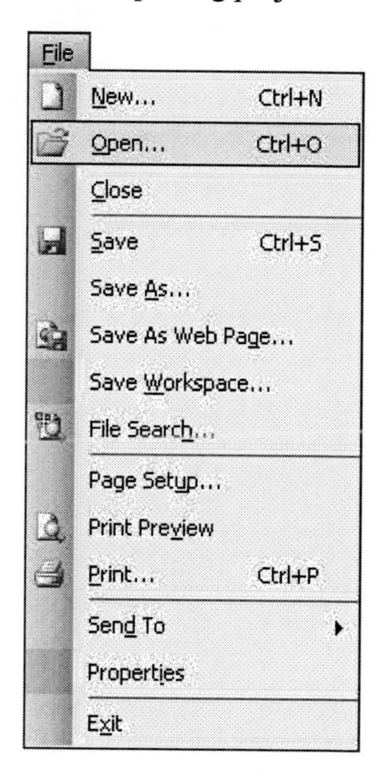

2-5: File

Edit (2-6): Includes commands for cutting, copying, pasting, and finding data as well as managing tasks.

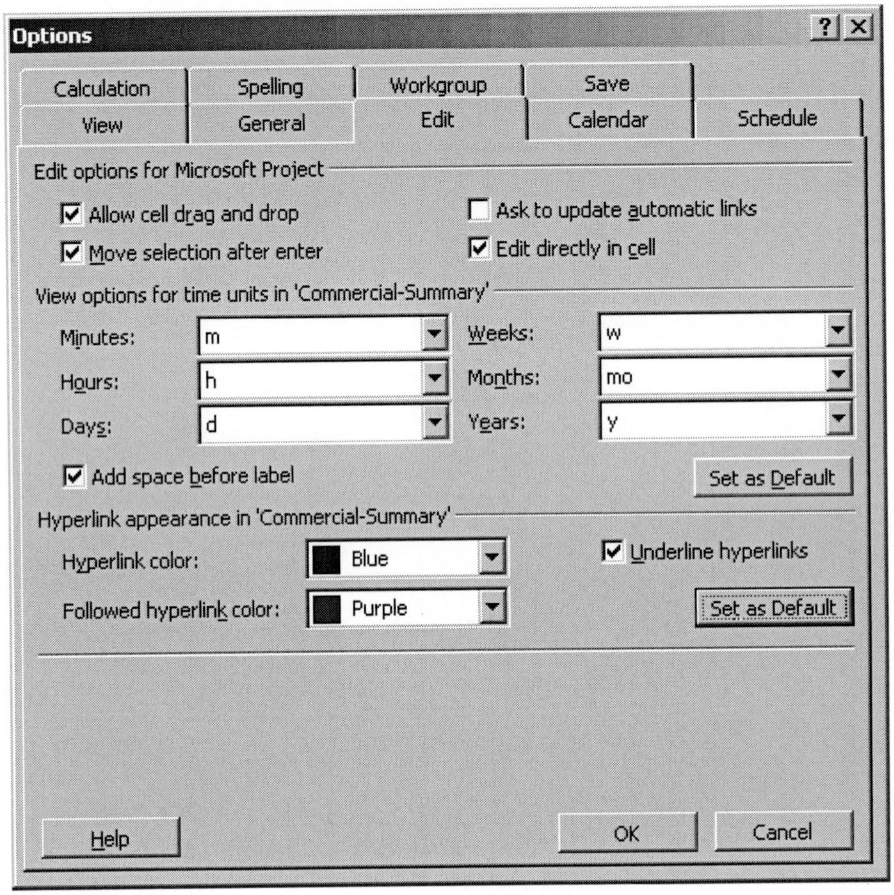

2-6: Edit

View (2-7): Offers various methods of viewing the schedule data including as a calendar, Gantt chart, network, tasks, and resources, depending on the version of Project used. The various views will be covered in greater detail below.

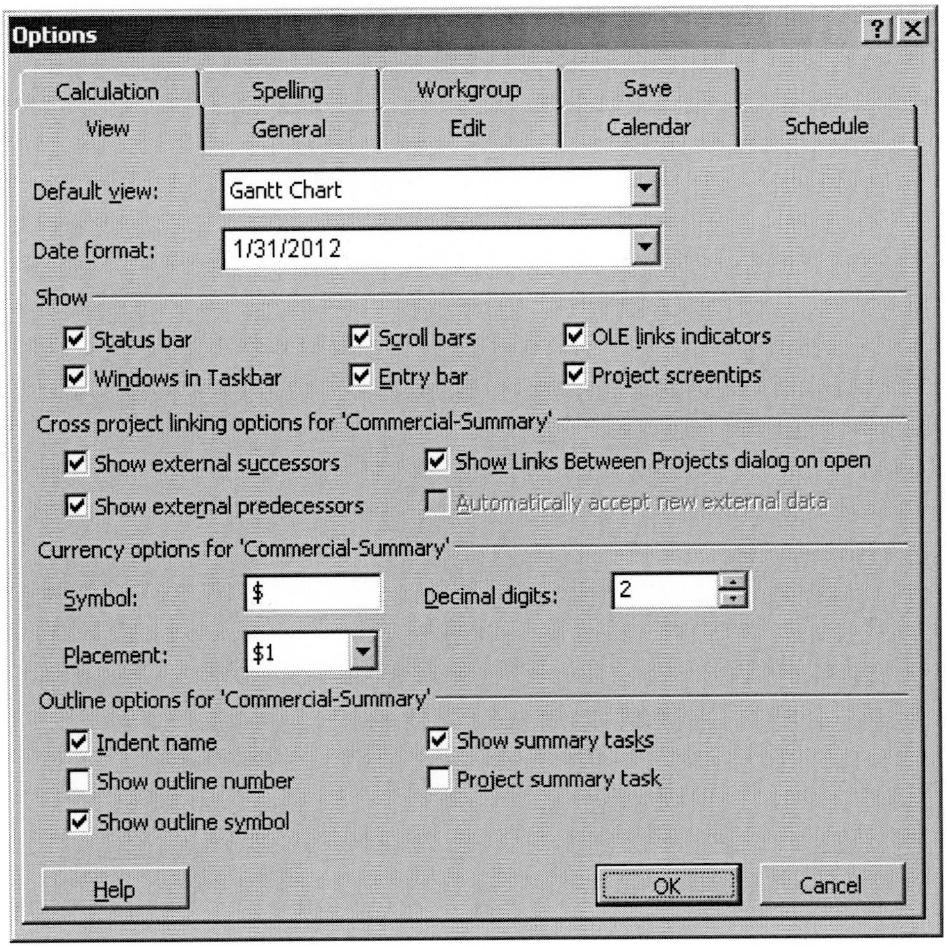

2-7: View

Insert (Figure 2-8): Allows you to place objects (tasks, resources, drawings, links, other projects) into a project plan.

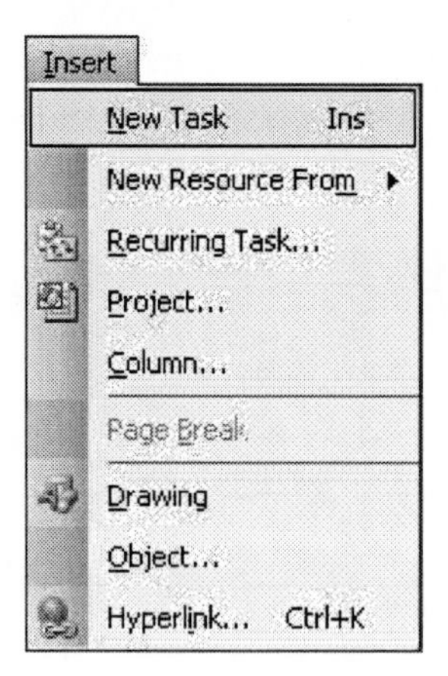

2-8: Insert

Format (Figure 2-9): Used to customize the type font, bars, and styles used in displaying project data.

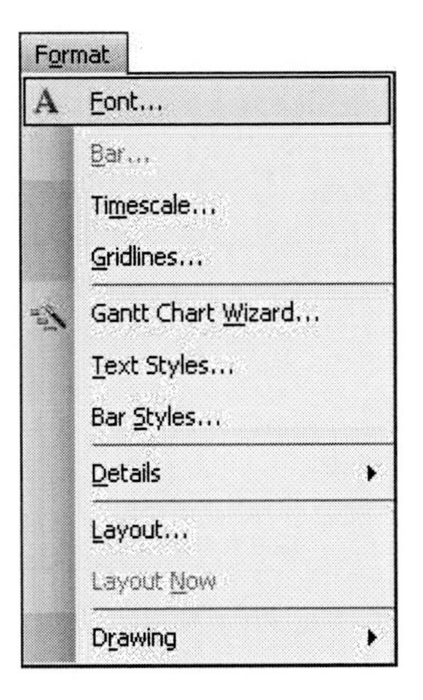

2-9: Format

Tools (Figure 2-10): Includes a variety of tools to improve efficiency such as changing work time, sharing resources, tracking, and customizing projects.

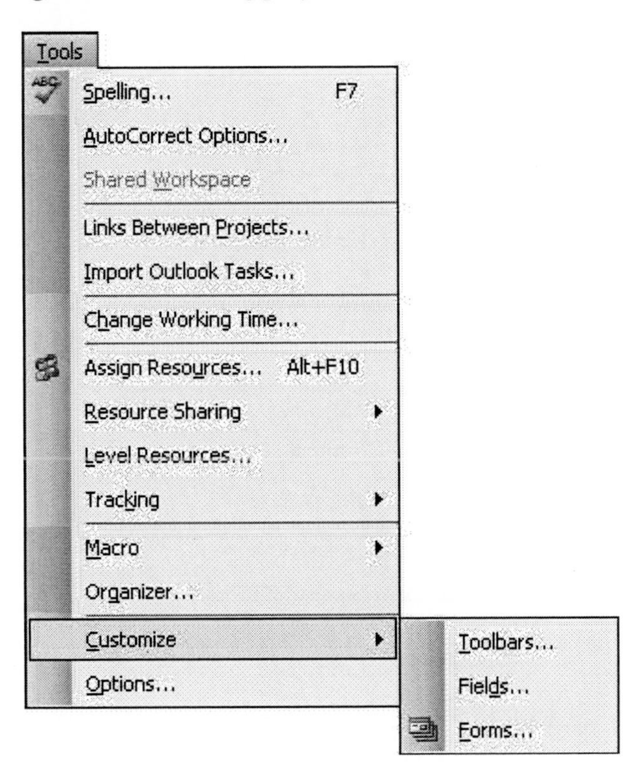

2-10: Tools

Project (Figure 2-11): Makes sorting and filtering data easier and documents project information.

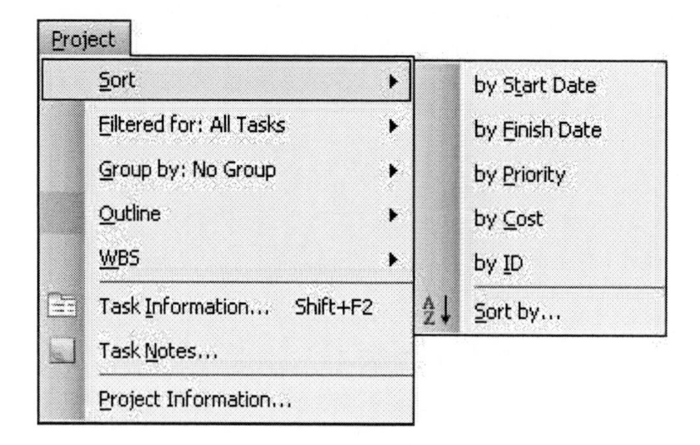

2-11: Project

Window (Figure 2-12): Allows you to customize the display of individual windows for easier viewing and manipulation of data.

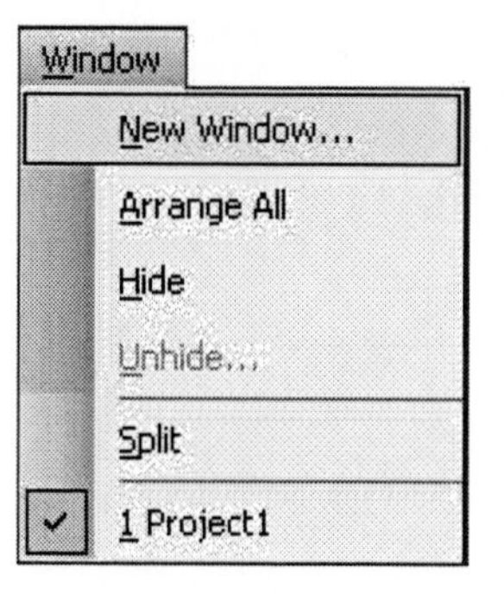

2-12: Window

Help (Figure 2-13): Offers assistance on specific components and topics within Microsoft Project.

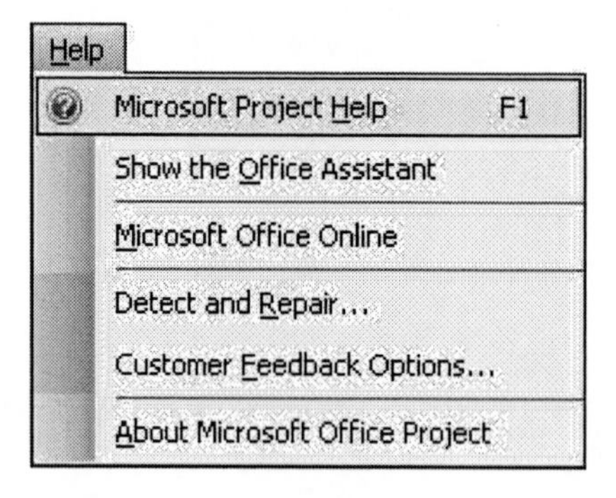

2-13: Help

Now that you've seen the structure of the Project interface menu, you will better understand how to enter and control data for specific construction and remodeling projects. Before creating a project, take a closer look at the different views available to you:

Calendar (Figure 2-14): Fits your project onto a standard monthly calendar to help you visualize the workload and the timeline.

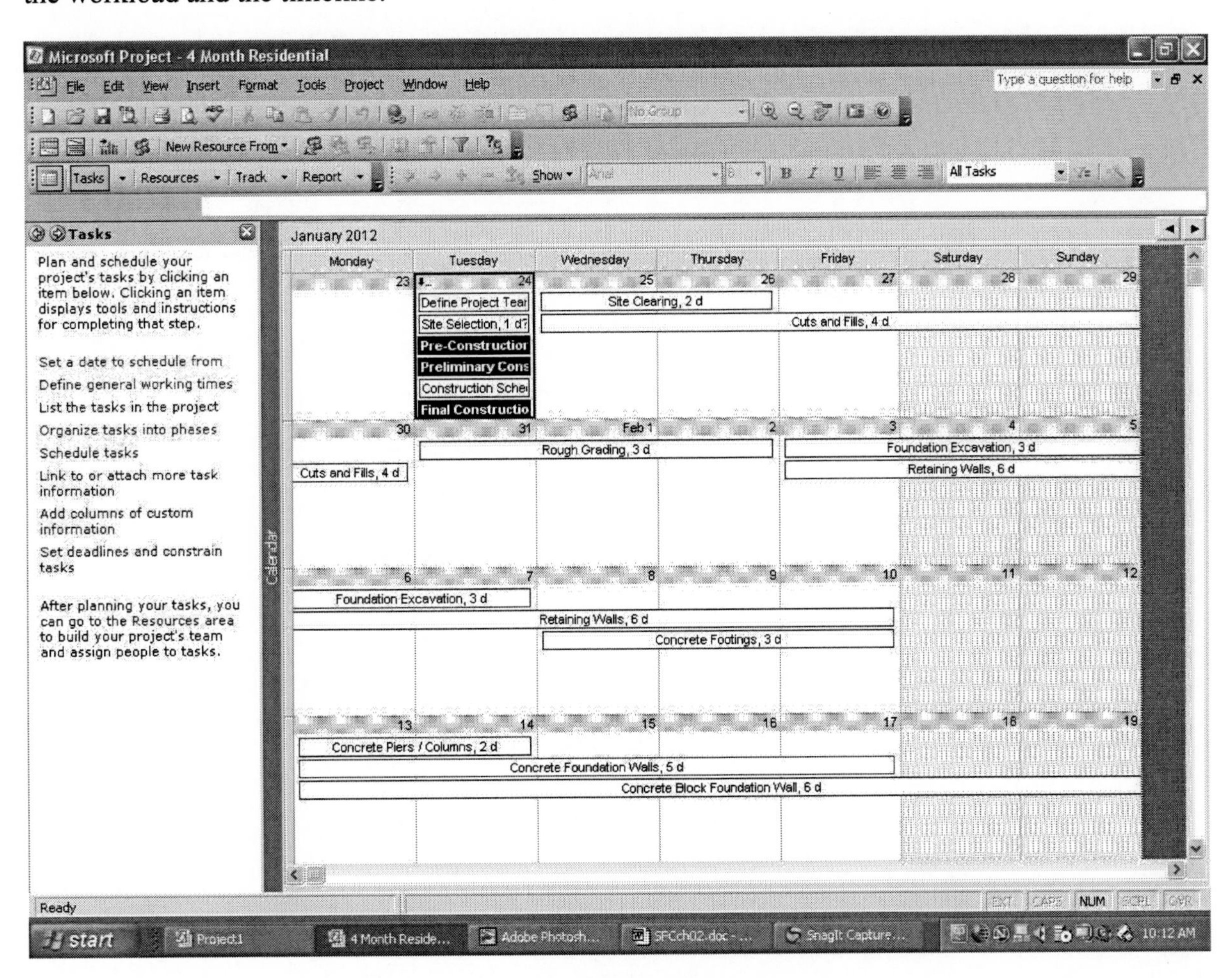

2-14: Calendar

Gantt Chart (Figure 2-15): Provides basic data about tasks such as Start, Finish, and Duration. In addition, the Detail *Gantt* helps you visualize critical and non-critical tasks and the Tracking *Gantt* allows visual comparison between the current plan and the baseline.

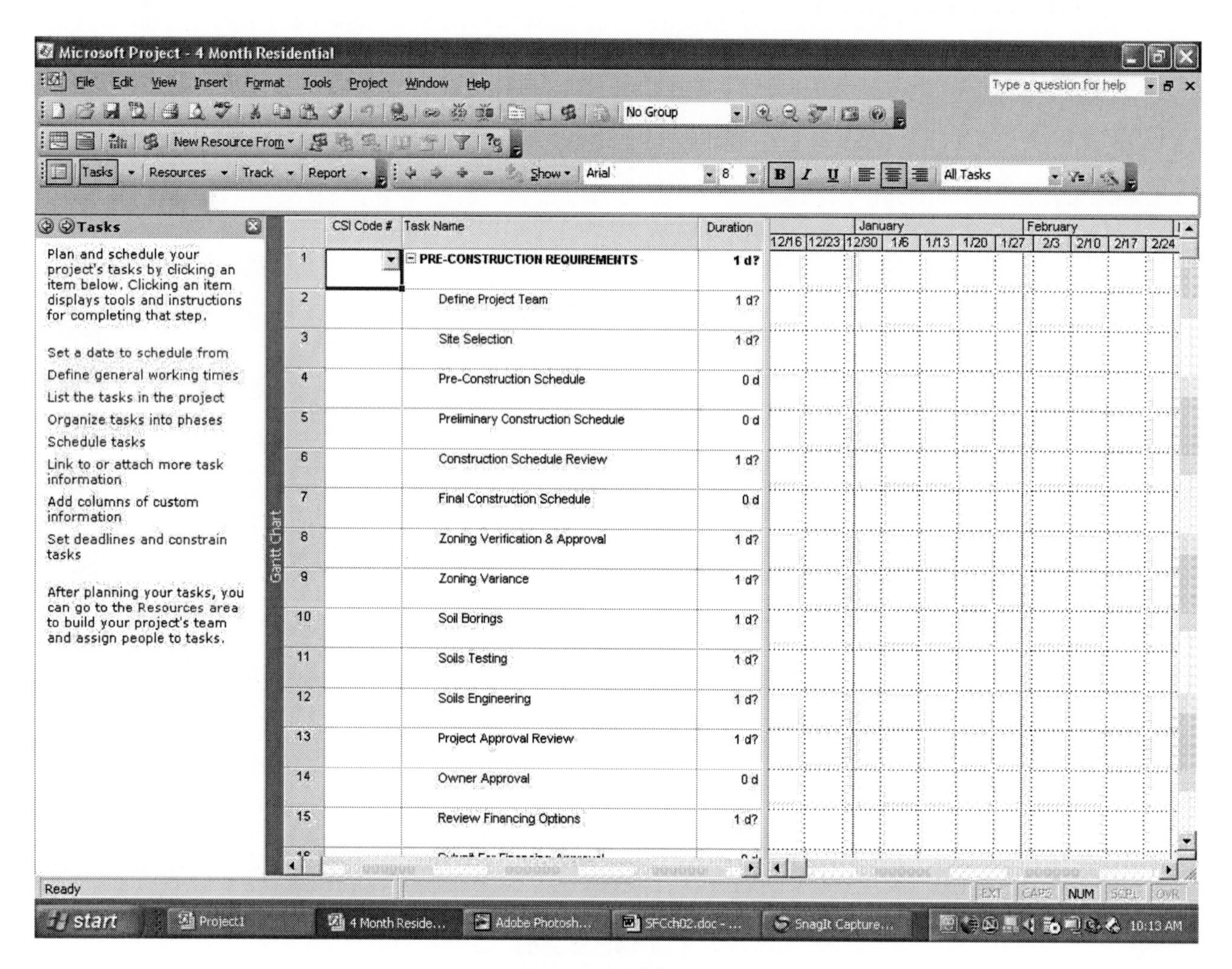

2-15: Gantt Chart

Network Diagram (Figure 2-16): Also known as the PERT (Program Evaluation and Review Technique), this view illustrates the relationship between tasks. It shows the project data based on relationships rather than on time. The Network view helps you make sure that the logic of your schedule follows the logic of the real world.

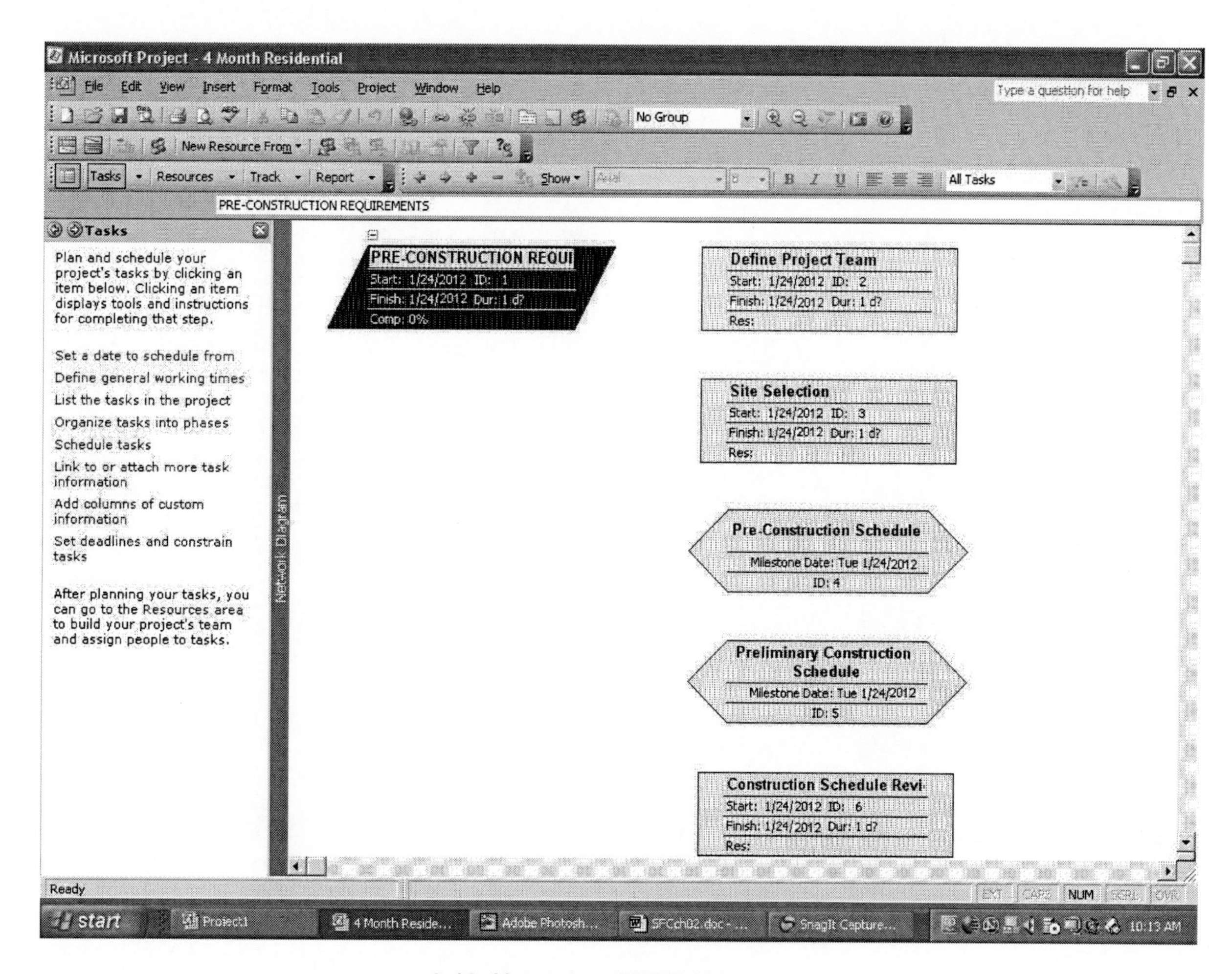

2-16: Network or PERT Chart

Task Usage (Figure 2-17): Helps you compare the specific tasks of a project, such as initiating electric service or installing sub-flooring, against a calendar.

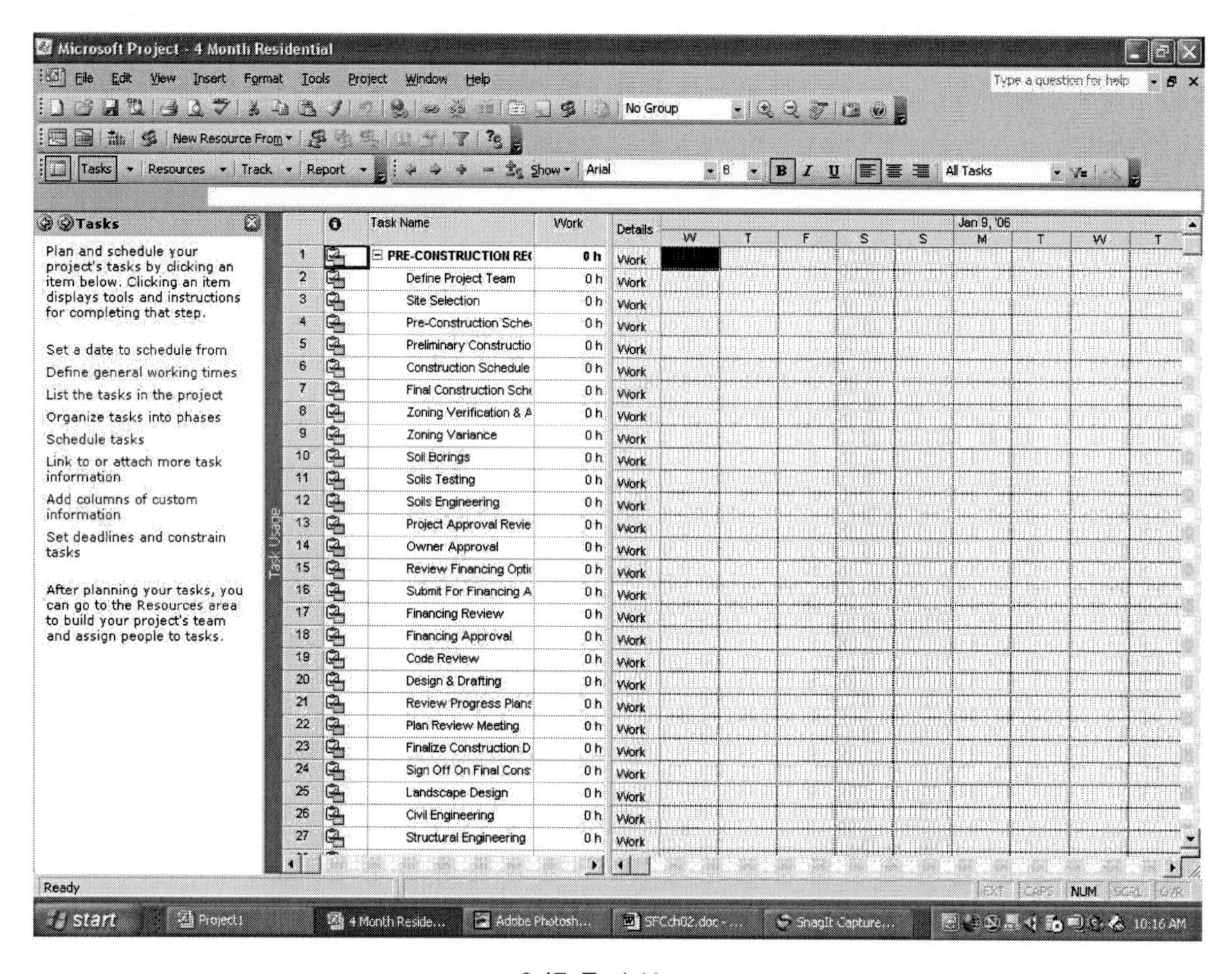

2-17: Task Usage

Resource Usage (Figure 2-18): Displays a comparison of specific resources against a calendar. A resource is any person, machine, equipment, or material used to complete the work on a task.

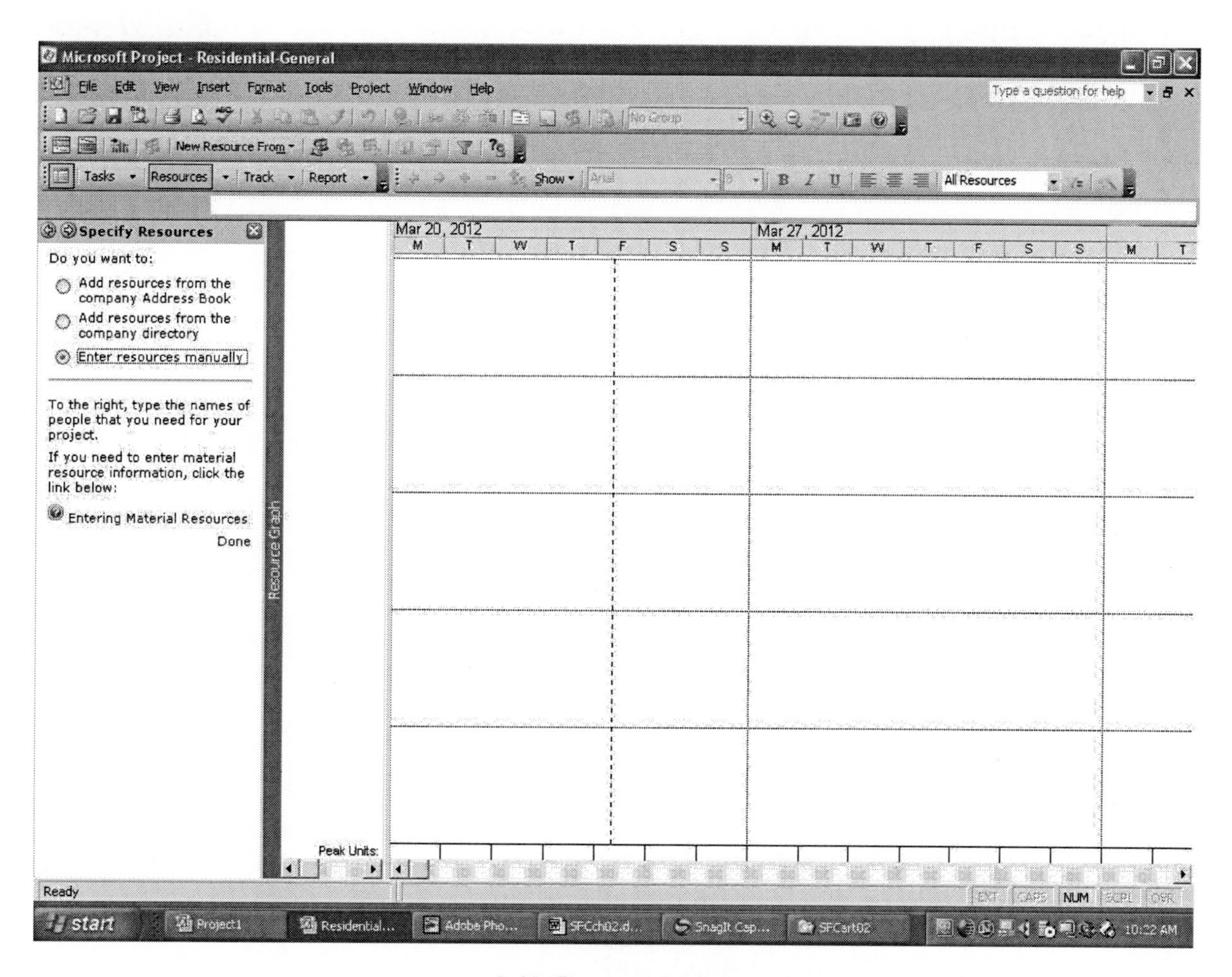

2-18: Resource Usage

Microsoft Project is quite versatile and customizable. In the beginning, it is best to use the default setup. Soon you will be using templates and customizing the interface to make the task of scheduling projects easier.

Creating a Project

It's time to create a schedule for a construction or remodeling project. You can begin with a blank project or you can start with a third-party template designed to reduce input time and effort. Let's assume that you are starting from scratch so that you can get the most from the learning process.

To create a new project:

1. **Open** the File menu and click New.
2. On the New Project task pane, **select** Blank Project. If you are using an existing template, instead select Templates, On My Computer. You also may go online to find templates that fit the project requirements.

To set the project dates:

1. Once the new project is created (above), **click** Tools, Customize, Toolbars.
2. **Check** the Analysis toolbar option.
3. **Click** Close.
4. **Click** Adjust Dates on the Analysis toolbar.
5. **Enter** a new start date for the project and click OK.

To define the project information (Figure 2-19):

1. **Click** Project, Project Information.
2. **Enter** the project's Start Date.
3. **Select** whether the project should be scheduled from the Start Date forward or from the Finish Date backward.
4. For now, leave the Status Date as NA. It will be covered under Tracking Progress later in this chapter.
5. **Define** the project Calendar as desired.
6. **Enter** a Priority for this project. This is useful when you have multiple projects at the same time.
7. **Click** the Statistics button to summarize project data.

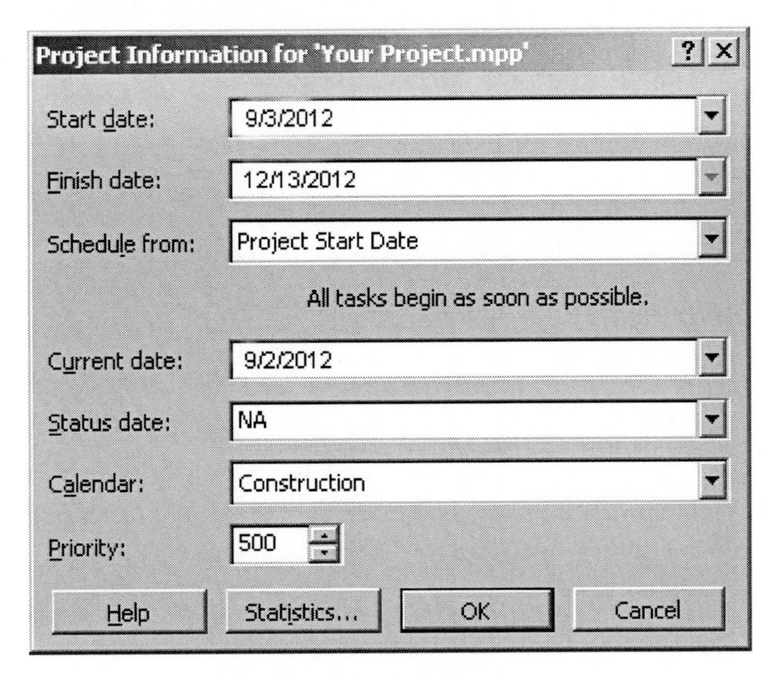

2-19: Project Information

To record the file properties:

1. **Click** File, Properties.
2. **Enter** the project Title. This is often a customer name.
3. **Enter** the project Subject.
4. **Enter** the project manager's name. This will be your own, though it could be that of the project lead.
5. **Enter** Keyword information about the project to make it easier to find.
6. **Click** OK.

To save a project:

1. **Click** File, Save.
2. **Select** a save location from the drop-down list.
3. **Enter** a file name; the Properties may suggest one based on the project title. The extension will be .MPP, although you can choose to save it as a Project Template (.MPT), database (MPD), or other format.
4. **Click** Tools, General Options to protect the file with a password if you wish, then click OK.
5. **Click** Save.

To open a saved project:

1. **Click** File, Open.
2. **Select** the project file; you may need to look on the drop-down list.
3. **Click** Open. Alternately, you can select Open Read-Only (can't change) or Open As Copy (can retain original and change copy).

To search for a project:

1. **Click** File, Open
2. On the Open dialog box toolbar **click** Tools, Search.
3. **Select** the Basic or Advanced tab and insert data to identify the file. A Basic search looks for text within a type of file, while an Advanced search can look across files for a variety of properties, conditions, or values.
4. **Click** Go.

To make a shortcut to My Places:

1. **Go** to the file location you want to add to the My Places bar.
2. **Click** Tools, Add to My Places.
3. **Resize** the My Places bar, if desired, by right-clicking on it and selecting Small Icons from the context menu.

Setting Project Options

Microsoft Project, as you can see, offers many options. The location from which you can control these options is called, logically, Options. You can access Options through Tools on the menu bar. Here you can establish preferences for view, edit, calendar, schedule, calculations, spelling, saving, interface, and security, as well as general or miscellaneous options.

To set the View options (Figure 2-20):

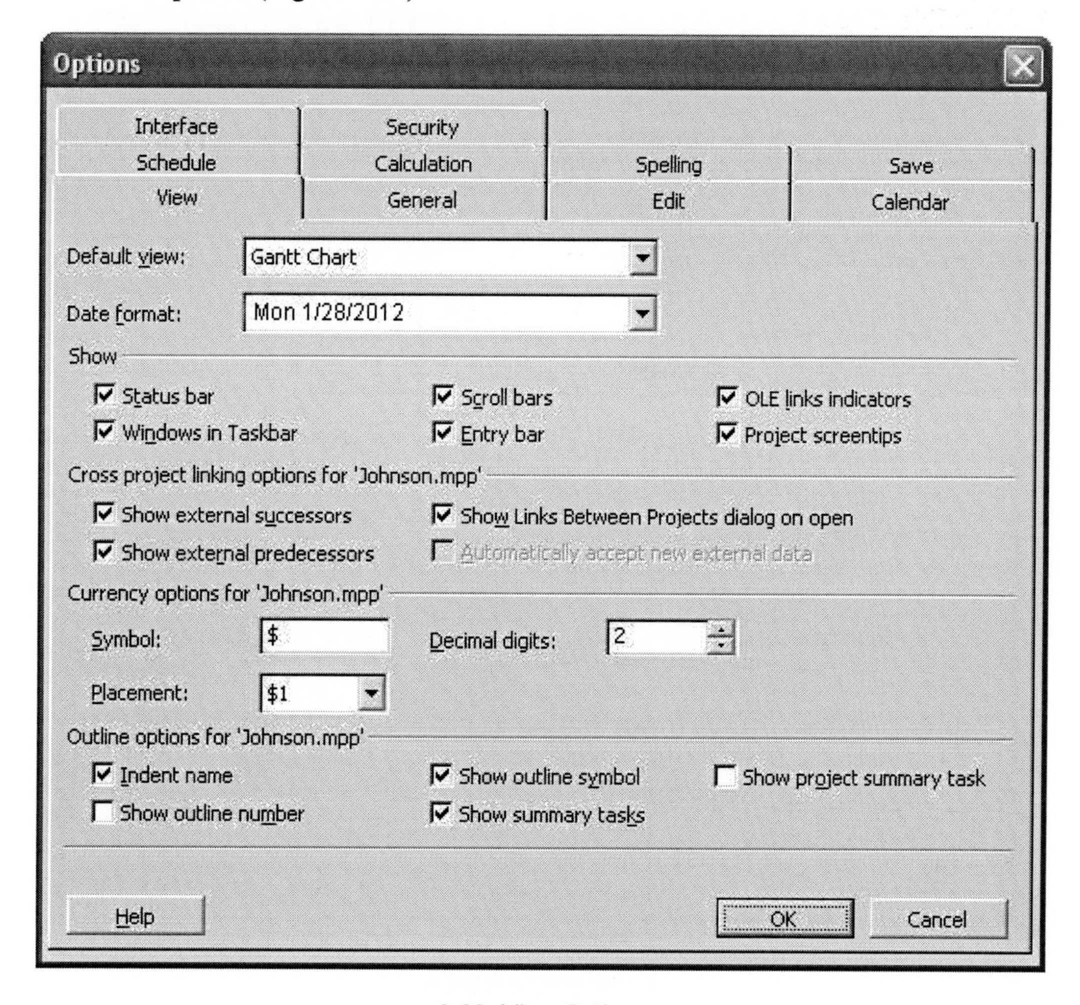

2-20: View Options

1. **Click** Tools, Options, and the View tab.
2. **Set** the Default View and Date Format as desired.
3. **Select** preferences for what to show (Status bar, Windows in Taskbar, etc.)
4. **Select** cross-project linking options for the selected file.
5. **Select** currency options or allow defaults.
6. **Select** Outline options.
7. As needed, **select** Help to learn more about what these options and defaults mean and how to select the most appropriate for your project.

General options is a good place to start to establish preferences for the overall look and action of Microsoft Project, including what shows up on the screen, how files are handled and saved, and how much assistance you wish to have. Once you have things as you want them, you can click the Set-As-Default tab so you don't have to change future project files.

To set General options (Figure 2-21):

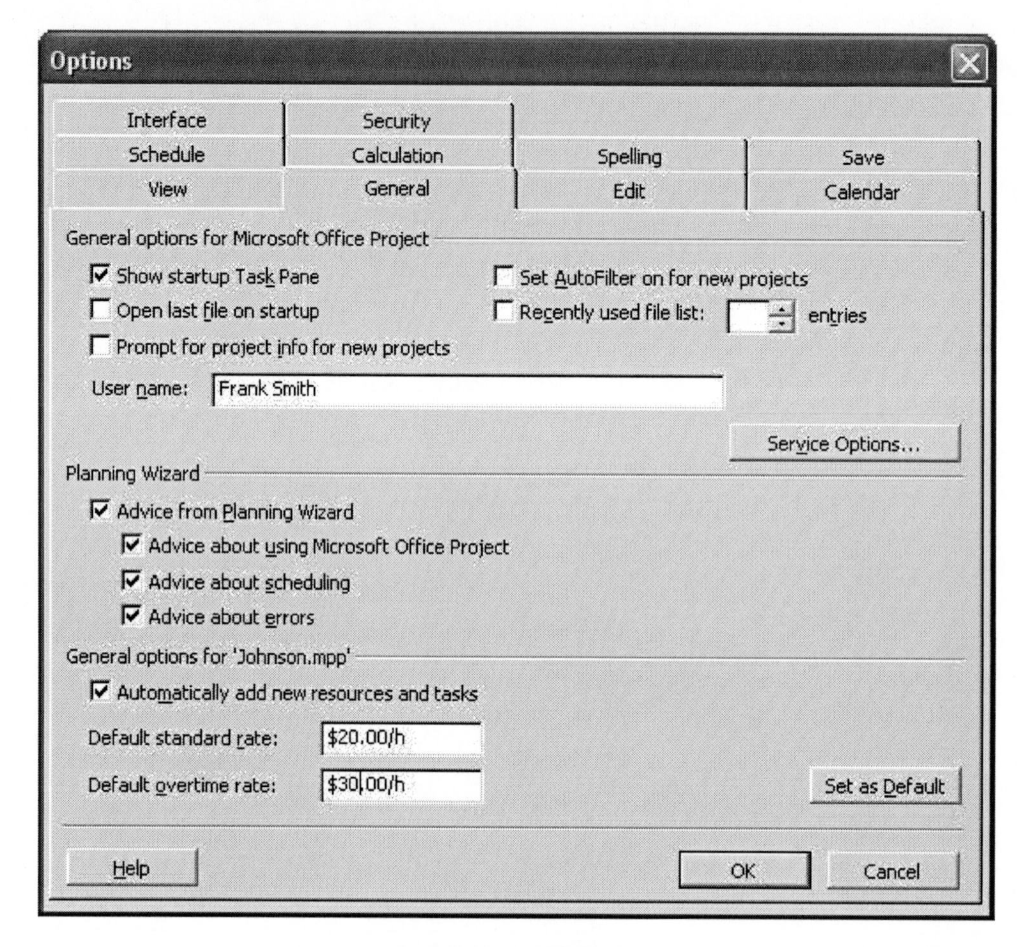

2-21: General Options

1. **Click** Tools, Options, and the General tab.

2. **Select** the General options for files and standard rates. The user name can be changed here.

3. **Select** the Planning Wizard options to include or exclude advice as needed.

As you do more file editing, you can set defaults and make your job easier by setting things up in the Edit options section.

To set Edit options (Figure 2-22):

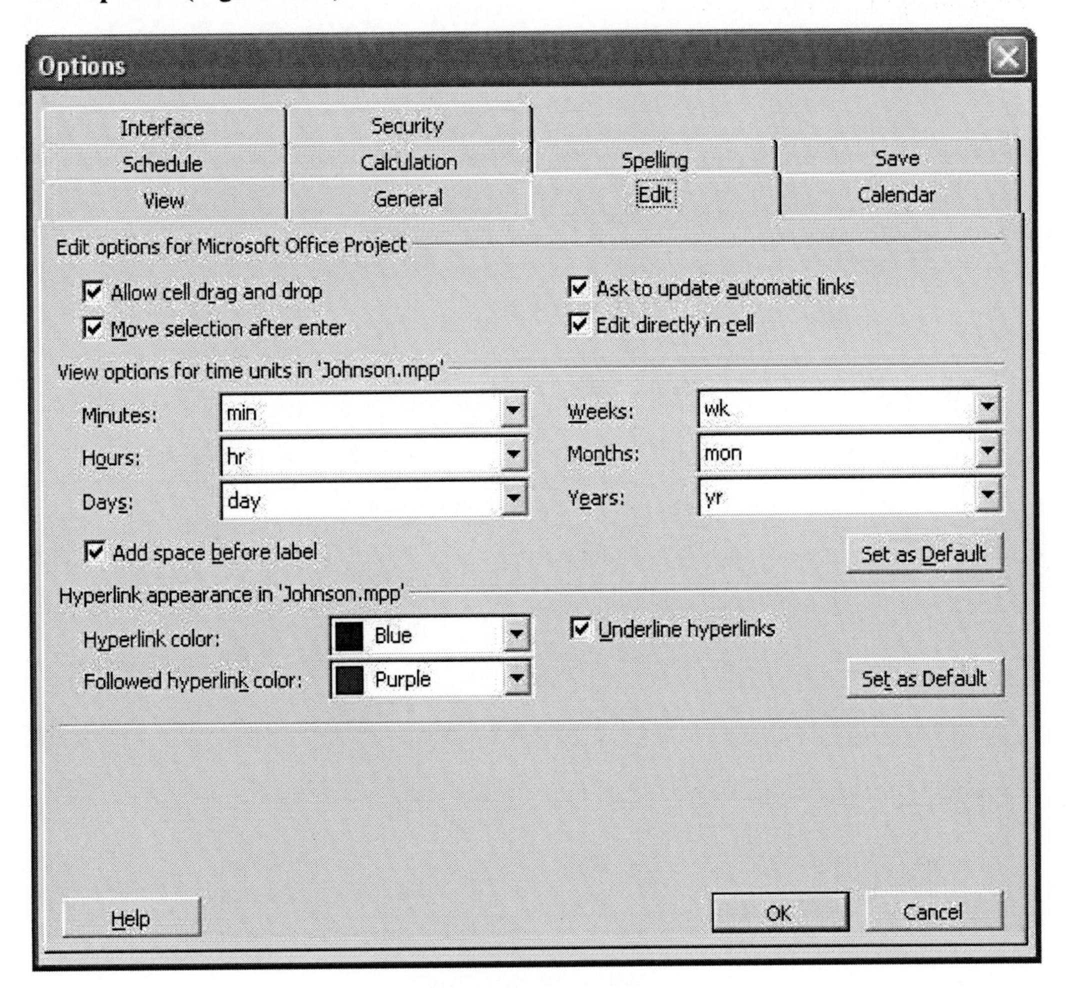

2-22: Edit Options

1. **Click** Tools, Options, and the Edit tab.

2. **Select** the editing options you prefer, checking the Help function to assist in selecting your preferences.

3. If you want your selected editing options to be applied to future projects, **click** the Set As Default button(s). If not, do not select this button and the choices will only be applied to the current project.

To set Calendar options (Figure 2-23):

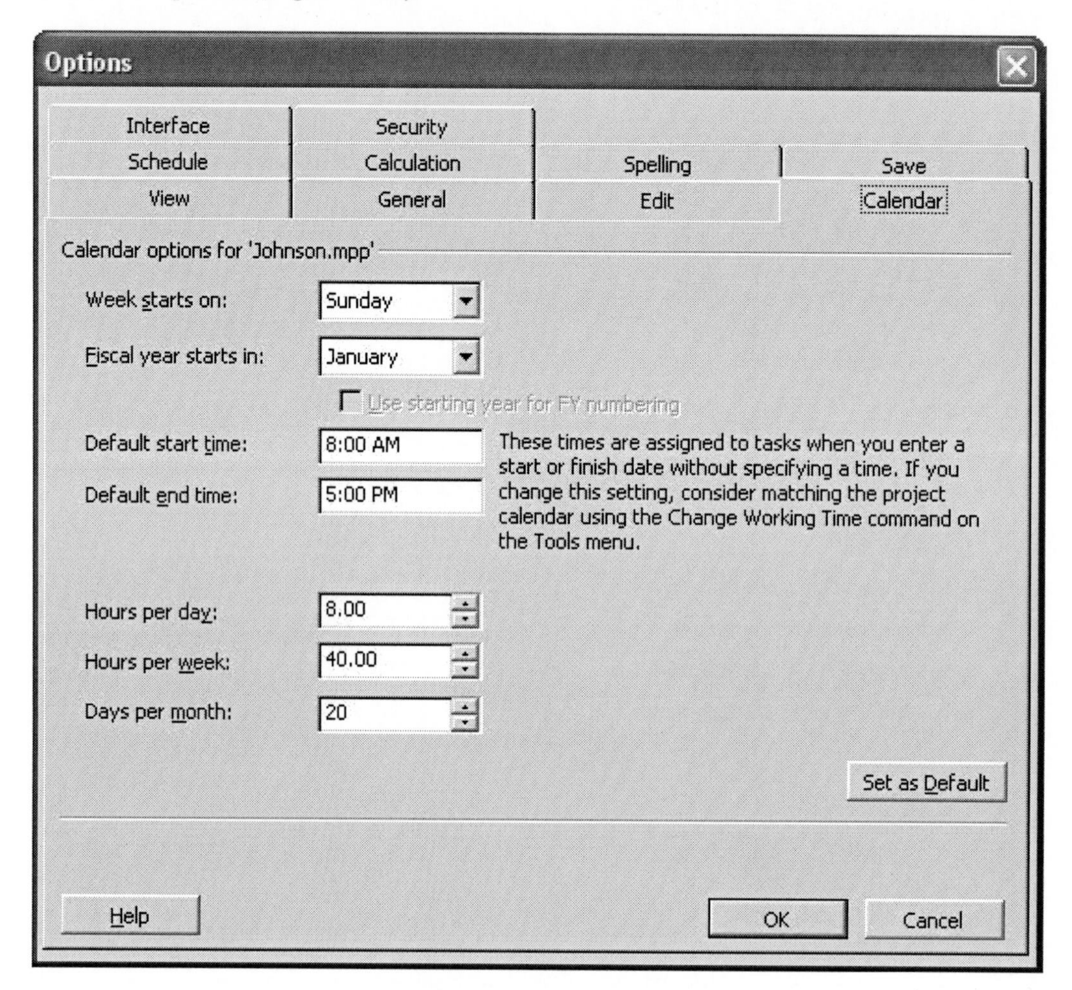

2-23: Calendar Options

1. **Click** Tools, Options, and the Calendar tab.

2. **Select** the day of the week on which you want the calendar for this project to start (Sunday, Monday, etc.)

3. **Select** the first month in your fiscal year (January, February, etc.).

4. **Select** the default start and end time of the normal work day (8:00 AM, 5:00 PM, etc.). You can modify the start and end times of specific days within the project file, but selecting the default will help you define what is typical and make entry easier.

5. **Select** the number of hours in the typical work day (8.00, etc.), hours per week (40.00, etc.), and work days per month (20, 21, etc.).

6. If the above data reflects your typical calendar for most projects, **click** the Set as Default button.

To set Schedule options (Figure 2-24):

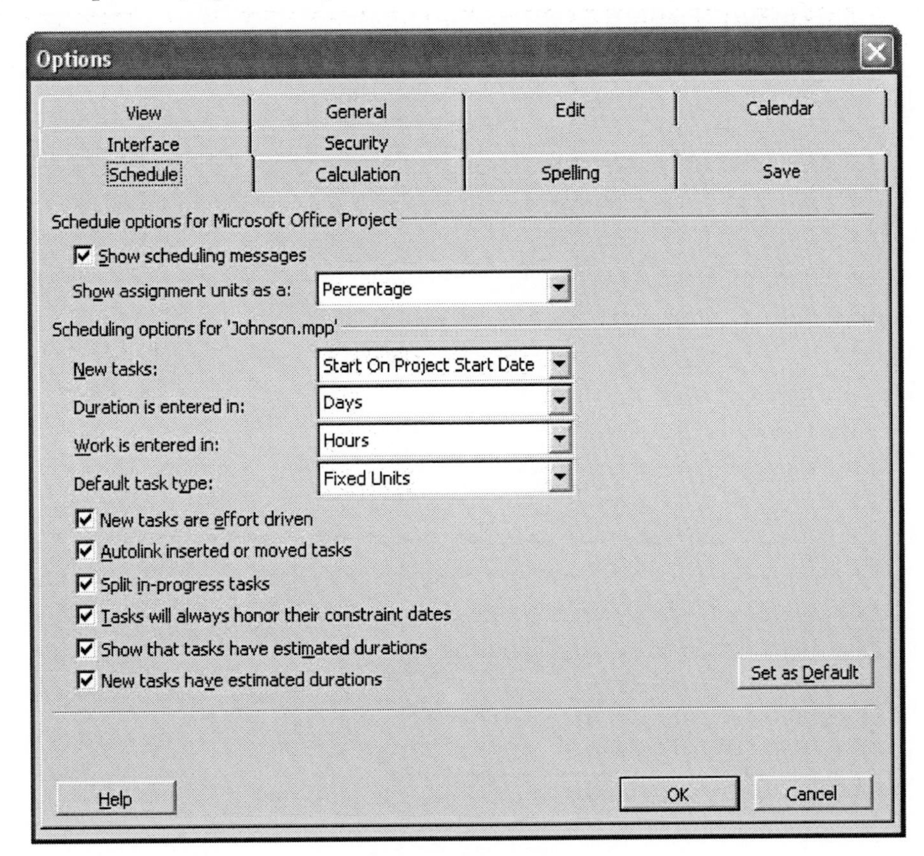

2-24: Schedule Options

1. **Click** Tools, Options, and the Schedule tab.
2. To allow Microsoft Project to notify you of schedule inconsistencies, **select** Show Scheduling Messages.
3. **Select** whether you want resource assignment units to be displayed as a percentage or a decimal.
4. **Select** whether you want new tasks to start on the project start date or on the current date. Default is Start on Project Start Date.
5. **Select** Minutes, Hours, Days, Weeks, or Months as the unit for project duration. Default is Days.
6. **Select** Minutes, Hours, Days, Weeks, or Months as the unit for project work. Default is Hours.
7. **Select** the Default Task Type for new tasks: Fixed Units, Fixed Work, or Fixed Duration. This feature allows you to select which component of the basic project formula (Duration = Work/Units) remains fixed and which can be adjusted. The default is Fixed Units.
8. **Select** Schedule checkbox as appropriate. Use the Help button to learn more. The default is all boxes checked.

To set Calculation options (Figure 2-25):

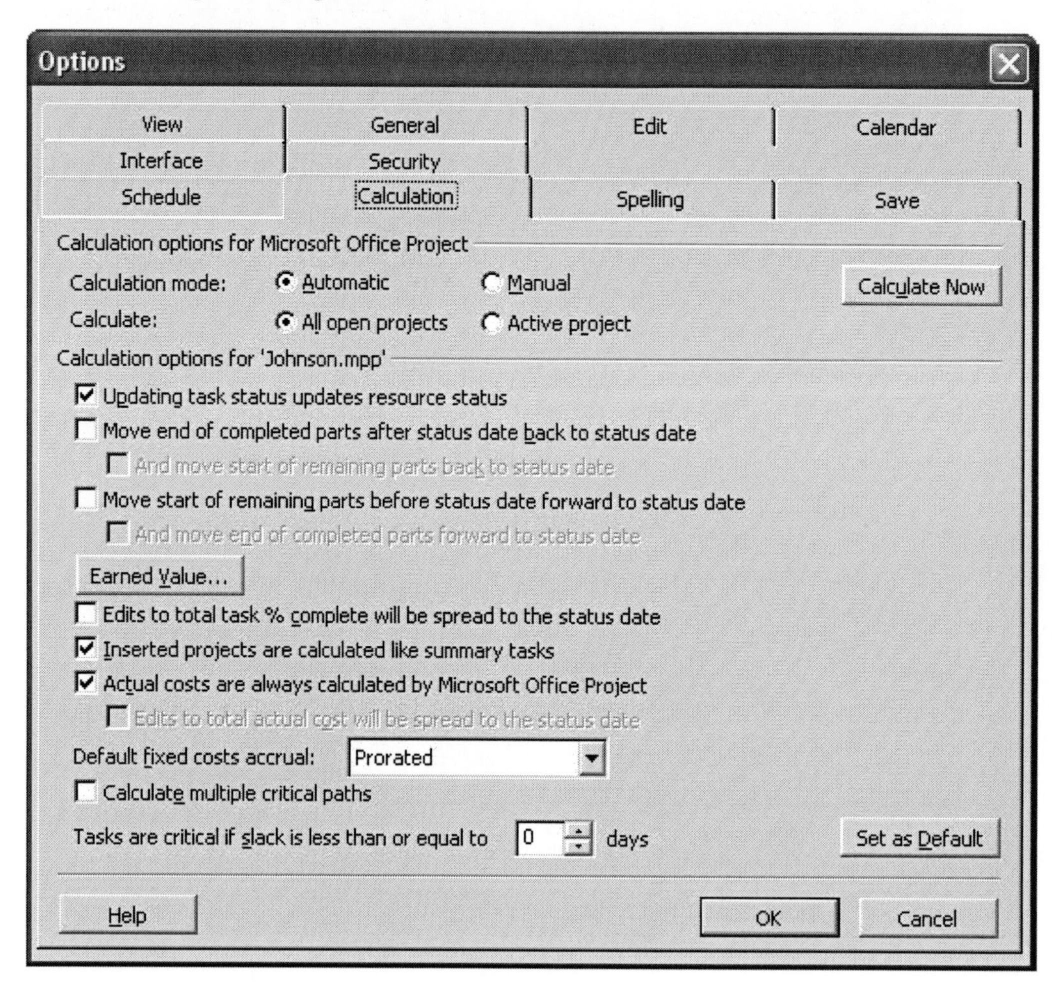

2-25: Calculation Options

1. **Click** Tools, Options, and the Calculation tab.

2. **Select** the calculation options for all projects, as desired. The defaults are Automatic and All open projects. Don't change the Calculation mode to Manual until you are experienced.

3. **Select** the Calculation options for the open project. The default is Up-dating task status updates resource status.

4. **Select** the Default fixed costs accrual as Start, Prorated, or End. The default is Prorated.

5. **Refer** to Help for explanations of additional check boxes. If desired, Set as Default so the selected Calculation settings are applied to all projects.

To set Spelling options (Figure 2-26):

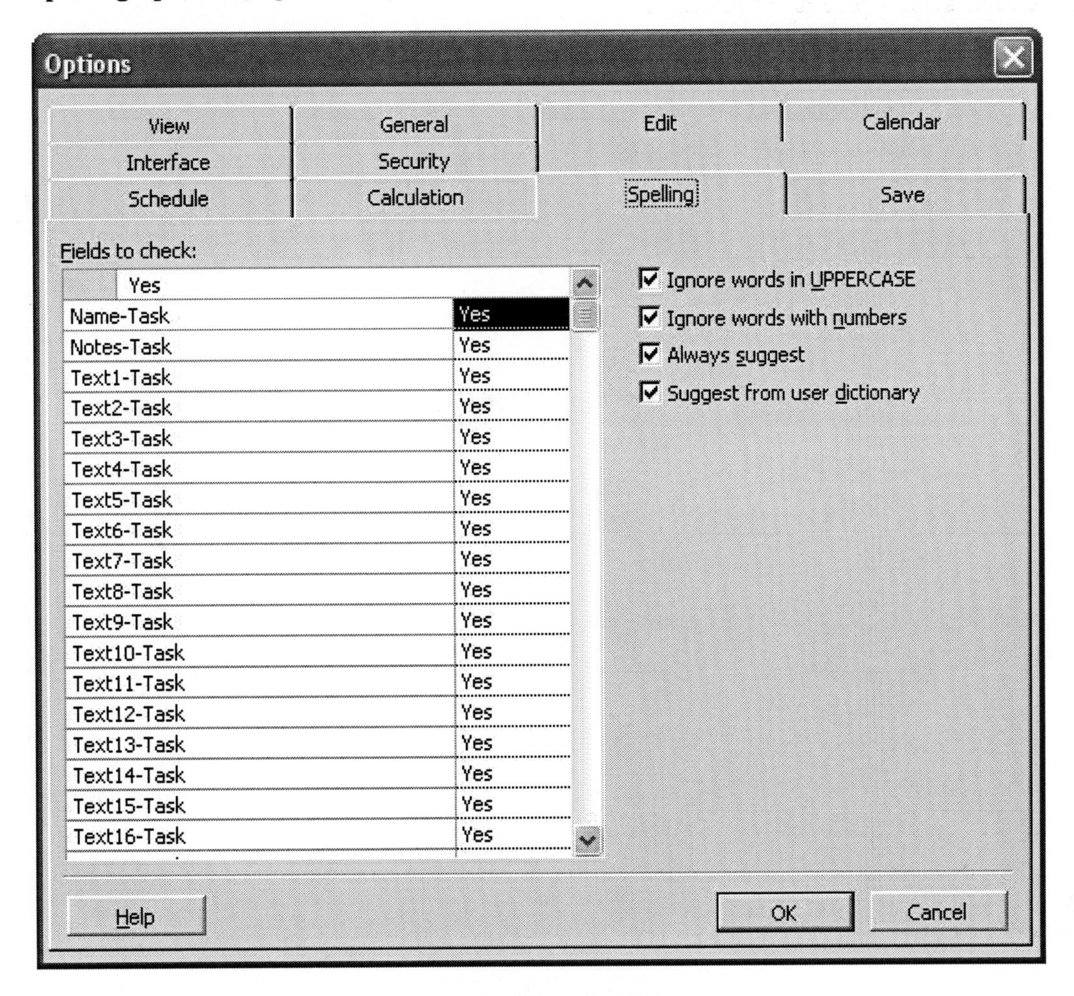

2-26: Spelling Options

1. **Click** Tools, Options, and the Spelling tab.
2. **Use** the Help feature to determine which functions you wish to change. In most cases, using the default settings is adequate.

To set Save options (Figure 2-27):

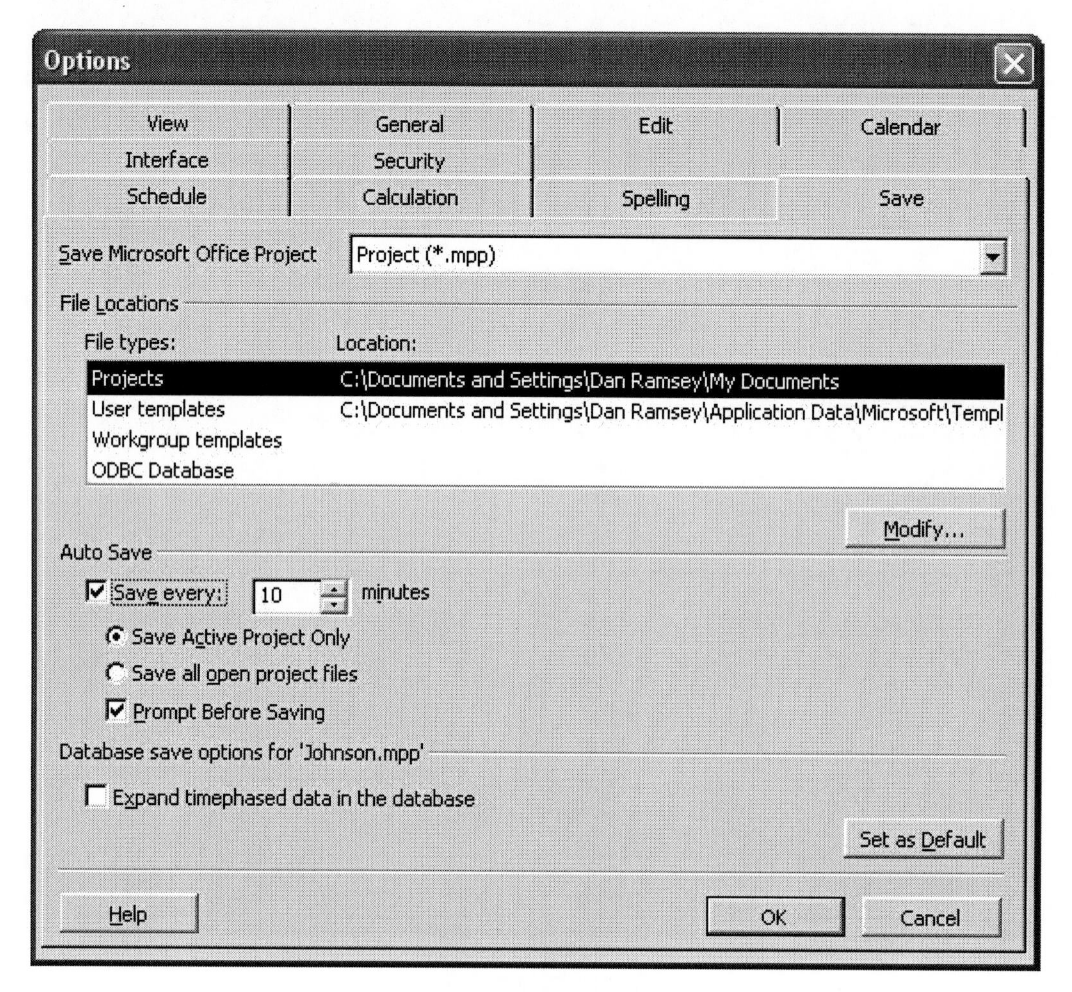

2-27: Save Options

1. **Click** Tools, Options, and the Save tab.

2. To change the location where files are saved, modify the File Locations path for Projects and User Templates. For advanced users, you also can change the path to Workgroup templates and ODBC Database files, covered later in this book.

3. **Select** whether you want to have Project automatically save open files and how often. Automatic save is preferred for new users who may forget to frequently save data; however Auto Save uses up memory and may lead to a program crash or failure.

To set Interface options (Figure 2-28):

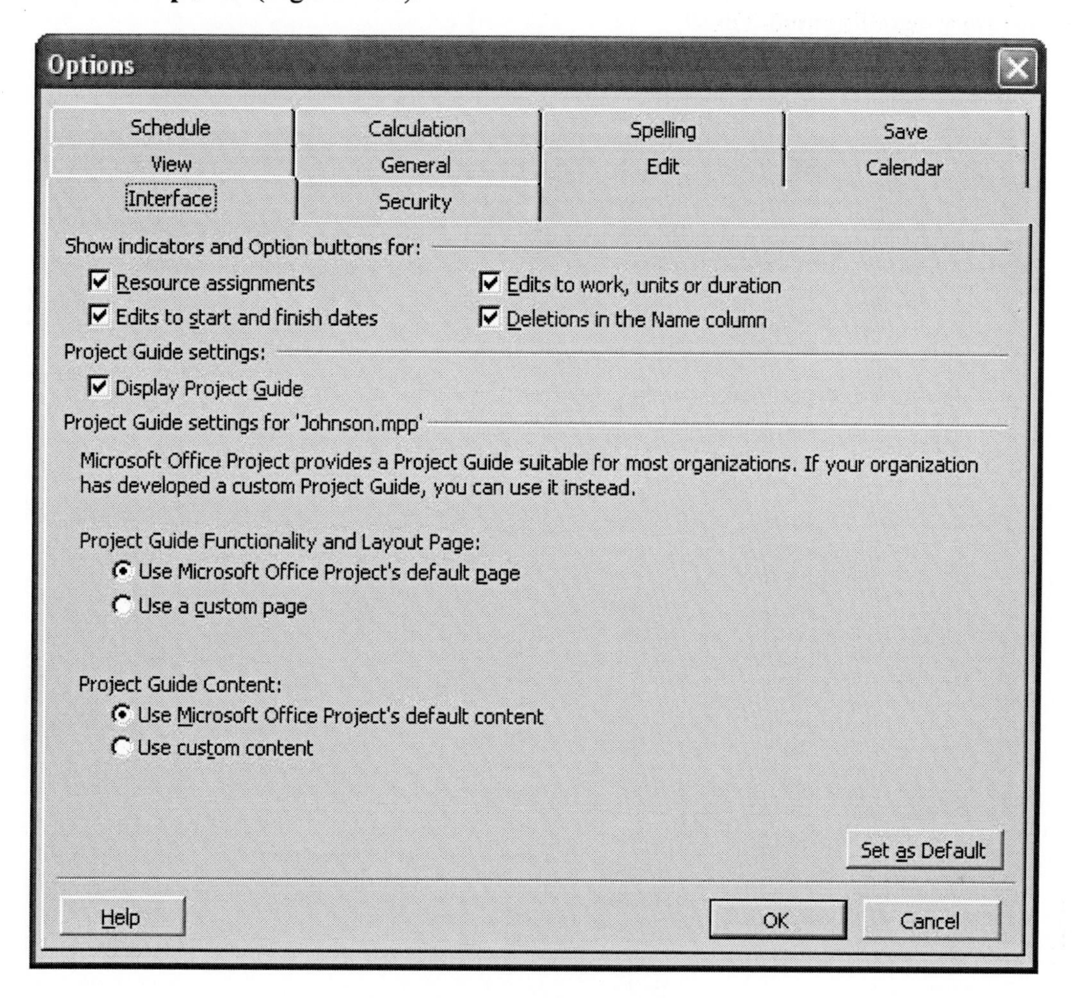

2-28: Interface Options

1. **Click** Tools, Options, and the Interface tab.

2. **Use** the Help feature to determine which functions you wish to change. In most cases, using the default settings is adequate.

To set Security options (Figure 2-29):

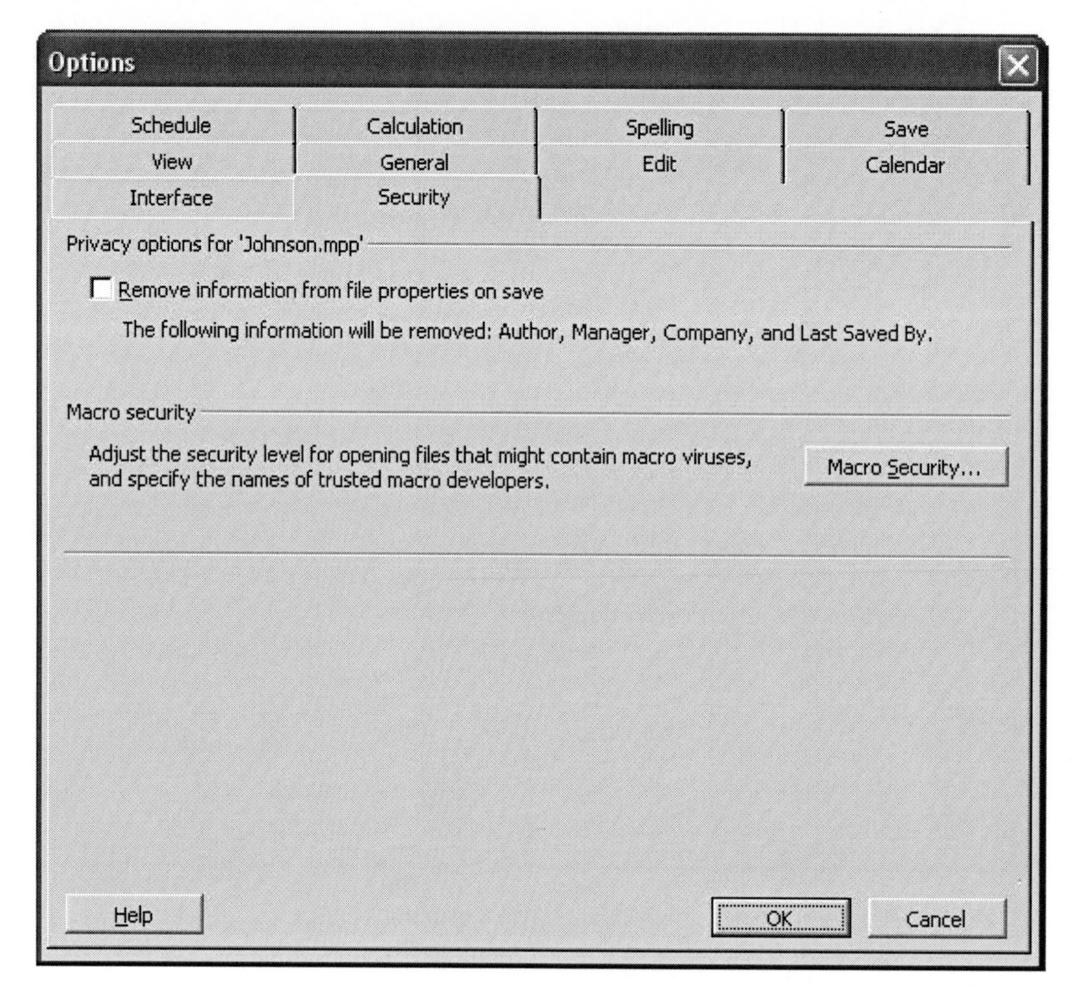

2-29: Security Options

1. **Click** Tools, Options, and the Security tab.
2. **Use** the Help feature to determine which functions you wish to change. In most cases, using the default settings is adequate. If your computer is connected to the Internet and not fully protected by a firewall, security issues are more important. Also, select higher security settings if other people use your computer.

Note: Macros are subroutines or small programs that can enhance productivity or damage your data, depending on the design intent. Most are written in VBA (Visual Basic for Applications) to perform tasks on data. By setting macro security on High you can stop unauthorized programs from accessing and potentially changing your Project data.

Starting a Project

You've customized and set up Microsoft Project to make data entry easier and files more secure. Now you are ready to begin entering tasks, which is called building a project. If you are using pre-built templates the job will be easier, as many of the tasks will already be built and your job will simply be to insert the starting or ending date and make some changes to the job tasks. However, enable you to learn more about using Microsoft Project, this chapter will assume that you don't have a pre-built template. Later chapters will assume that you have either input the tasks or are using a template with the tasks already entered.

To start a new project:

1. **Click** on View and select Calendar, Gantt Chart, Network Diagram, Task Usage, Tracking Gantt, or one of the Resource views, as appropriate. Most users prefer the Gantt Chart view to start.

2. **Select** Tasks on the Project Guide toolbar.

3. **Read** the instructions on the sidebar titled Tasks (Figure 2-30).

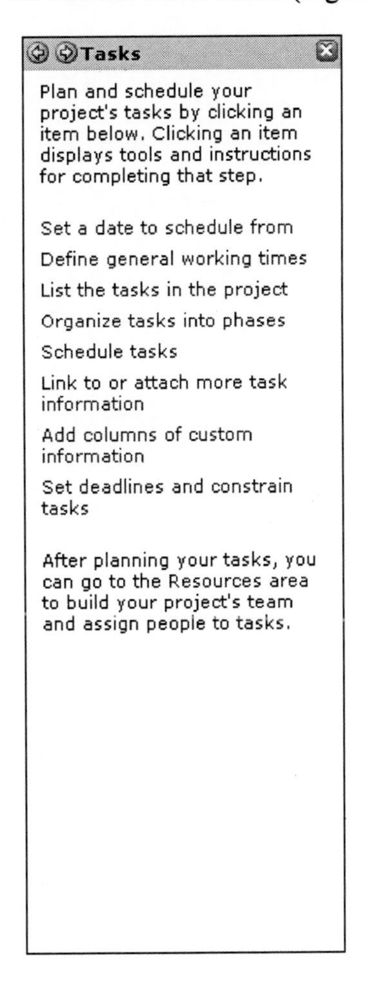

2-30: Tasks Sidebar

4. **Select** the first choice on the Tasks menu: Set a date to schedule from.

5. **Follow** the instructions on the sidebar, entering the starting date. When done, select Done and the sidebar will return to the Tasks topic.

6. **Select** the second choice on the Tasks menu: Define general working times.

7. **Continue** entering tasks and data as directed.

8. Once done entering tasks, **select** Resources (Figure 2-31) on the Project Guide toolbar or click on the words Resources area on the sidebar.

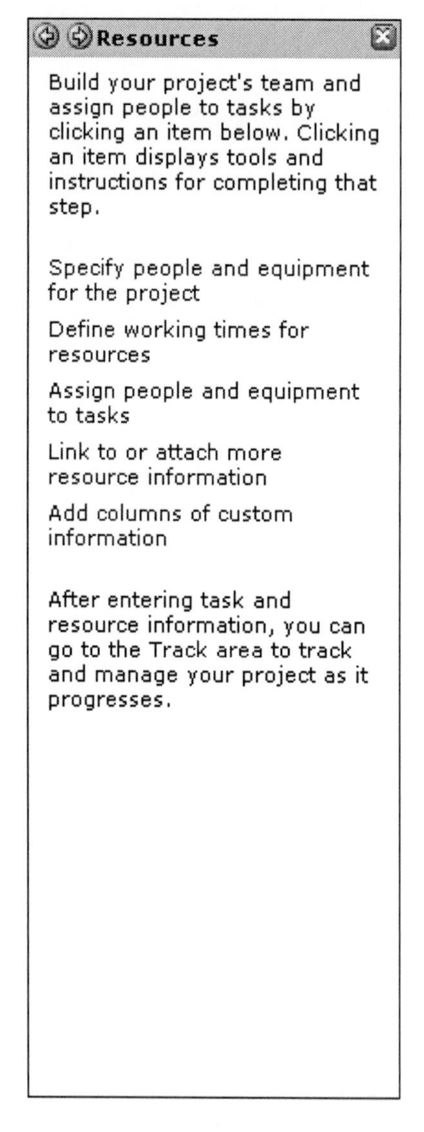

2-31: Resources Sidebar

In the same way, you can enter the project Resources, Track (Figure 2-32), and Report (Figure 2-33) systems by simply following the instructions on the sidebar. Even if you're using a pre-existing template for your project, you'll need to go through each of these four components of the Project Guide bar to make sure the structure is accurate for your specific project.

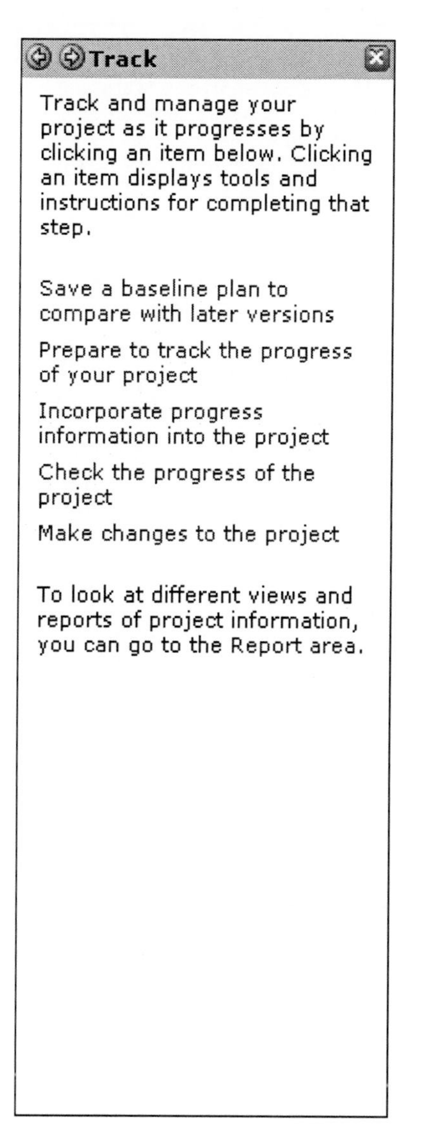

2-32: Track System

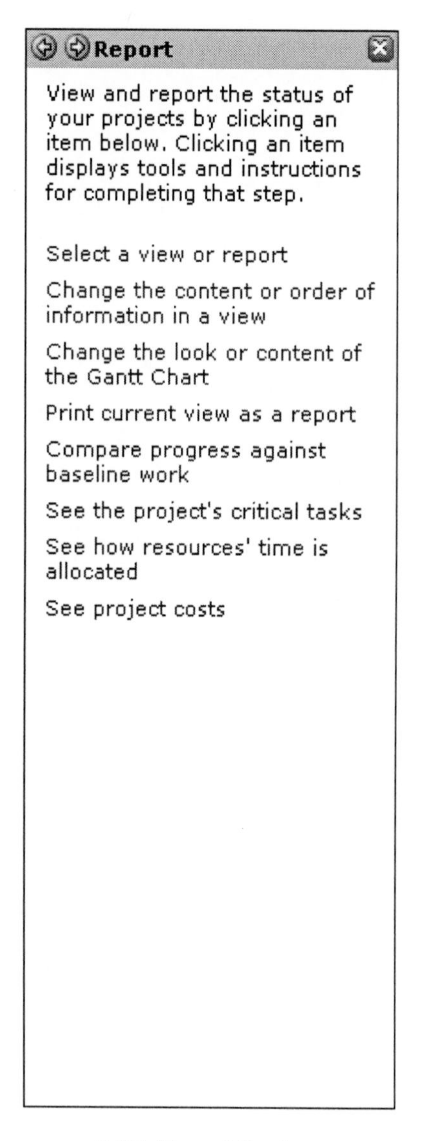

2-33: Report System

Also note that at any time you can change how you view the project without impacting the project data. That is, you can switch from a Gantt Chart view to a Calendar or even a Task Usage view without upsetting the data. Just select the view you want under the View menu.

That's an overview of how projects are started using Microsoft Project. Now let's look at the specifics of entering tasks, identifying resources, customizing and formatting, tracking progress, and reporting on projects.

Entering Tasks

To enter Tasks:

1. **Select** a blank Task Name field and enter the task name.
2. **Enter** the Duration of the task.
3. **Enter** the Start date.
4. **Enter** the Finish date.
5. **Enter** the Predecessors, or tasks that have to be completed before this task is started.
6. **Enter** the Resource Names. Resources are defined using the Resources button on the Project Guide toolbar.

Note: that the results of the task entry are shown on the calendar adjacent to the table or spreadsheet. Also note that you can use the mouse to move or reorder tasks on the table as well as timeline bars on the calendar. In addition, you can select a task and use the right mouse button to cut, copy, paste, or perform other actions. This feature is especially handy if you are using a pre-designed template to develop your project schedule.

The properties of any task can be viewed and changed easily. Simply select the Task ID number on the far left to highlight the entire task, and then double-click on the task. A small screen will appear with six tabs: General, Predecessors, Resources, Advanced, Notes, and Custom Fields. Selecting a tab offers numerous options. For example, the General tab allows you to edit the task Name, Duration, and whether the Duration is Estimated (a question mark identifies estimated durations). You also may indicate the Percent complete and assign a Priority number. You can change the Start and Finish dates as well as Hide task bar and Roll up Gantt bar to summary. The other tabs have similar customized characteristics for each specific Task on a project table.

Predecessors tell the project table which tasks must come before others. Successors indicate which tasks must come after specific tasks. Use the Predecessor tab to indicate the hierarchy. Usually, predecessors and successors are identified by their task numbers. Task 134, for example, may require that Tasks 152 and 166 be completed first. The tasks don't have to be sequentially numbered as long as Project knows which tasks come before or after the selected task.

Note: that you can enter all the Tasks for a specific job without assigning Predecessors, Resources, or even dates and durations. Just make sure that each Task gets these assignments before finishing the project schedule.

There are a couple more points to make before leaving Tasks. The first relates to summary tasks. You can use Microsoft Project to establish primary or summary tasks that incorporate other tasks or subtasks. For example, you can set up a Concrete Foundation summary task that has subtasks of concrete footings, concrete piers/columns, and concrete foundation walls, each with its own duration, predecessor(s), resources, and notes. Summary tasks can then be collapsed to hide their subtasks and make viewing the overall project easier.

To define summary tasks:

1. **Select** the Task on the table.
2. **Double-click** the task to open the Task Information screen.
3. **Select** the General tab.
4. **Select** Roll up Gantt bar to summary.
5. On the table, indent all subtasks below the summary task. Hint: Use the left arrow on the formatting toolbar to indent tasks.

Summary tasks will then calculate the duration of all subtasks to offer a total duration for the summary task.

The second useful action for entering tasks is to create milestones for a project. A milestone is an important point within a specific project. It could be substantial completion, the punchlist, or the owner moving in. For other projects it could be finishing framing or passing the electrical inspection.

To create milestones:

1. **Select** the Task you want to make into a milestone.
2. **Enter** a Duration of 0 days for the task.
3. **Open** Task Information and the Advanced tab.
4. **Check** the box Mark task as milestone.

A diamond will appear on the Gantt chart to indicate that the task is a milestone.

If you only have one project and don't have to worry about time or resource limits, that would be the end of making an accurate schedule. However, in the real world tasks have to be completed in a specific order and often must be completed by a specific date. These are called dependencies and constraints. A dependency is the order of tasks; one task must precede another. A constraint is a time limit; a task must be completed by a specified date or cannot be started until a specific date when materials arrive. Fortunately, Microsoft Project can handle both dependencies and constraints.

There are four types of dependencies:

- Finish-to-Start (FS): the successor task cannot start until the predecessor task is finished.
- Start-to-Start (SS): the successor task cannot start until the predecessor is started.
- Finish-to-Finish (FF): the successor task cannot be finished until the predecessor is finished.
- Start-to-Finish (SF): the successor task cannot be finished until the successor has started.

To create a Finish-to-Start dependency:

1. **Select** the predecessor task.
2. **Hold-Down** the Ctrl key and select the successor task.
3. **Click** the Link Tasks toolbar button (on the Standard toolbar) to create a link between the tasks.

To change the link to type other than Finish-to-Start (the most popular), edit the link. Alternately, you can use the mouse to select and link dependencies.

Constraints are useful if you have tasks that must be completed by specific dates. However, overusing them can limit your schedule and make your job as scheduler more difficult, so use them sparingly.

There are eight types of constraints:

- As Soon As Possible (ASAP): for the earliest start dates available.
- As Late As Possible (ALAP): for the latest finish dates available.
- Must Start On: (MSO): to start a task on a specific date.
- Must Finish On: (MFO): to finish a task on a specific date.
- Finish No Earlier Than (FNET): to finish later, but not earlier than specific date.
- Finish No Later Than (FNLT): to finish earlier, but not later-than specific date.
- Start No Earlier Than (SNET): to start later, but not earlier-than specific date.
- Start No Later Than (SNLT): to start earlier, but not later-than specific date.

To create a constraint on a task:

1. **Select** the task to be constrained.
2. **Double-click** on the task to open the Task Information window.
3. **Select** the Advanced tab.
4. **Modify** the Constraint task Deadline (Figure 2-34), Constraint type, and Constraint date as appropriate.

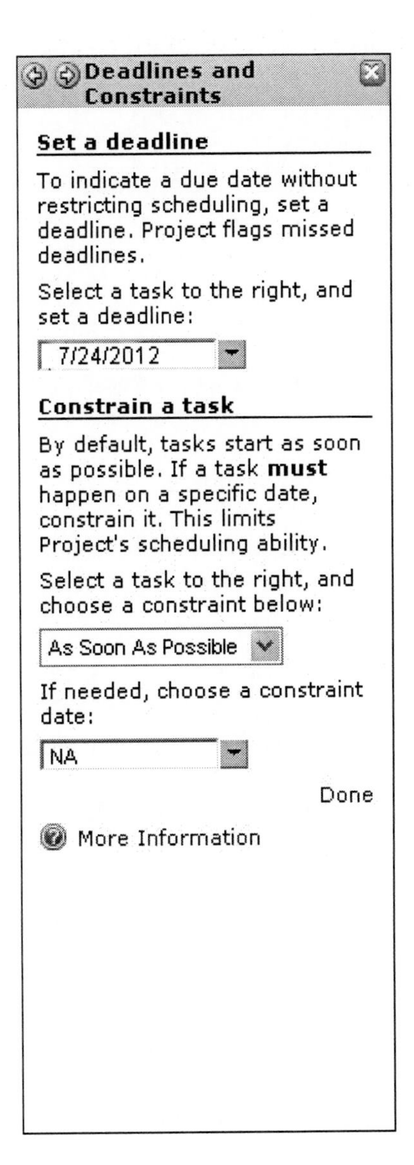

2-34: Deadline

Identifying Resources

Resources are the people and things that will be part of the project. Microsoft Project calls them work and material. Work resources have calendars that define their working and nonworking days and times. Material resources have no calendar and are considered always available.

To create work resources:

1. **Click** View, Resource Sheet.
2. **Enter** a name into the first available Resource Name field. Use first and last name for employees and business name for subcontractors.
3. **Specify** a Max Units value for the resource.
4. If you wish to track costs, **enter** a Rate for the Resource.

To create material resources:

1. **Click** View, Resource Sheet.
2. **Enter** a description or label into the first available Resource Name field.
3. **Change** the Type field from Work (default) to Material.

Note: The descriptions refer to the generic template that comes with Microsoft Project. If you have purchased an after-market template specific to construction, columns and fields may be different.

You can select and edit each of the work or material resources on the Resource Sheet. Simply double-click on the resource you want to edit, and then select the tab you want to edit. The tabs include:

- General - resource name, email, initials, group, code, type, and resource availability.
- Working Time (if Work Resource) - base calendar, working time for selected dates, set selected dates to, from.
- Costs - cost rate tables, cost accrued (start, prorated, end).
- Notes - allows you to write and format notes regarding the resource.
- Custom Field - allows you to design custom fields for resources.

Each of the Resource Information tabs includes a Details button that can be used to expand information regarding the resource. Most importantly, each tab has a Help button to assist you in understanding and using each feature. Also, any of the field names can be modified. Double-click on the header to open a Column Definition screen that allows you to change the field name and title, align the title and data (left, center, right), set the column width, and check the box to allow Header Text Wrapping. Alternately, you can select Best Fit and let the program determine the appropriate column width.

You can modify resource availability to reflect the real world. For example, if you have a framer who will be on vacation for two weeks in October, you can modify Resource Information, General, Resource Availability to exclude specific dates for that resource. You also can manage a work resource using the Working Time tab under Resource Information. The Costs tab can be used to modify estimated resource costs effective on specific dates, such as an announced increase for specific materials or a raise promised an employee on a specific date.

Once the majority of work and material resources have been defined, you can assign them to specific projects or jobs. In most cases, you won't have to repeat the process of defining resources. You'll just select and assign your defined resources to new projects.

To assign resources:

1. **Click** View, Gantt Chart.
2. **Click** Tools, Assign Resources (shortcut: Alt+F10).
3. **Select** the task to which you want to assign a resource.
4. **Select** the resource you want to assign.
5. **Click** the Assign button.
6. If necessary, edit the Units value.
7. **Check** resources for the selected task.
8. **Click** Close.

You have now assigned work or material resources to specific projects or jobs. You can filter the assignments to display all resources or just specific groups of resources such as Material or Work.

Once you have resources assigned the way you want them, you can graph the assignments. The three categories are Remaining Availability, Work, and Assignment Work.

To graph assigned resources:

1. **Click** Tools, Assign Resources.
2. **Click** Graphs.

Microsoft Project offers many ways of looking at the same data. Another method of assigning resources is by using the Task Form. Try this:

1. **Click** View, Resource Sheet.
2. **Select** a resource in the Resource Name column.
3. **Click** Window, Split.

In the bottom of the window will appear a Resource Form that can be used to assign resources for specific tasks.

To edit Assignment Information:

1. **Click** View, Track Usage.
2. **Select** a task in the Task Name column.
3. **Double-click** on the Assignment below the task name.
4. **Edit** the General, Tracking, or Notes tabs as needed.

What can you do if you over-allocate resources -- schedule more resources than you actually have? Fortunately, Microsoft Project can help you identify over-allocations and correct them. An over-allocation happens when the Assignment Units for a resource are greater than the Max Units.

To identify over-allocated resources:

1. **Click** View, Toolbars, and Resource Management.
2. **Select** the Resource Allocation View icon on the Resource Management toolbar.
3. **Select** the Go to Next Over-allocation icon on the Resource Management toolbar.
4. **Edit** the resource allocation so there are no conflicts. Use the Help feature for specific assistance.

To view and edit resources graphically:

1. **Click** View, Task Usage.
2. **Click** Window, Split to show the lower pane with the Task Form.
3. **Select** any field in the Task Form.
4. **Click** View, Resource Graph.

Customizing and Formatting

Project data is much easier to understand and use if it is presented in a friendly format with terminology that you commonly use. Fortunately, Microsoft Project is customizable. You can create custom fields, create values and set them to defaults, enter custom field formulas, import custom fields, redefine the graphical indicators, and make other customizations.

To create custom fields (Figure 2-35):

1. **Click** Tools, Customize, Fields...

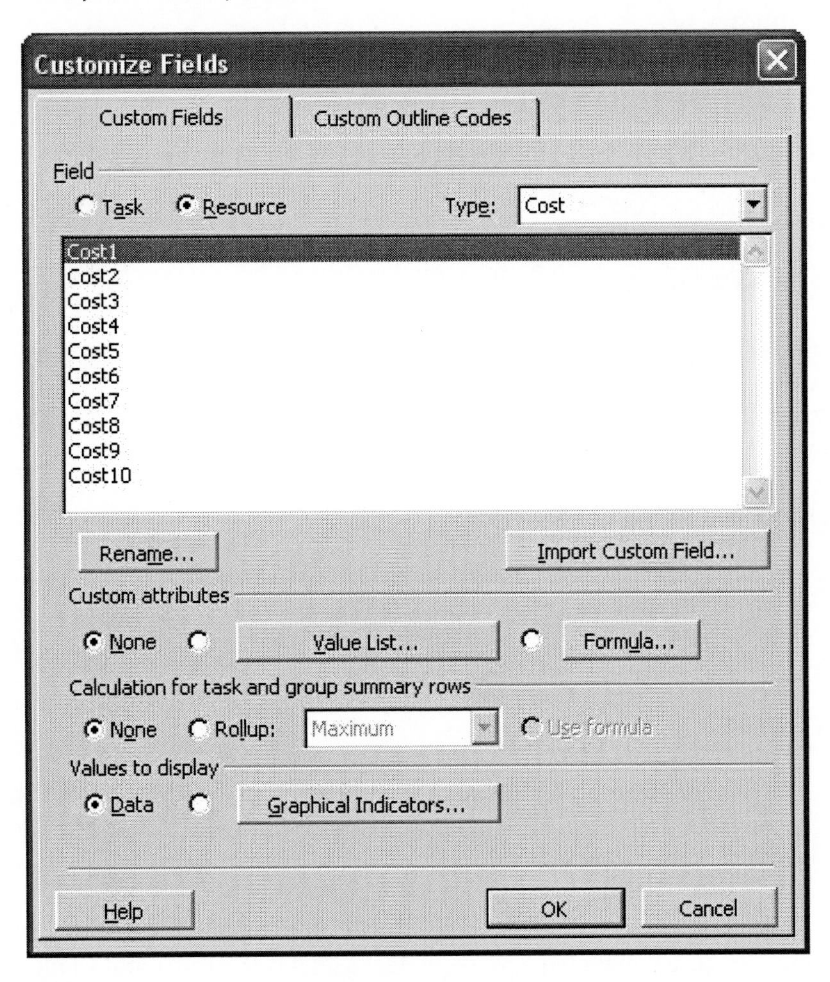

2-35: Customize Fields

2. **Select** the Custom Fields tab.
3. **Select** the field to customize: Task or Resource.
4. **Select** the field Type: Cost, Date, Duration, Finish, Flag, Number, Start or Text.
5. **Rename** the Field.
6. **Select** Custom attributes from the Value List or Formula.
7. **Select** Calculation for task and group summary rows: Rollup or Use Formula.
8. **Select** Values to Display: Data or Graphical Indicators.

It is recommended that you don't attempt to customize fields until you have developed experience with Microsoft Project and know better what you want the program to do for you. You can then follow the Microsoft Project Help (F1) screens to learn more about customizing fields. For now, be aware that Project is quite customizable. You also can customize how you see project data, which is called views. You can add and remove fields in existing views, modify tables, create data filters that limit what you see, group data, build bar charts, and customize text. As with customizing fields, these tools are for advanced and experienced Project users, but you should be aware of what you can do when ready.

To create new tables:

1. **Click** View, Table, More Tables.
2. **Select** Task for a new task table or Resource for a new resource table.
3. **Select** the type of table.
4. **Click** the New button.
5. **Enter** a table name.
6. **Define** at least one column for the table. You may add, remove, and rename columns as desired.
7. **Select** data and format for the table.
8. **Select** Organizer to define where the table will appear in views and what it will look like.

Tracking Progress

The primary purpose of learning and using a scheduling tool like Microsoft Project is to be able to track progress on specific jobs and make sure everything is running as efficiently -- and profitably -- as possible. Schedulers call this following the critical path.

A task is critical to a project if missing its finish date causes the entire project to slip. The critical path is the 'longest connected' chain of tasks that are all critical. Tracking job progress typically means tracking its critical path and making sure that it continues to stay on schedule. Microsoft Project can help you do this.

To view job's critical path:

1. **Click** View, Tracking Gantt.
2. **View** the job's critical path, in red.

A baseline is a view of several task fields in time. You can set a baseline at the beginning of a project and then compare it to your most current baseline. A baseline can include the Cost, Start, Finish, Duration, and Work fields.

To save a baseline:

1. **Click** Tools, Tracking, Save Baseline.
2. **Select** Baseline.
3. **Choose** Entire Project.
4. **Click** OK.

You also can select specific tasks such as plumbing and save the baseline for them by replacing Step 3 with Choose Selected Tasks and identifying the tasks you wish to include in the limited baseline.

To view and compare baselines: Click View, Tracking Gantt.

To view and compare multiple Gantt charts:

1. **Click** View, More Views.
2. **Select** Multiple Baseline Gantt.

To manage projects, you need to periodically update the status of tasks in Microsoft Project.

To update task status:

1. **Click** View, Table, and Tracking.
2. **Select** the task to be updated,
3. **Enter** the task's Actual Start or Actual Finish date if different than the estimated date.
4. **Revise** the Percent Complete as appropriate.
5. **Continue** updating the status of other tasks on the project.

You also can use the Update Tasks button on the Tracking toolbar.

If jobs get behind schedule (as they often do), you can move unfinished tasks to start after the status date, called rescheduling remaining work. Once you have updated the project status date you can use the Reschedule Work button on the Tracking toolbar to update the work schedule.

Reporting on Projects

Depending on the size of your business, the number of jobs you schedule, and the number of clients and others you need to report to, the reporting process can be relatively easy or quite difficult. Microsoft Project can make the job easier no matter what.

Reporting on jobs simply means selecting and printing views or snapshots as reports. First you look at the view on the screen to make sure the columns are of the best width to view the data. Columns can be resized with the mouse. Then printing options are set up to print the report on paper. You can see what the report will look like on paper by selecting File, Print Preview. Pages that won't quite fit on the page can be scaled using Page Setup on the File or Print menu.

Report pages can have headers and footers added by selecting File, Page Setup.

Reports are easier to read if the Gantt chart symbols, colors, and shades are clarified. You can include a legend on the reports by selecting File, Page Setup, Legend tab, Legend on.

Alternately, you can save the report as a Project file (.MPP), template (.MPT), database (.MPD), web page (.HTML), Microsoft Access database (.MDB), Microsoft Excel spreadsheet (.XLS), tab-delimited text (.TXT), comma-delimited text (.CSV), or XML (.XML) format. The report file can then be saved to a disk or emailed as an attachment to the client, supplier, and subcontractor.

Chapter Three

Other Scheduling Systems

Using Other Scheduling Systems

Not everyone needs or wants to use Microsoft Project. Some prefer software that is written specifically for construction scheduling. Others would rather use estimating software that has a scheduling component. Still others want a program that can handle all primary business tasks for contractors: estimating, scheduling, documentation, management, and finances. A few prefer the manual method.

It's a good thing that you have choices. This chapter offers an overview of those choices, showing you what other software programs can do for you. You'll then have the knowledge to make the wisest investment of time, money, and effort.

Tip: The software market is constantly changing. Today's products will be updated or outmoded within a few years. Select systems from reputable developers, get adequate training, and keep the systems up-to-date.

Selecting Scheduling Software

There are various software programs that are designed and developed specifically for construction and remodeling job scheduling. The primary reason to choose one of these instead of Microsoft Project is that they have features that are more useful to contractors than the generic Project program offers. Though aftermarket construction templates can furnish many of these tools, some users still prefer to work with a product that is designed for, and often by, contractors. This book does not recommend a specific program but offers you an overview so you can decide which should be considered further. It also gives you contact information for these products.

VirtualBoss

VirtualBoss (Figures 3-1 through 3-4) is a job scheduling and task management program designed for contractors by contractors. It operates on any Windows-based PC including desktops, laptops, and Pocket PCs, both as a stand-alone system or on a computer network. The various tasks that *VirtualBoss* performs include scheduling jobs, managing and tracking service calls, managing and tracking punch lists, organizing contacts (customers, contractors, suppliers, etc.), and tracking worker's compensation and liability insurance. It is designed to let you know the exact status of various projects and task lists at any time.

VirtualBoss follows critical path management (CPM) methods, allowing lag and lead times within schedules. The developers say that the learning curve on their product is 1 to 2 hours. Schedules and related documents can be quickly emailed or faxed to job participants as needed.

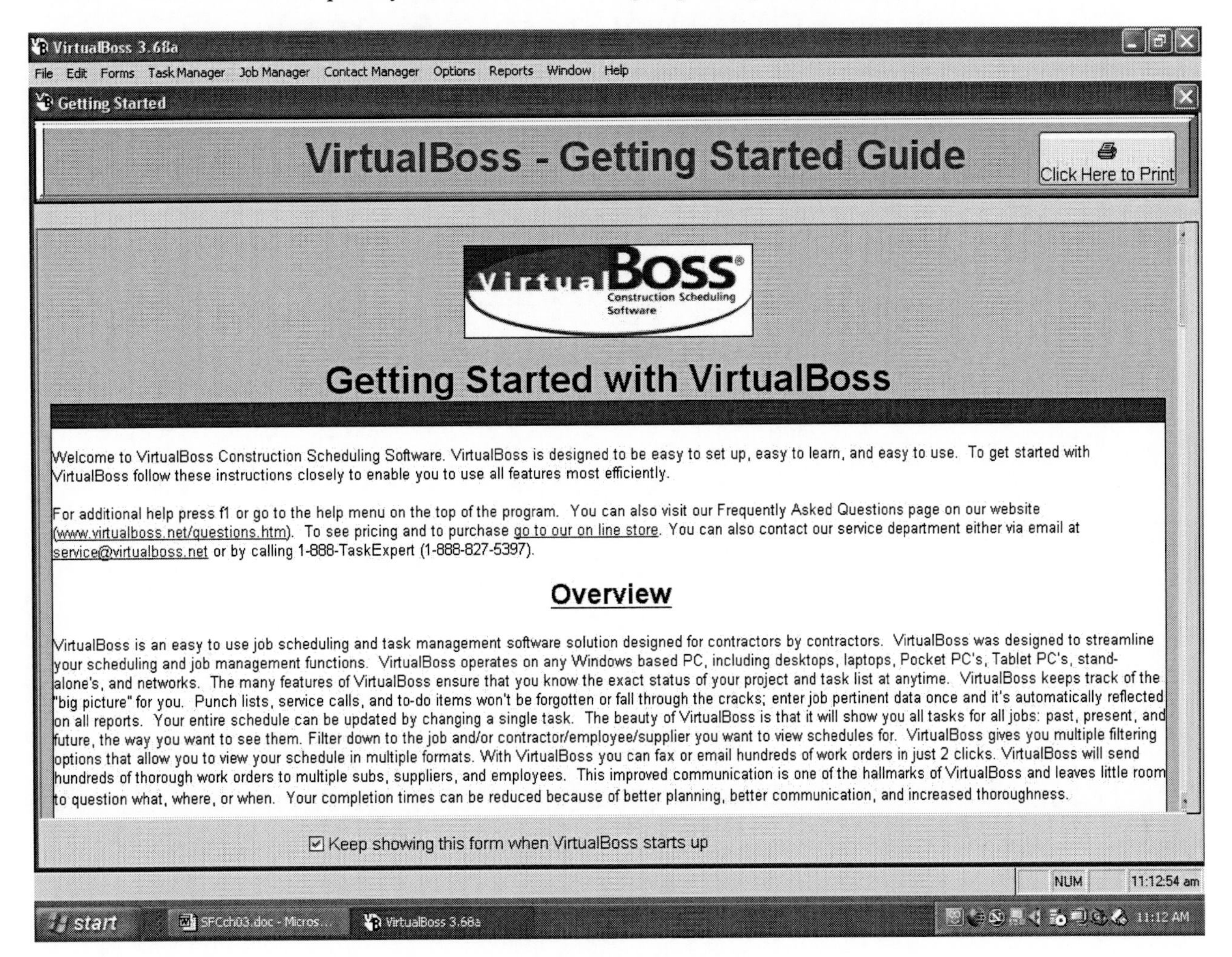

3-1: VirtualBoss Overview

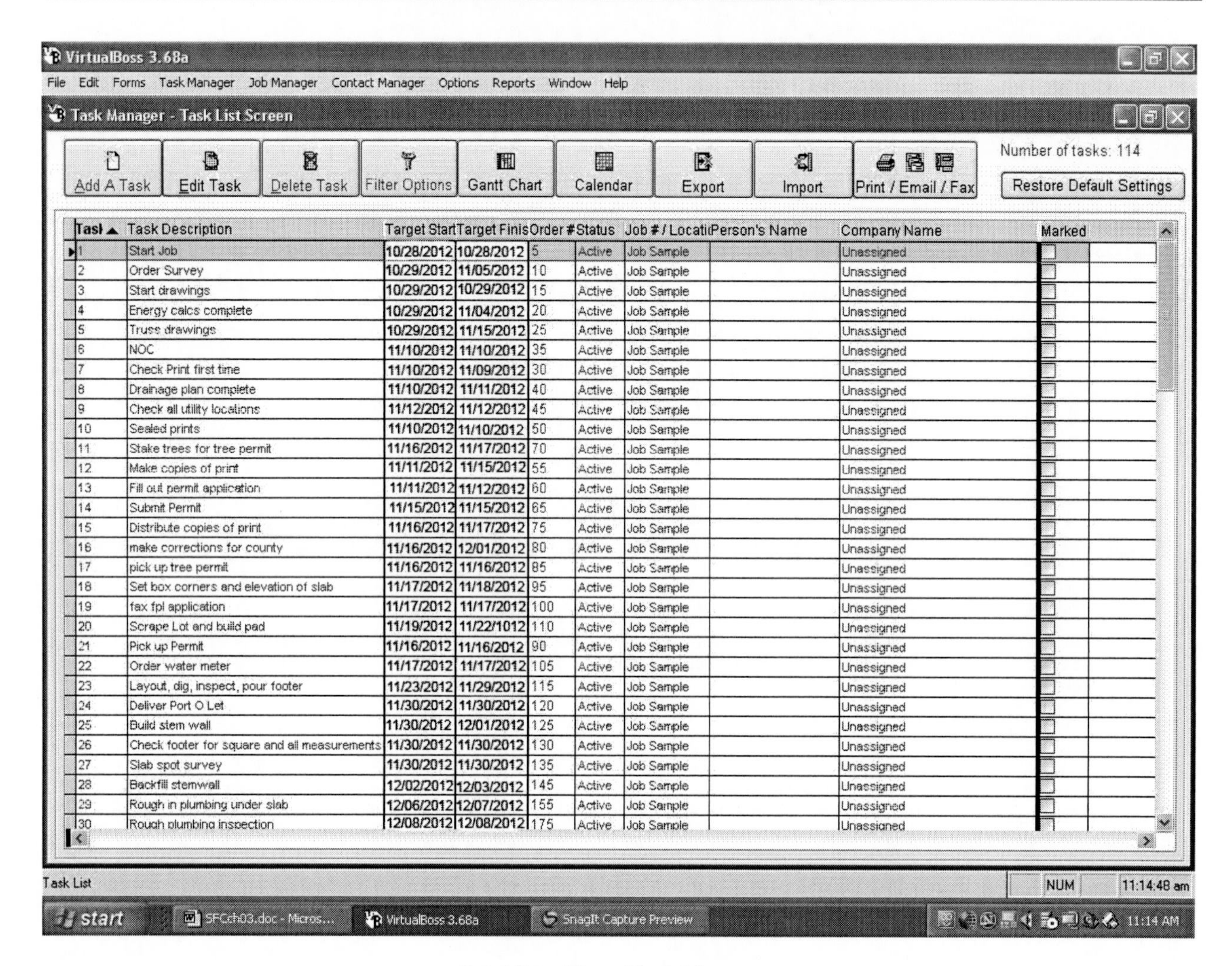

Tasl	Task Description	Target Start	Target Finis	Order #	Status	Job # / Locati	Person's Name	Company Name	Marked
1	Start Job	10/28/2012	10/28/2012	5	Active	Job Sample		Unassigned	
2	Order Survey	10/29/2012	11/05/2012	10	Active	Job Sample		Unassigned	
3	Start drawings	10/29/2012	10/29/2012	15	Active	Job Sample		Unassigned	
4	Energy calcs complete	10/29/2012	11/04/2012	20	Active	Job Sample		Unassigned	
5	Truss drawings	10/29/2012	11/15/2012	25	Active	Job Sample		Unassigned	
6	NOC	11/10/2012	11/10/2012	35	Active	Job Sample		Unassigned	
7	Check Print first time	11/10/2012	11/09/2012	30	Active	Job Sample		Unassigned	
8	Drainage plan complete	11/10/2012	11/11/2012	40	Active	Job Sample		Unassigned	
9	Check all utility locations	11/12/2012	11/12/2012	45	Active	Job Sample		Unassigned	
10	Sealed prints	11/10/2012	11/10/2012	50	Active	Job Sample		Unassigned	
11	Stake trees for tree permit	11/16/2012	11/17/2012	70	Active	Job Sample		Unassigned	
12	Make copies of print	11/11/2012	11/15/2012	55	Active	Job Sample		Unassigned	
13	Fill out permit application	11/11/2012	11/12/2012	60	Active	Job Sample		Unassigned	
14	Submit Permit	11/15/2012	11/15/2012	65	Active	Job Sample		Unassigned	
15	Distribute copies of print	11/16/2012	11/17/2012	75	Active	Job Sample		Unassigned	
16	make corrections for county	11/16/2012	12/01/2012	80	Active	Job Sample		Unassigned	
17	pick up tree permit	11/16/2012	11/16/2012	85	Active	Job Sample		Unassigned	
18	Set box corners and elevation of slab	11/17/2012	11/18/2012	95	Active	Job Sample		Unassigned	
19	fax fpl application	11/17/2012	11/17/2012	100	Active	Job Sample		Unassigned	
20	Scrape Lot and build pad	11/19/2012	11/22/1012	110	Active	Job Sample		Unassigned	
21	Pick up Permit	11/16/2012	11/16/2012	90	Active	Job Sample		Unassigned	
22	Order water meter	11/17/2012	11/17/2012	105	Active	Job Sample		Unassigned	
23	Layout, dig, inspect, pour footer	11/23/2012	11/29/2012	115	Active	Job Sample		Unassigned	
24	Deliver Port O Let	11/30/2012	11/30/2012	120	Active	Job Sample		Unassigned	
25	Build stem wall	11/30/2012	12/01/2012	125	Active	Job Sample		Unassigned	
26	Check footer for square and all measurements	11/30/2012	11/30/2012	130	Active	Job Sample		Unassigned	
27	Slab spot survey	11/30/2012	11/30/2012	135	Active	Job Sample		Unassigned	
28	Backfill stemwall	12/02/2012	12/03/2012	145	Active	Job Sample		Unassigned	
29	Rough in plumbing under slab	12/06/2012	12/07/2012	155	Active	Job Sample		Unassigned	
30	Rough plumbing inspection	12/08/2012	12/08/2012	175	Active	Job Sample		Unassigned	

3-2: VirtualBoss Task Manager

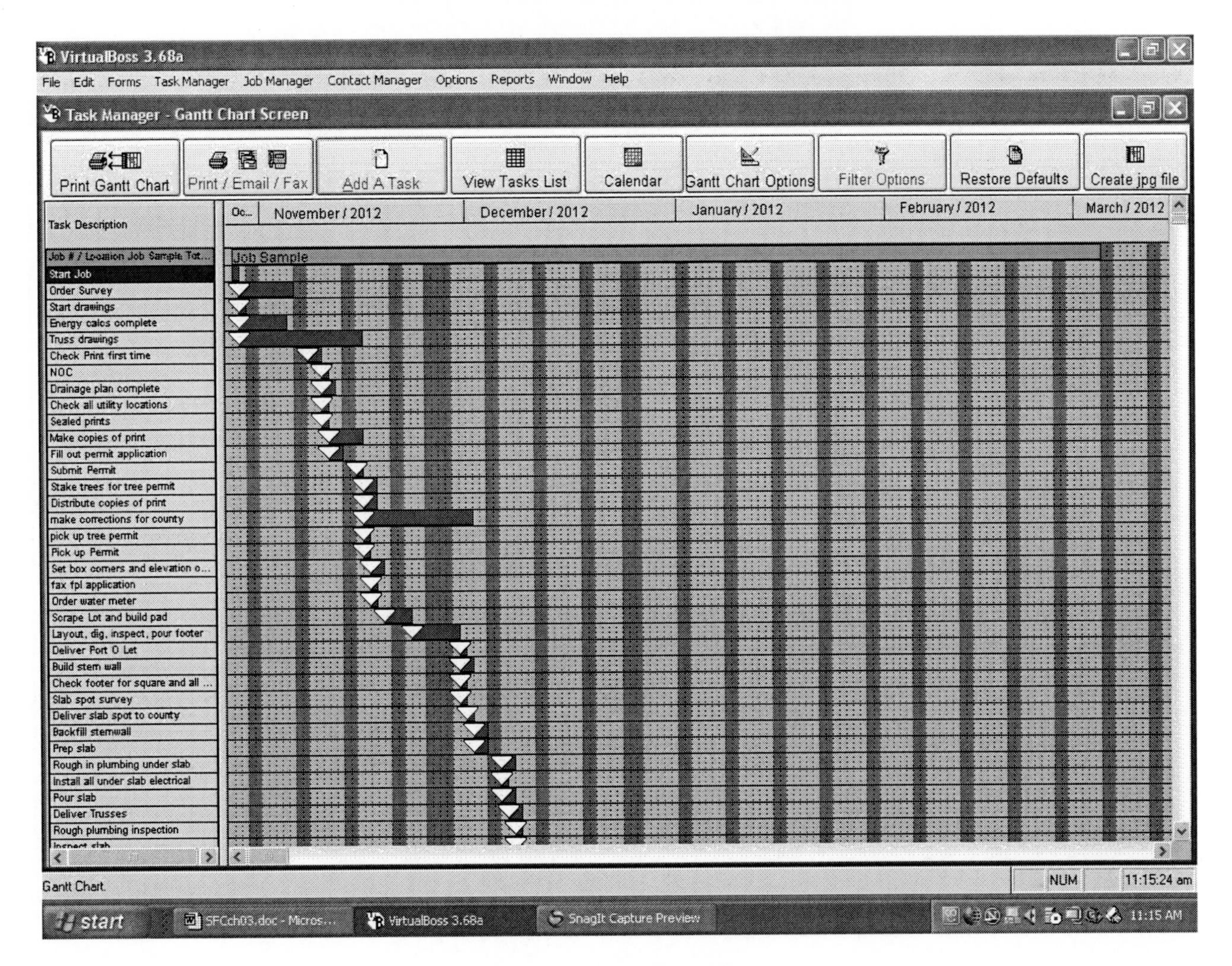

3-3: VirtualBoss Gantt Chart

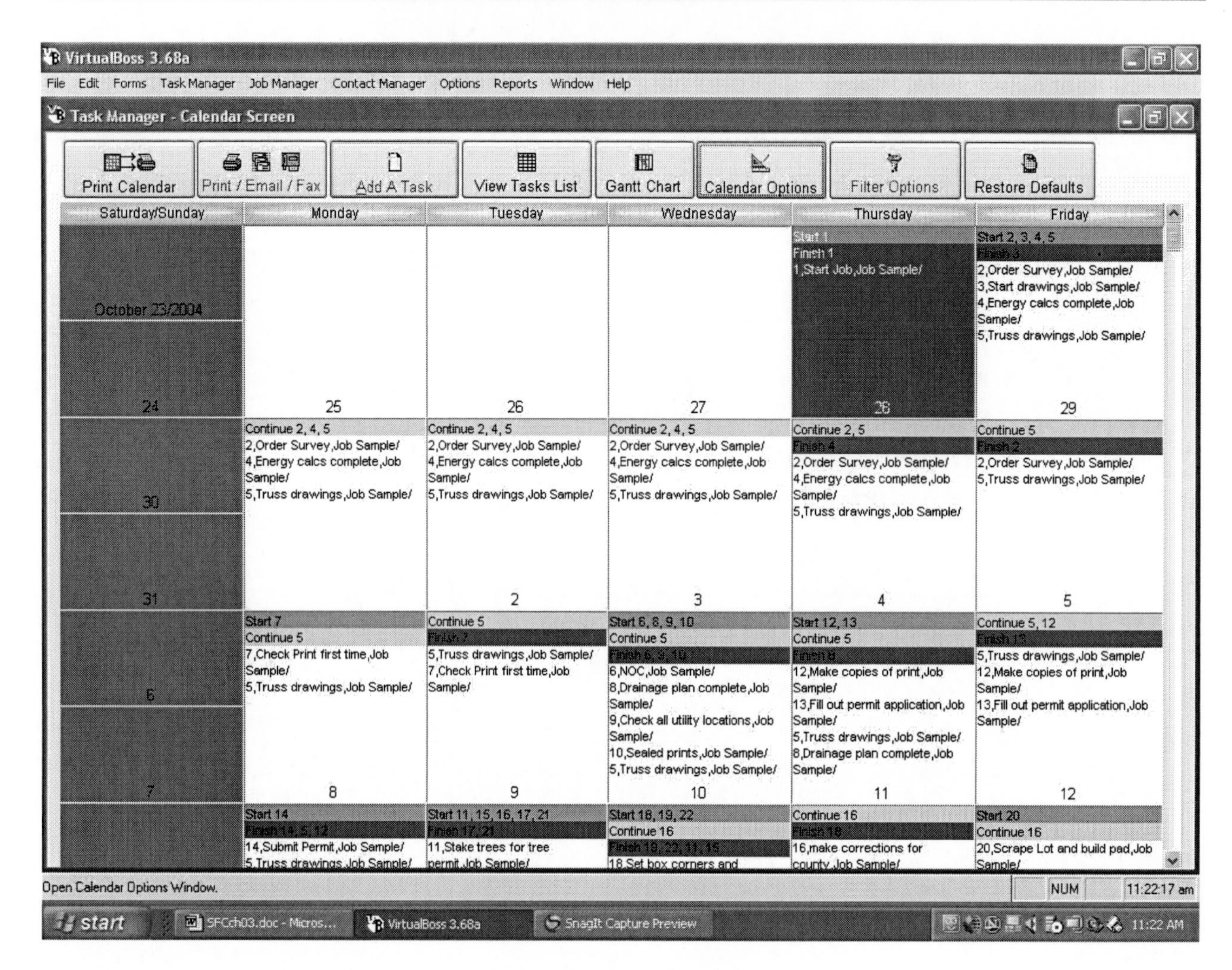

3-4: VirtualBoss Calendar

VirtualBoss builds schedules from which it tracks to-do lists, punch list items, service items, scheduling items, and other components. First, enter all contact information in the Contact Manager using the New button and the five tabs: Address, Profile, Notes, Work Order Distribution, and Captions. Next, enter the specific job into the Job Manager. Then you will assign tasks to the job using the Task Manager, including job number and location, client information, status, target start date, and notes.

One advantage *VirtualBoss* has over Microsoft Project is that it allows you to keep all your jobs in a single database rather than in separate files. In addition, *VirtualBoss* allows you to print, fax, or email work orders and schedules directly from the program to keep others updated.

According to *VirtualBoss,* there are more than 18,000 of their programs installed worldwide. A free demo version is available online. A single workstation (1 PC) license is about $400 and upgrades are $200 annually, including ongoing technical support. An add-on is available to import/export with QuickBooks accounting software for $100. Network versions are also available. Contact *VirtualBoss* Software (308 Louisiana Ave., Perrysburg OH 43551, 888-827-5397; www.virtualboss.net) for additional information and pricing.

Primavera

Primavera Contractor (Figures 3-5 through 3-7) is a planning and scheduling program designed specifically for the construction industry. It also makes versions for other industries including engineering, financial, power and energy, high tech, aerospace and defense, and consulting. This means you can easily share files with engineers, subs, and others who also use Primavera software products.

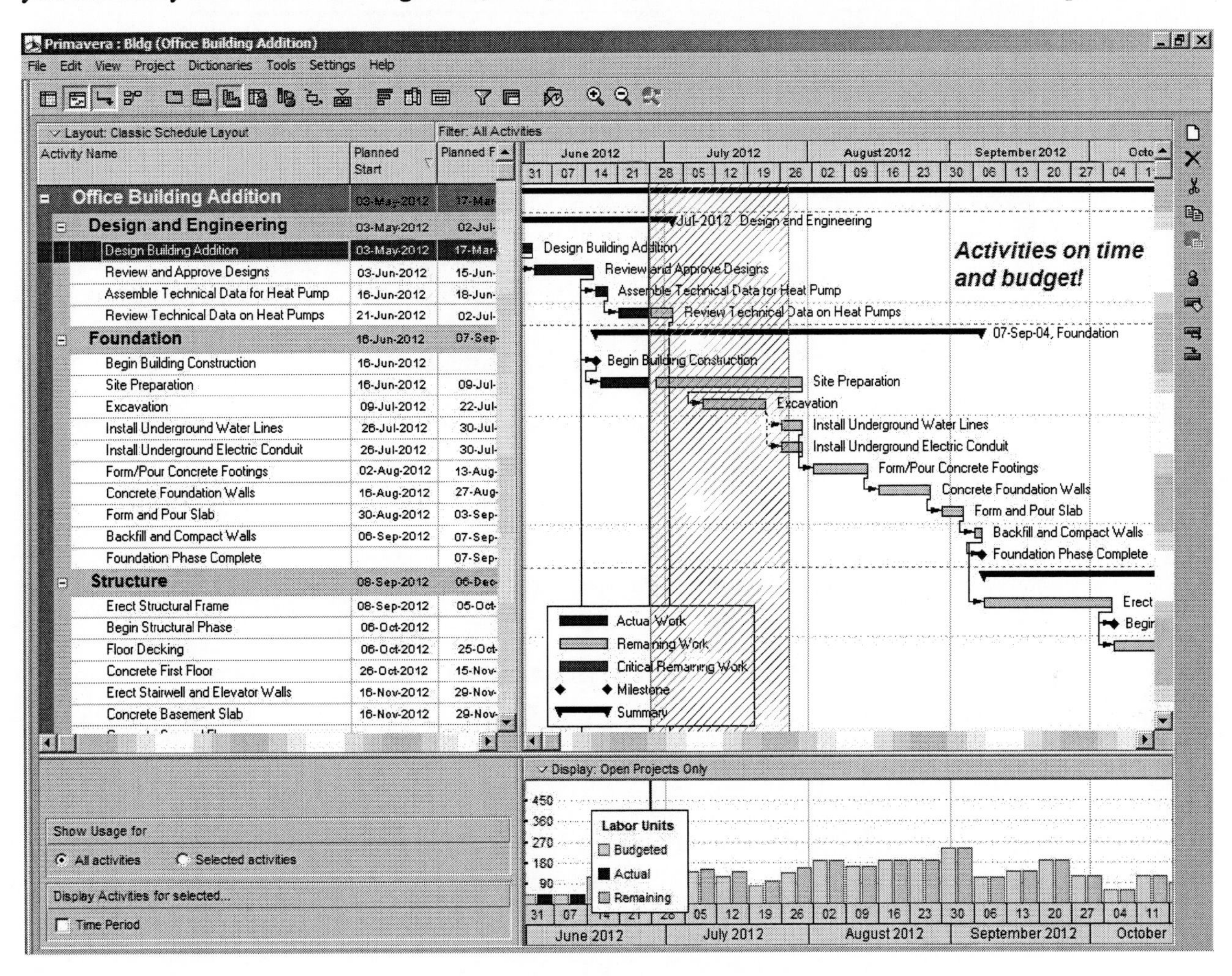

3-5: Primavera Classic Schedule Layout

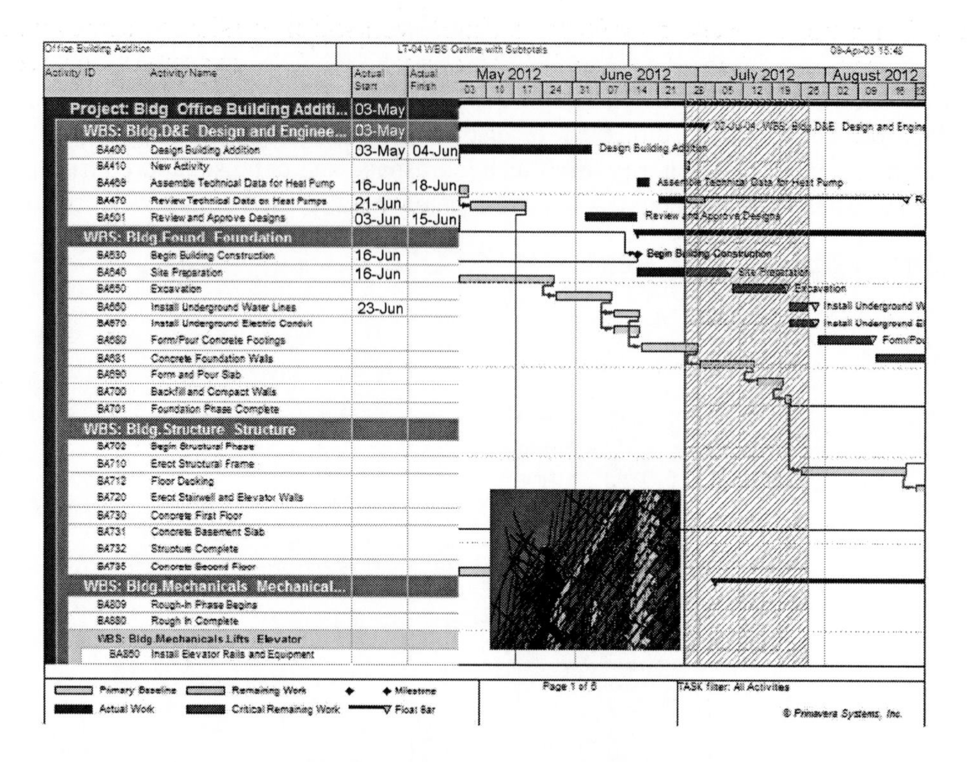

3-6: Primavera Detailed Schedule

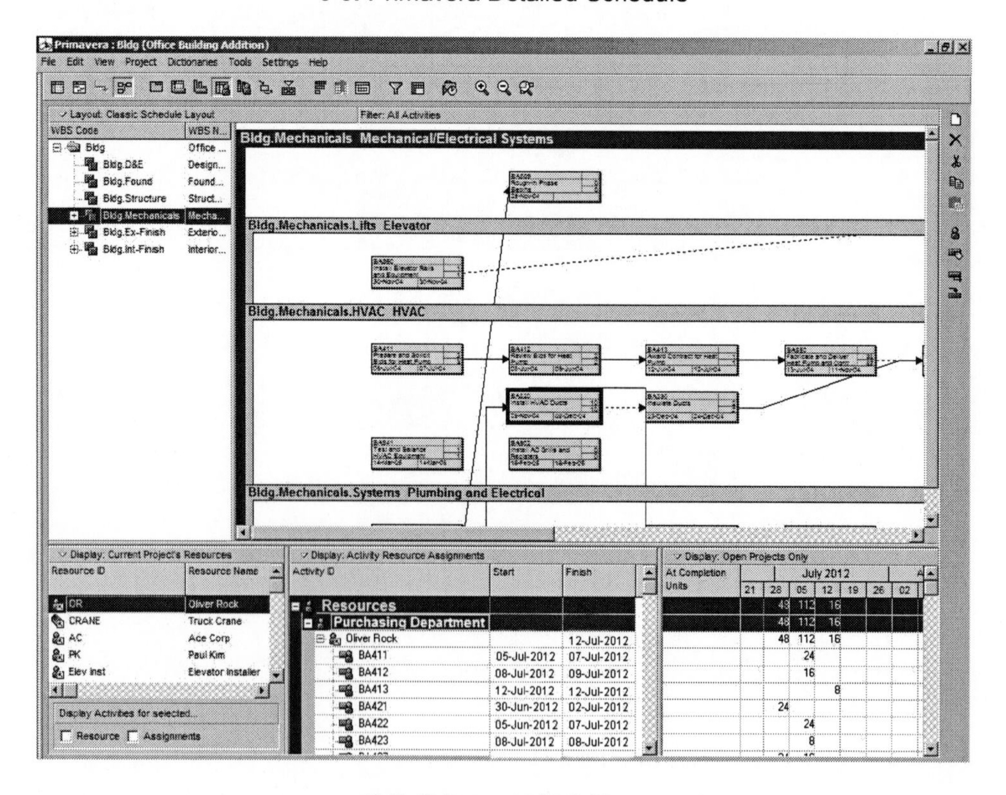

3-7: Primavera Activities

Primavera Contractor is priced at $500 for a single workstation version. It comes with 90 days of maintenance and support; additional support is $140 a year or $300 for two years. Contact Primavera (3 Bala Plaza West, Suite 700, Bala Cynwyd PA 19004; www.primavera.com) for additional information and pricing. Primavera also offers SureTrak Project Manager software and other technology tools. For additional "construction scheduling software" use www.google.com or www.a9.com to search the Internet.

Opting for Integrated Software

There are also software programs that offer many related construction functions, including scheduling, in an integrated system. The advantage of these is that data is shared between the programs making it easier to look at the entire project, not just the schedule. If you are planning to buy and learn an estimating program anyway, it makes sense to consider an integrated software system for contractors. The disadvantages of integrated software are price and training. You may find that an integrated one-system-does-it-all program has more features than you may need. In addition, the initial training on the system may require days or even weeks rather than hours. Let's take a look at some of the more popular integrated software systems for small contractor businesses.

Construction Office

Construction Office is an integrated software package produced by UDA Technologies. There are three products available in stand-alone and network versions. The Builder/Remodeler version includes estimating, contracts, residential specifications, document management, mobile (Pocket PC) integration, and add-ons for scheduling, QuickBooks integration, and Microsoft Project integration. There also is a Professional version with scheduling included for integrated project management. A Developer edition offers tools for property developers. Construction Office's scheduling component includes many of the same features as Microsoft Project (see Chapter 2) that are customized specifically to the construction and remodeling trades. Construction Office Builder/Remodeler is priced beginning at $300 and the Professional version starts at $500, each with add-ons available. Contact UDA Technologies (P.O. Box 3355, Auburn AL 36831, 800-700-8321; www.udatechnologies.com) for additional information and specific pricing.

The American Contractor

The American Contractor by Maxwell Systems is an extensive software program offering numerous integrated components for construction and remodeling contractors. The components include project management, job cost tracking, general ledger, financial reports, accounts payable and receivable, job invoice printing, work orders, service agreements, payroll, bid writing, subcontractor control, inventory, project scheduling, and many other functions.

American Contractor comes in three versions, each with increasingly more components and features: Express, Essential, and Enterprise. The project scheduling component is only included in the Essential and Enterprise products.

Because American Contractor has numerous components and configurations, pricing and support are quoted on specific systems. Contact The American Contractor (5200 Soquel Ave. Suite 201, Santa Cruz CA 95062, 800-333-8435) for additional information and pricing.

Goldenseal

Goldenseal (Figures 3-8 through 3-10) is a job estimating program that also produces job schedules and assists in project management. Though estimating isn't the topic of this book, it is crucial for many home construction and remodeling contractors who must bid competitively to stay in business.

Goldenseal offers numerous starting points for project estimating and management. You can set up prospects or customer accounts, estimates or assemblies, project accounts or logs, material accounts or purchases, employee accounts or labor hours, real estate accounts or leases, and checking accounts or transactions. For scheduling a job you will start with Project Account. However, you may need to set up other components that serve the schedule, such as Customer Accounts, Employee Accounts, and Material Accounts.

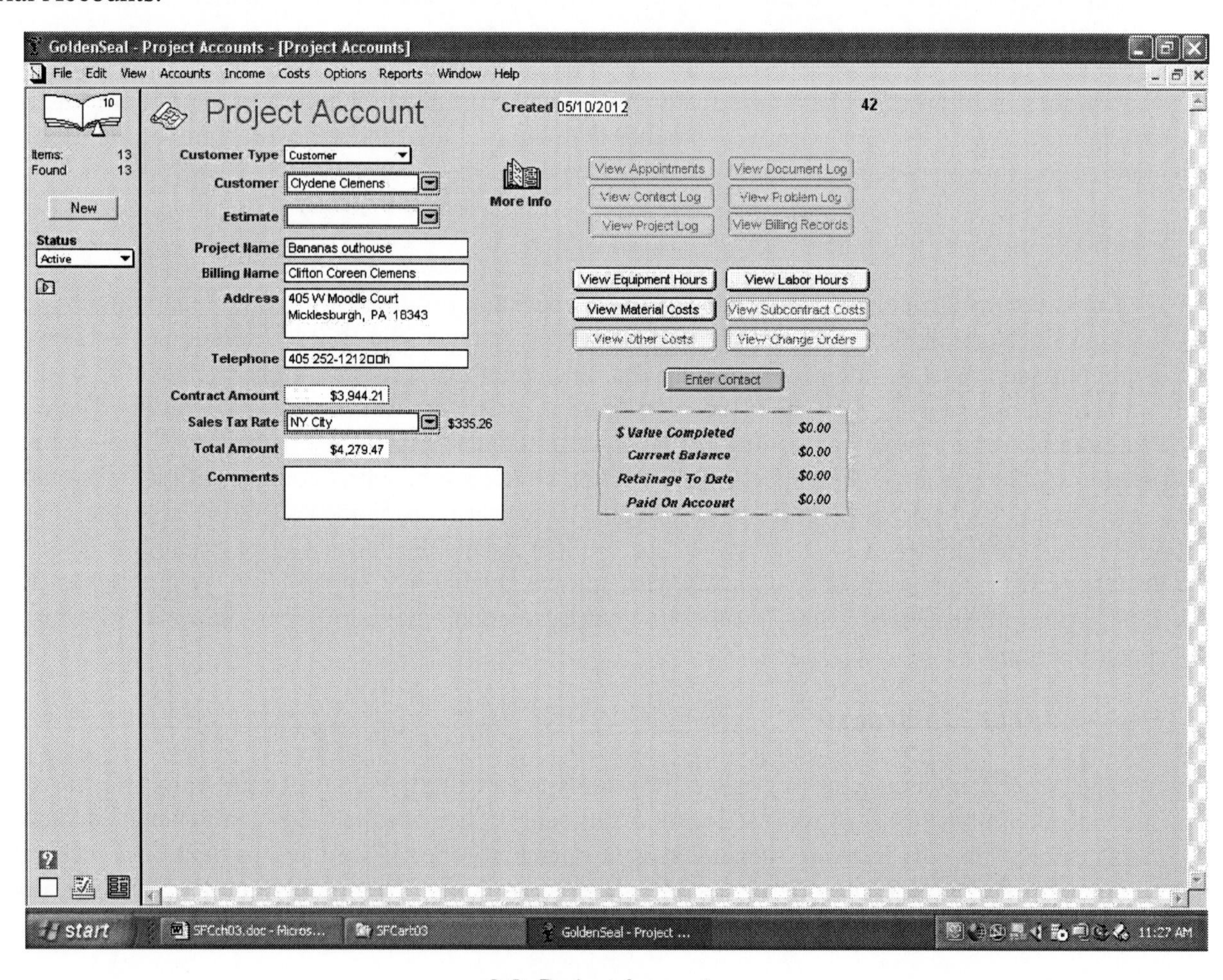

3-8: Project Account

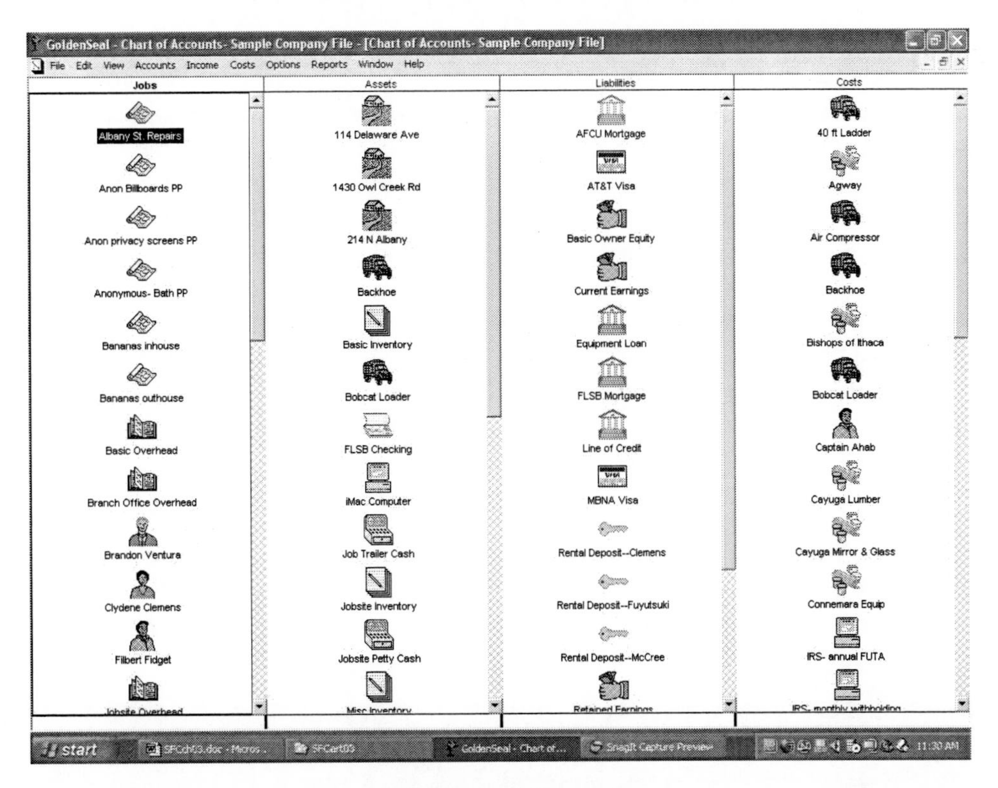

3-9: Chart of Accounts

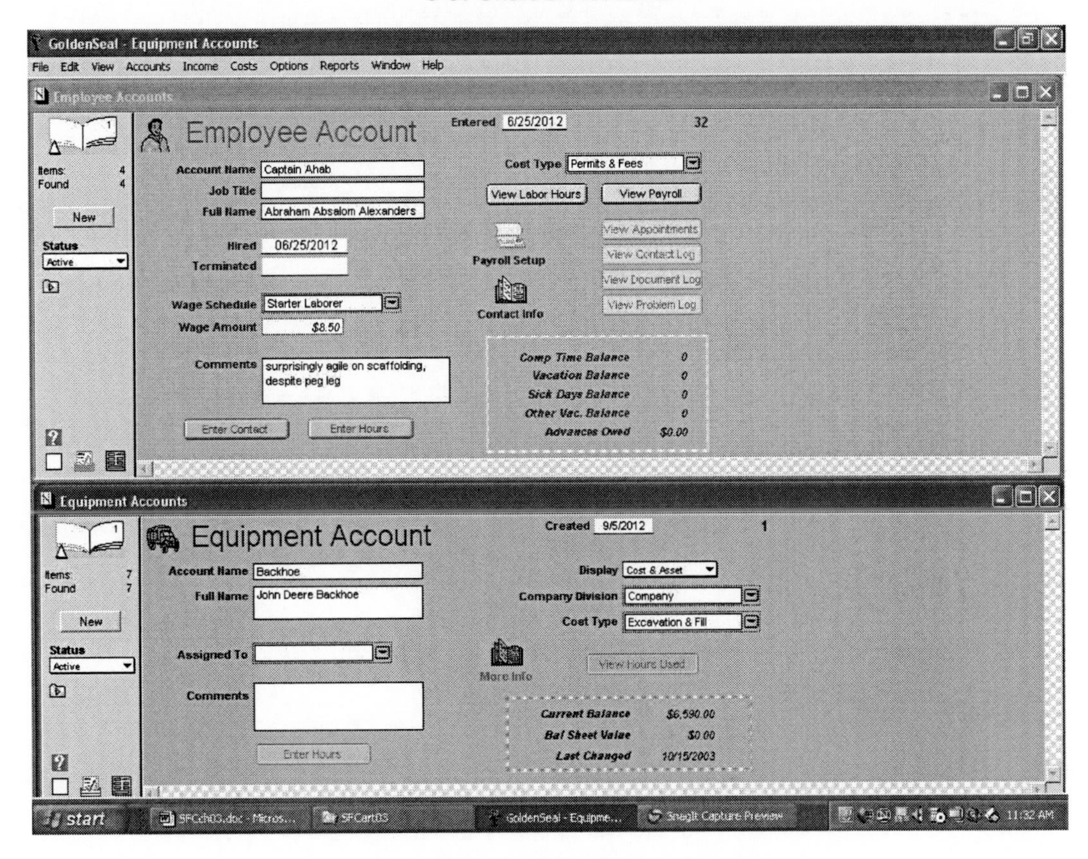

3-10: Employee and Equipment Accounts

A Project Account setup requires customer type (customer, equipment, investment, overhead, and real estate), customer name, estimate name, project name, billing name and address, contract amount, and comments. Once set up, you can view material costs, labor costs, and subcontractor costs from the Project Account. You can also change the project status: active, inactive, pending, completed, or closed. You can also select the job cost setup and the billing setup. To show a job schedule using Goldenseal, select Reports, Estimates, Schedule. Then select an estimate in the Estimate field and click on the Update button. You can change the dates, update, and produce a new schedule in seconds.

Goldenseal estimating software is priced at $300 and the complete system (with accounting, job costs, payroll, and project billing) is $500. Network systems are available at additional charge. It comes with a 180-page manual and some onscreen help. Contact Turtle Creek Software (116 S. Cayuga St., Ithaca NY 14850, 607-272-1008; www.turtlesoft.com) for additional information and pricing.

WinEstimator

WinEstimator (Figures 3-11 through 3-13), as the name implies, is an estimating tool written for Microsoft Windows operating system. Instead of having a scheduling module or function, WinEstimator integrates (imports and exports) with Microsoft Project (Chapter 2) and with Primavera (above) products.

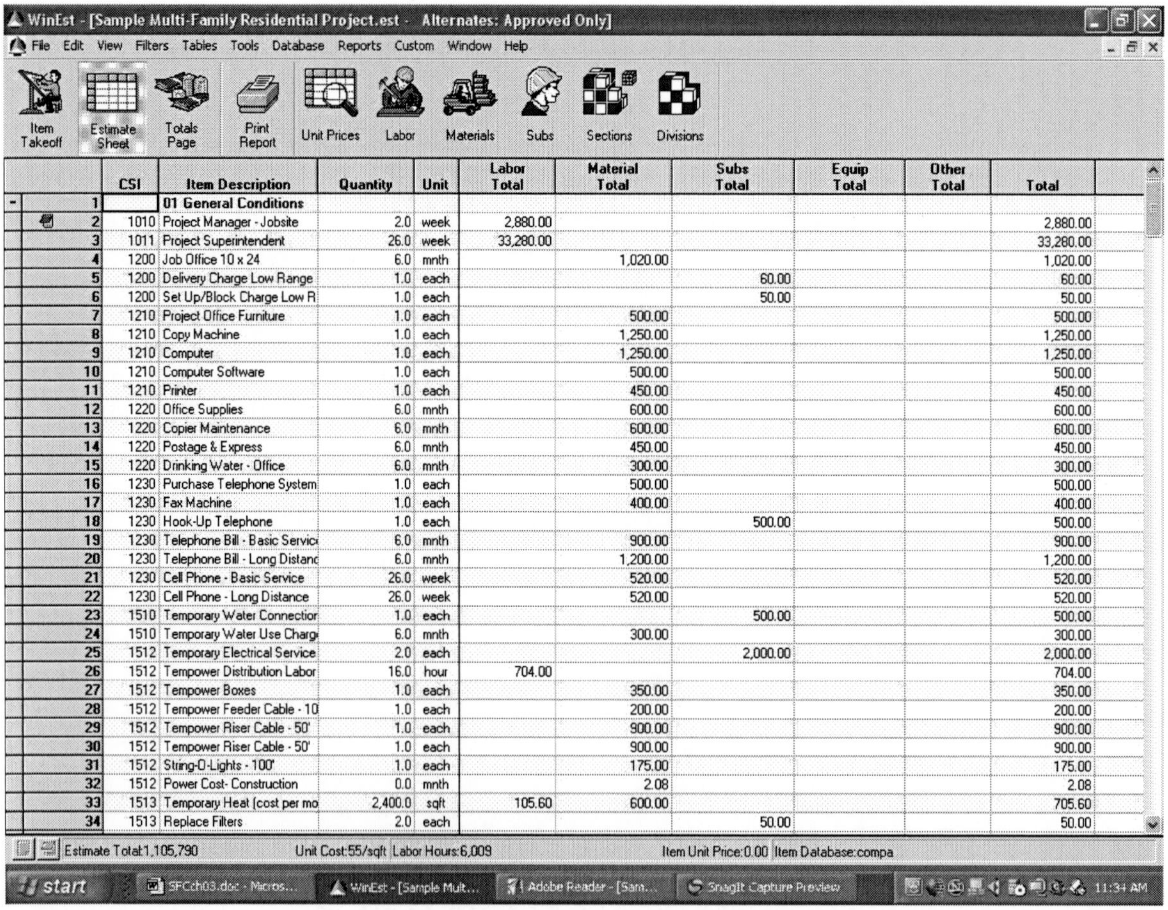

WinEst - [Sample Multi-Family Residential Project.est - Alternates: Approved Only]

File Edit View Filters Tables Tools Database Reports Custom Window Help

Item Takeoff | Estimate Sheet | Totals Page | Print Report | Unit Prices | Labor | Materials | Subs | Sections | Divisions

	CSI	Item Description	Quantity	Unit	Labor Total	Material Total	Subs Total	Equip Total	Other Total	Total
1		01 General Conditions								
2	1010	Project Manager - Jobsite	2.0	week	2,880.00					2,880.00
3	1011	Project Superintendent	26.0	week	33,280.00					33,280.00
4	1200	Job Office 10 x 24	6.0	mnth		1,020.00				1,020.00
5	1200	Delivery Charge Low Range	1.0	each			60.00			60.00
6	1200	Set Up/Block Charge Low R	1.0	each			50.00			50.00
7	1210	Project Office Furniture	1.0	each		500.00				500.00
8	1210	Copy Machine	1.0	each		1,250.00				1,250.00
9	1210	Computer	1.0	each		1,250.00				1,250.00
10	1210	Computer Software	1.0	each		500.00				500.00
11	1210	Printer	1.0	each		450.00				450.00
12	1220	Office Supplies	6.0	mnth		600.00				600.00
13	1220	Copier Maintenance	6.0	mnth		600.00				600.00
14	1220	Postage & Express	6.0	mnth		450.00				450.00
15	1220	Drinking Water - Office	6.0	mnth		300.00				300.00
16	1230	Purchase Telephone System	1.0	each		500.00				500.00
17	1230	Fax Machine	1.0	each		400.00				400.00
18	1230	Hook-Up Telephone	1.0	each			500.00			500.00
19	1230	Telephone Bill - Basic Servic	6.0	mnth		900.00				900.00
20	1230	Telephone Bill - Long Distanc	6.0	mnth		1,200.00				1,200.00
21	1230	Cell Phone - Basic Service	26.0	week		520.00				520.00
22	1230	Cell Phone - Long Distance	26.0	week		520.00				520.00
23	1510	Temporary Water Connectior	1.0	each			500.00			500.00
24	1510	Temporary Water Use Charg	6.0	mnth		300.00				300.00
25	1512	Temporary Electrical Service	2.0	each			2,000.00			2,000.00
26	1512	Tempower Distribution Labor	16.0	hour	704.00					704.00
27	1512	Tempower Boxes	1.0	each		350.00				350.00
28	1512	Tempower Feeder Cable - 10	1.0	each		200.00				200.00
29	1512	Tempower Riser Cable - 50'	1.0	each		900.00				900.00
30	1512	Tempower Riser Cable - 50'	1.0	each		900.00				900.00
31	1512	String-O-Lights - 100'	1.0	each		175.00				175.00
32	1512	Power Cost- Construction	0.0	mnth		2.08				2.08
33	1513	Temporary Heat (cost per mo	2,400.0	sqft	105.60	600.00				705.60
34	1513	Replace Filters	2.0	each			50.00			50.00

Estimate Total:1,105,790 | Unit Cost:55/sqft | Labor Hours:6,009 | Item Unit Price:0.00 | Item Database:compa

start | SFCch03.doc - Micros... | WinEst - [Sample Mult... | Adobe Reader - [Sam... | SnagIt Capture Preview | 11:34 AM

3-11: WinEstimator Estimate Sheet

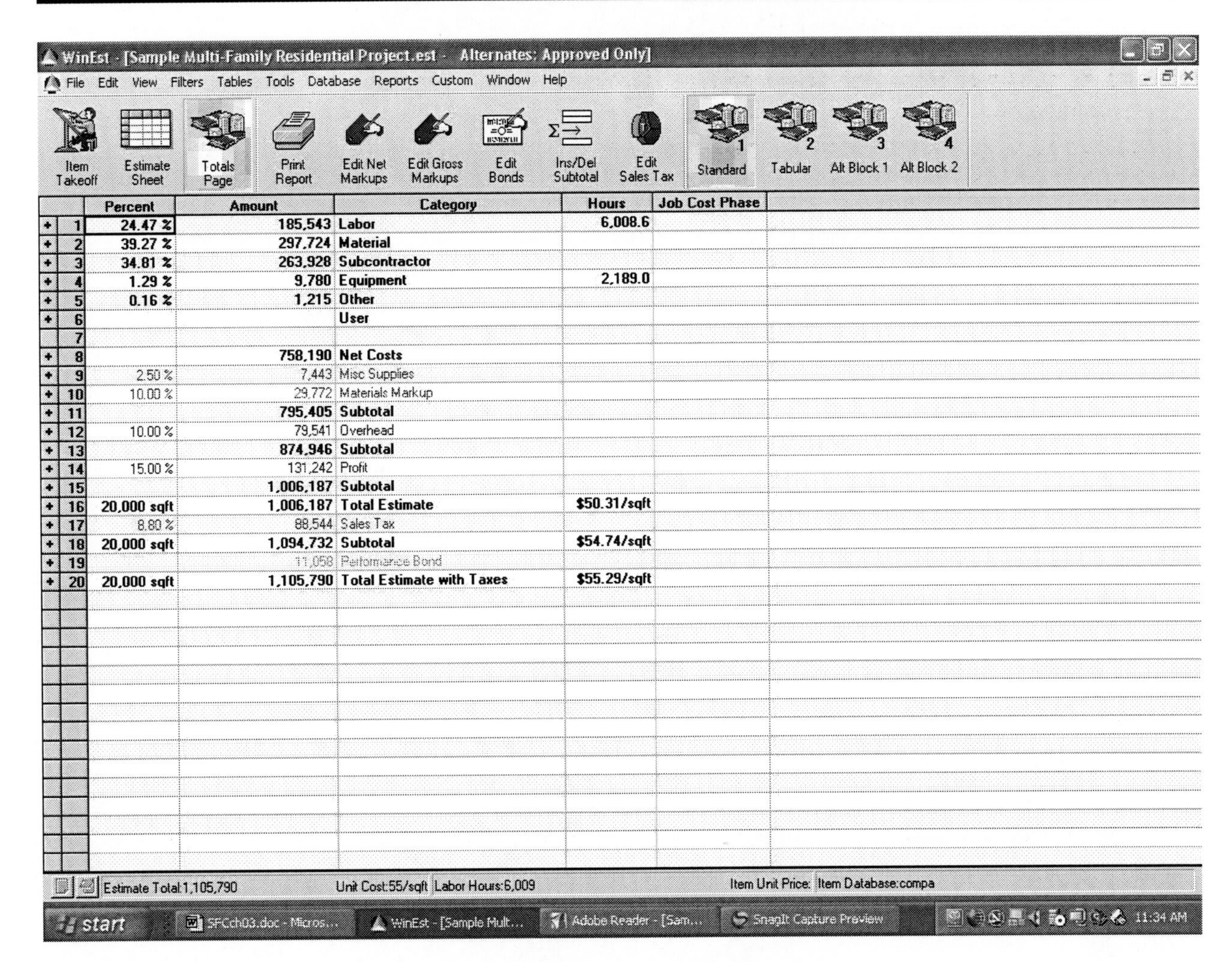

	Percent	Amount	Category	Hours	Job Cost Phase
1	24.47 %	185,543	Labor	6,008.6	
2	39.27 %	297,724	Material		
3	34.81 %	263,928	Subcontractor		
4	1.29 %	9,780	Equipment	2,189.0	
5	0.16 %	1,215	Other		
6			User		
7					
8		758,190	Net Costs		
9	2.50 %	7,443	Misc Supplies		
10	10.00 %	29,772	Materials Markup		
11		795,405	Subtotal		
12	10.00 %	79,541	Overhead		
13		874,946	Subtotal		
14	15.00 %	131,242	Profit		
15		1,006,187	Subtotal		
16	20,000 sqft	1,006,187	Total Estimate	$50.31/sqft	
17	8.80 %	88,544	Sales Tax		
18	20,000 sqft	1,094,732	Subtotal	$54.74/sqft	
19		11,058	Performance Bond		
20	20,000 sqft	1,105,790	Total Estimate with Taxes	$55.29/sqft	

3-12: WinEstimator Totals Page

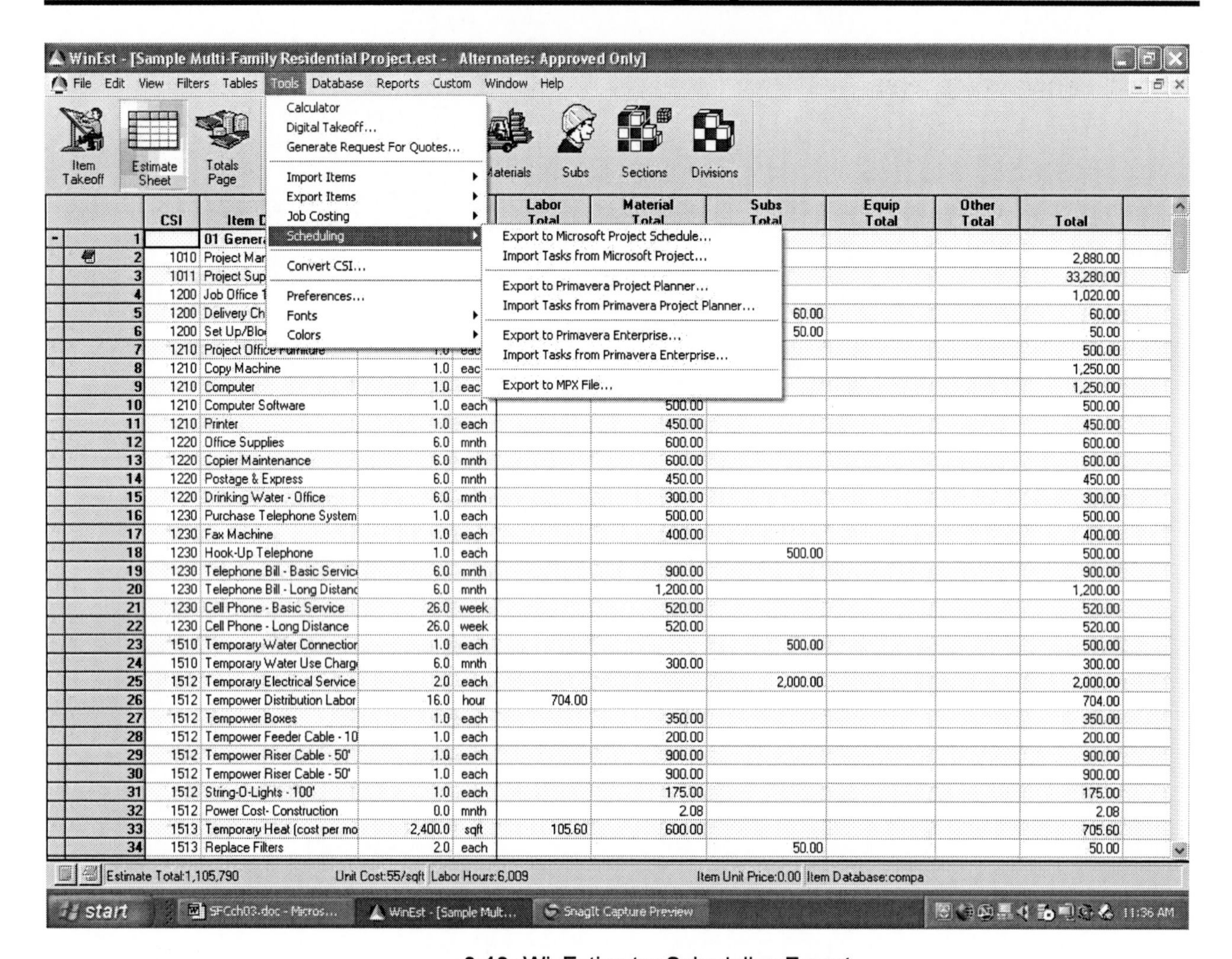

3-13: WinEstimator Scheduling Export

Contact WinEstimator Inc. (P.O. Box 1208, Kent WA 98035, 800-950-2374, www.winest.com) for additional information, compatibilities, and pricing.

For additional "construction software" use www.google.com or www.a9.com to search the Internet.

National Estimators

Craftsman Books produces and publishes National Estimator books annually. Though not specific to scheduling, they offer powerful tools for computerizing one of the most difficult tasks that contractors face: estimating project costs and time. Each book includes tables for quickly estimating labor and materials for common projects. In addition, each book includes a CD-Rom with all data from the book to develop, print, and export accurate construction and remodeling estimates.

National Construction Estimator covers labor and material costs, man-hours, and city cost adjustments for all residential, commercial, and industrial construction.

National Home Improvement Estimator (Figure 3-14) includes labor costs, material costs, city cost modifiers, and selling prices for all home improvement projects.

National Repair & Remodeling Estimator covers current labor and material prices for all residential repair and remodeling work.

National Renovation & Insurance Repair Estimator includes labor and material costs for all insurance repair and renovation work.

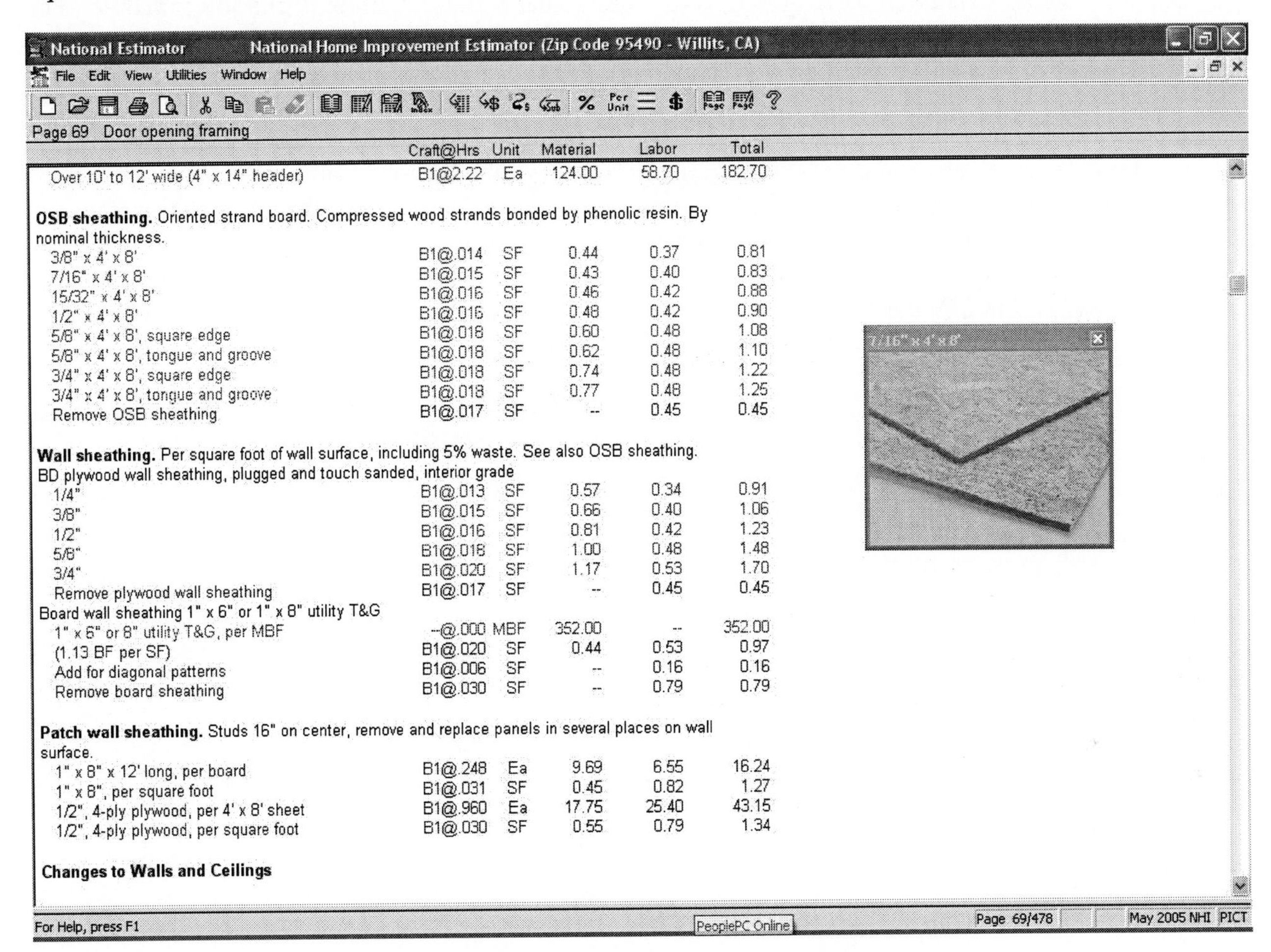

	Craft@Hrs	Unit	Material	Labor	Total
Over 10' to 12' wide (4" x 14" header)	B1@2.22	Ea	124.00	58.70	182.70
OSB sheathing. Oriented strand board. Compressed wood strands bonded by phenolic resin. By nominal thickness.					
3/8" x 4' x 8'	B1@.014	SF	0.44	0.37	0.81
7/16" x 4' x 8'	B1@.015	SF	0.43	0.40	0.83
15/32" x 4' x 8'	B1@.016	SF	0.46	0.42	0.88
1/2" x 4' x 8'	B1@.016	SF	0.48	0.42	0.90
5/8" x 4' x 8', square edge	B1@.018	SF	0.60	0.48	1.08
5/8" x 4' x 8', tongue and groove	B1@.018	SF	0.62	0.48	1.10
3/4" x 4' x 8', square edge	B1@.018	SF	0.74	0.48	1.22
3/4" x 4' x 8', tongue and groove	B1@.018	SF	0.77	0.48	1.25
Remove OSB sheathing	B1@.017	SF	--	0.45	0.45
Wall sheathing. Per square foot of wall surface, including 5% waste. See also OSB sheathing.					
BD plywood wall sheathing, plugged and touch sanded, interior grade					
1/4"	B1@.013	SF	0.57	0.34	0.91
3/8"	B1@.015	SF	0.66	0.40	1.06
1/2"	B1@.016	SF	0.81	0.42	1.23
5/8"	B1@.018	SF	1.00	0.48	1.48
3/4"	B1@.020	SF	1.17	0.53	1.70
Remove plywood wall sheathing	B1@.017	SF	--	0.45	0.45
Board wall sheathing 1" x 6" or 1" x 8" utility T&G					
1" x 6" or 8" utility T&G, per MBF	--@.000	MBF	352.00	--	352.00
(1.13 BF per SF)	B1@.020	SF	0.44	0.53	0.97
Add for diagonal patterns	B1@.006	SF	--	0.16	0.16
Remove board sheathing	B1@.030	SF	--	0.79	0.79
Patch wall sheathing. Studs 16" on center, remove and replace panels in several places on wall surface.					
1" x 8" x 12' long, per board	B1@.248	Ea	9.69	6.55	16.24
1" x 8", per square foot	B1@.031	SF	0.45	0.82	1.27
1/2", 4-ply plywood, per 4' x 8' sheet	B1@.960	Ea	17.75	25.40	43.15
1/2", 4-ply plywood, per square foot	B1@.030	SF	0.55	0.79	1.34
Changes to Walls and Ceilings					

3-14: National Estimators Home Improvement Estimator

Each of these estimating tools is updated annually with the latest costs and data. National Estimator books with CDs are available from the Craftsman Book Company (6058 Corte del Cedro, Carlsbad CA 92018, 800-829-8123; www.craftsman-book.com).

Considering Other Scheduling Options

The previous pages show some of the major scheduling and integrated software systems for smaller construction and remodeling contractors. You have other options as well, including using graphic programs to develop Gantt and PERT charts and other visual representations of the job schedule. You can also use Microsoft Excel or other spreadsheet programs to schedule and manage jobs, or you can simply use paper or a calendar using proven scheduling methods to keep jobs on track.

SmartDraw (Figure 3-15) is a drawing tool that can produce flow charts, graphs, organization (org) charts, calendars, timelines, and other visual tools for tracking jobs. Graphics are especially useful if you have a job, such as constructing a specific house plan, where you want to see the task relationships but in which the schedule, once started, usually doesn't change or slip. In addition, you can produce floor plans and other graphics using SmartDraw. Contact SmartDraw (10085 Carroll Canyon Rd. #200, San Diego CA 92131, 858-549-0314; www.smartdraw.com) for additional information and pricing. A 30-day demonstration version of SmartDraw is available online.

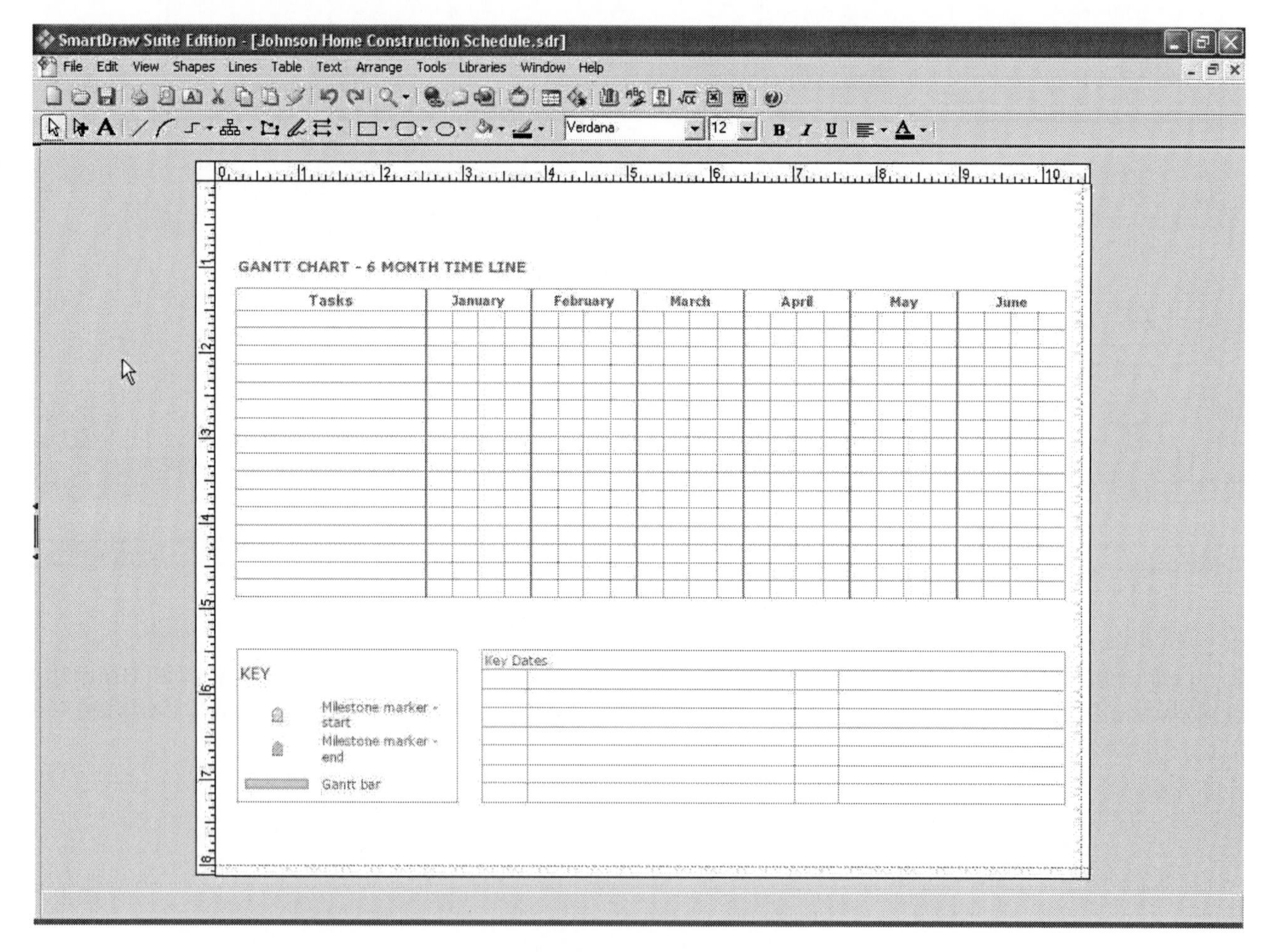

3-15: SmartDraw Gantt Chart

Other Scheduling Systems

You can also develop job schedules using Microsoft Excel (Figure 3-16), the most popular spreadsheet software program and part of the Microsoft Office group. It requires manual updating and isn't as easy to set up or modify as programs designed specifically for scheduling, but if you already have Excel you can set up a single sheet on a project to track the schedule. Make sure you use the Date format for cells in which dates are inserted.

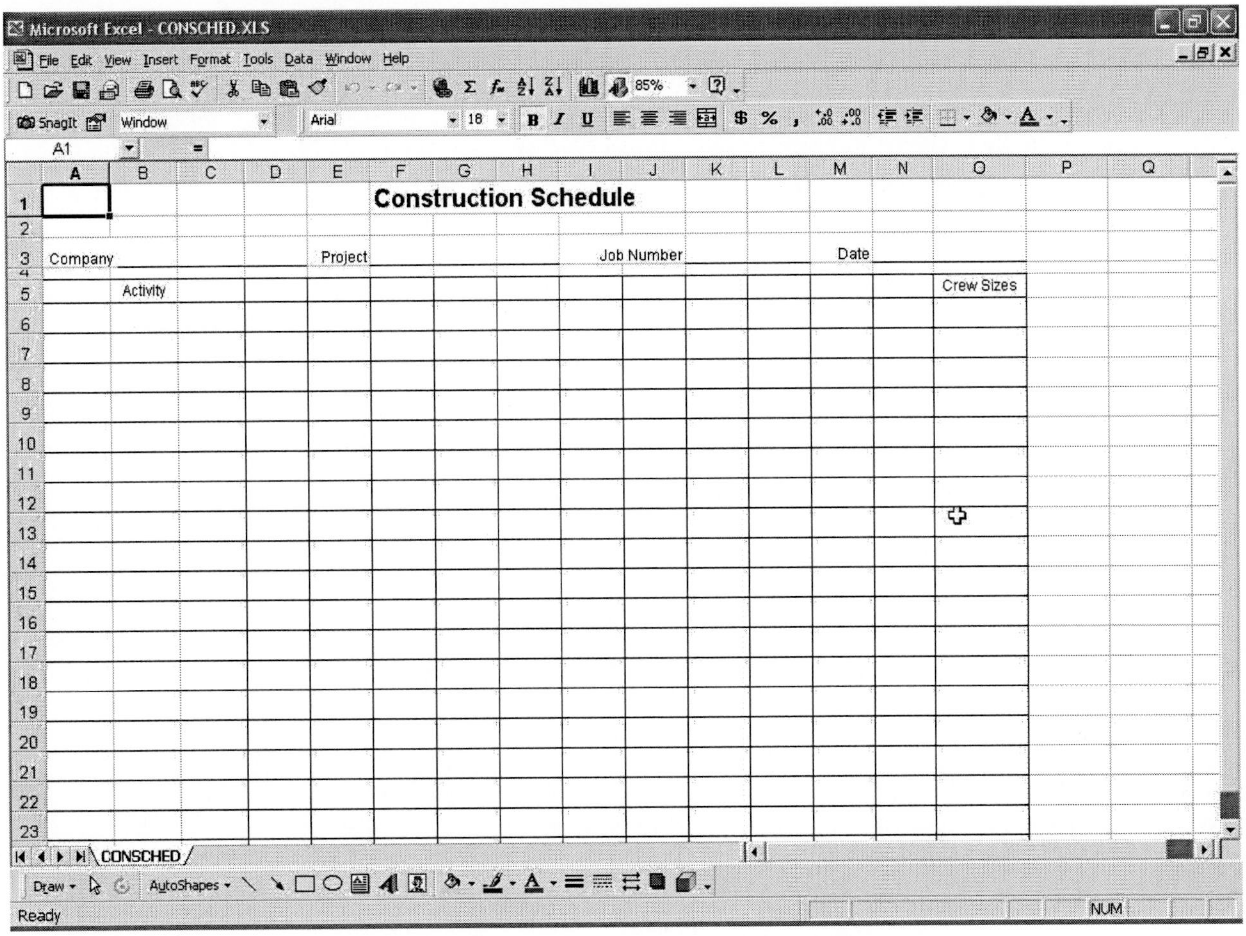

3-16: Microsoft Excel Construction Schedule

Then there's paper. You can use preprinted forms and plain paper to produce and update simple job schedules. Now that you know how job scheduling works (Chapter 1) and how popular software programs handle scheduling, you can design your own rudimentary system for managing jobs. Consider all jobs as related tasks with goals, clients, resources, starting and ending dates, and remember that everything is subject to change! For many small contractors, using Microsoft Project and add-on templates makes the most sense, at least until your business grows and you need additional or more focused software. Fortunately, most other scheduling and business software programs allow Project files to be imported.

The next chapter shows you how to use Microsoft Project templates to profit from scheduling.

Chapter Four

Using Project Templates

Using Project Templates

A template is a standard file that you can reuse so you don't have to create a new file every time. A template file includes all basic task, resource, calendar, view, and report information for a specific type of job. You can edit and modify the output schedule to suit the requirements of the current project. All new project files should be created through the use of templates. Using templates for creating project files will add consistency and save a tremendous amount of time. It also will give you added control over scheduling that you cannot get from a manual system.

The purpose of this book is to empower you with a system that makes using Microsoft Project and other scheduling tools easier and more effective. All of the customizations included in the templates are described in this book. By using the templates as a basis for your schedules, modifications, and additions, you will create a process of communication that will help keep yourself and team members focused and give you the ability to adjust your actions and keep your projects on track.

Adopting and using these templates will allow you to share, integrate, and use project information with anyone using the same system. It will also set a familiar expectation for the people who are viewing and working with the same information. Effective communication lies in the ability to convey information in a consistent and understandable format.

Templates are your framework to guide planning they serve as a checklist to ensure that you use and do not miss any information. In construction we deal with similar tasks, resources and information across all projects. Therefore, to be effective and efficient you need to have a system to work within every project. One of the greatest benefits of Microsoft Project is the ability to consolidate projects. To benefit from viewing and reporting across different projects, it is critical that they are formatted in a similar way so that you can filter, view, and access information correctly.

You can create scheduling templates at whatever level you need. By using templates for your project schedules, you will save input time and benefit from the customizations consistent across all files created through the use of your templates.

Working With the Project Calendar

The Project Calendar set as the default in all of our Template schedules is a custom calendar named Construction. This calendar has all major holidays marked as nonworking and has a standard workweek of Monday through Friday from 8:00 am until 5:00 pm.

The working hours are less significant because you are working with fixed duration tasks. The reason these are fixed duration tasks is because they have deadlines that need to be maintained and must be scheduled to manage time rather than resource productivity. It is each contractor's responsibility to manage resources to meet the scheduled task durations. Fixed duration scheduling allows you to adjust the amount of resources or working time for a task to be completed within the scheduled time. The task items within the schedules are set as windows of time to allow for normal delays and downtime.

Understanding Tables and Views

The Microsoft Project Gantt chart is divided into tables and views (Figure 4-1).

Tables **Views**

CSI Code #	Task Name	Duration

April				May		
4/1	4/8	4/15	4/22	4/29	5/6	5/13

4-1: Gantt Chart Tables and Views

Tables (Figure 4-2) are the spreadsheet cells that contain all of your task and resource information and data (as shown below) and are usually aligned to the left of the scheduling window on the Gantt chart.

Tables

CSI Code #	Task Name	Duration
03300	⊟ **Concrete Slabs**	**36 d**
03310	Concrete Slab Prep. @ Office	2 d
03310	Concrete Slabs @ Office	2 d
03310	Concrete Slab Prep. @ Manufacturing	5 d
03310	Concrete Slabs @ Manufacturing	5 d

4-2: Tables

There are four main tables that we will be working with; the Entry Table, Tracking Table, Schedule of Values Table, and Master Entry Table.

Views (Figure 4-3) display project information in a particular format usually in a date or calendar format at the right of the Gantt chart screen. The detail that a view is displays is dependent on the active table. All views in this book and these templates show gridlines separating the rows and columns with the start date at the left of the task bar, the finish date to the right of the task bar, and the task name at the top of the task bar. This is easy to read as you navigate from the task list in the table to the view part of the screen.

Views

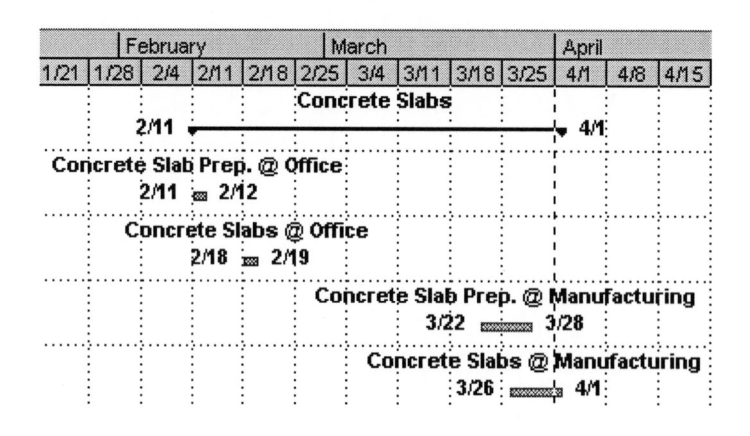

4-3: Views

As you plan and track a schedule, it's useful to look at different combinations of information. By changing the table applied to a sheet view or a Gantt Chart view, you can change the fields of information displayed in that view. If the tables or views provided within Microsoft Project or our custom templates do not meet your needs, you can create new tables and views or modify an existing table or view. Note that when you save a project, the new or modified table with the active view is saved with the project. You can reset the default table fields by inserting the necessary columns as described in the appendices of this book.

Understanding Task Relationships

A task dependency is the relationship between two tasks in which one task depends on the start or finish of another task in order to begin or end. The task that depends on another task is the successor, and the task it depends on is the predecessor. When you create a link between two tasks, Microsoft Project (Figure 4-4) calculates the successor's start date and finish date based on the predecessor's start date or finish date, the dependency type, and the successor's duration. In Microsoft Project, you can link predecessors and successors with four types of task dependencies. The nature of the dependency between the predecessor and successor determines the task dependency type to use.

Task dependency	Example	Description
Finish-to-start (FS)	A → B	Task (B) cannot start until task (A) finishes. For example, if you have two tasks Task A "Pour Foundation" and Task B "Backfill," "Backfill" can't start until "Pour Foundation" finishes. This is the most common type of dependency.
Start-to-start (SS)	A → B	Task (B) cannot start until task (A) starts. You have two tasks Task A "Foundation Excavation" and Task B "Pour Foundation", "Pour Foundation" can't begin until "Excavation" begins.
Finish-to-finish (FF)	A → B	Task (B) cannot finish until task (A) finishes. For example, if you have two tasks, Task A "Install Doors" and Task B "Install Hardware", "Install Hardware" can't finish until "Install Doors" finishes.
Start-to-finish (SF)	A → B	Task (B) cannot finish until task (A) starts. This dependency type can be used for a closely linked successor task which is dependent on the predecessor. If a related task Task B needs to finish before Task A but you want the task to be scheduled by Task B you can create an SF dependency between the task you want scheduled. Task A (Erect Panels) is linked and scheduled normally. Task B (Fabricate Panels) is dependent and closely linked to Task A (Erect Panels).

4-4: Task Relationships

Understanding Lead Time and Lag Time

It is very important to learn and master task linking and tasks dependencies. The task relationships and dependencies are the foundation of your planning. Lead time and Lag time (Figure 4-5) will assist you in shortening and lengthening your schedule.

You will often need to add more complex relationships than simple finish-to-start and start-to-start dependencies to accurately schedule your project. You can enter lag time to represent a delay between the finish of the predecessor and the start of the successor task. For example, if you have two tasks, "Concrete Slabs" and "Metal Studs," you need a delay between the finish of "Concrete Slabs" and the start of "Metal Studs" to allow the concrete to cure. You can enter lag time as a duration, such as 1d, or as a percentage of the predecessor's duration, such as 25%. For example, if the predecessor has a four-day duration, entering 1d or 25% would result in a one-day delay between the tasks. Note that the lag or lead time entered is based on the active calendar, such as a 5-day work week if you enter a 6 day lag, the time would exclude the weekend time and have a calendar duration of 8 days. If the successor task has a continuous time delay, such as in a typical 7 day concrete cure time, you would enter 7ED (elapsed days). In this case the delay would include the weekend and have a continuous delay, ignoring the calendar working time.

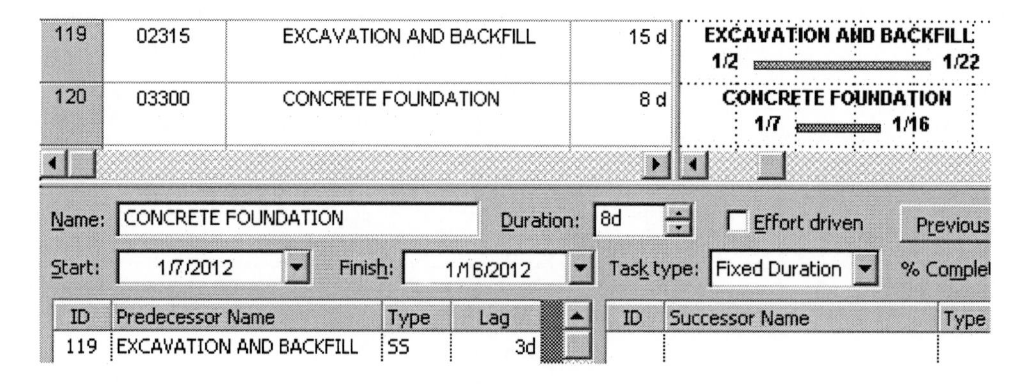

4-5: Lead and Lag Time

Lead time is an overlap of two tasks, when the successor starts before the predecessor finishes. Adding lead time can be useful if you want to give a successor task a head start. Lead time is entered as negative lag, such as -1d or -25%. You typically will use lead time when there are Finish to Start relationships.

When using Start to Start relationships, for example, for the tasks "Foundation Excavation" and "Pour Foundation Walls," use lag time with a 'start-to-start' relationship to begin "Pour Foundation Walls" when "Foundation Excavation" is half done (Figure 4-6).

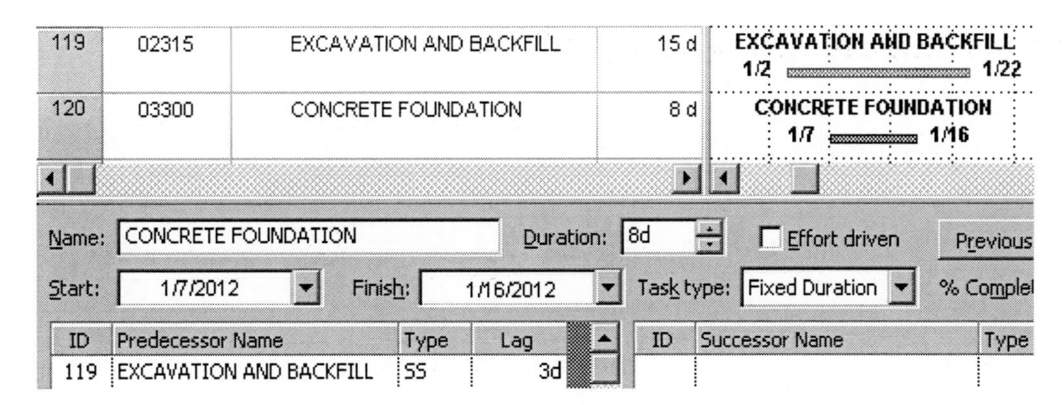

4-6: Lag Time Example

Understanding Constraints

A constraint (Figure 4-7) is a restriction or limitation that is set on the start or finish date of a task. For example, you can specify that a task must start on a particular date or finish no later than a particular date. When a new task is added to a project that is scheduled from the start date, Microsoft Project automatically assigns the As Soon As Possible constraint. Conversely, when a new task is added to a project that is scheduled from the finish date, Microsoft Project automatically assigns the As Late As Possible constraint.

Constraint Type	Scheduling Impact	Description
As Soon As Possible (ASAP)	Flexible	With this constraint, Microsoft Project schedules the task as early as it can, given other scheduling parameters. No additional date restrictions are put on the task. This is the default constraint for projects scheduled from the start date.
As Late As Possible (ALAP)	Flexible	With this constraint, Microsoft Project schedules the task as late as it can, given other scheduling parameters. No additional date restrictions are put on the task. This is the default constraint for projects scheduled from the finish date.
Finish No Later Than (FNLT)	Moderate	This constraint indicates the latest possible date that this task can be completed. It can be finished on or before the specified date. For projects scheduled from the finish date, this constraint is applied when you type a finish date for a task.
Start No Later Than (SNLT)	Moderate	This constraint indicates the latest possible date that this task can begin. It can start on or before the specified date. For projects scheduled from the finish date, this constraint is applied when you type a start date for a task.
Finish No Earlier Than (FNET)	Moderate	This constraint indicates the earliest possible date that this task can be completed. It cannot finish any time before the specified date. For projects scheduled from the start date, this constraint is applied when you type a finish date for a task.
Start No Earlier Than (SNET)	Moderate	This constraint indicates the earliest possible date that this task can begin. It cannot start any time before the specified date. For projects scheduled from the start date, this constraint is applied when you type a start date for a task.
Must Start On (MSO)	Inflexible	This constraint indicates the exact date on which a task must begin. Other scheduling parameters such as task dependencies, lead or lag time, resource leveling, and delay become secondary to this requirement.
Must Finish On (MFO)	Inflexible	This constraint indicates the exact date on which a task must be completed. Other scheduling parameters such as task dependencies, lead or lag time, resource leveling, and delay become secondary to this requirement.

4-7: Constraints

Apply constraints to tasks only when real-world time constraints make it absolutely necessary because applying constraints restricts Microsoft Project's ability to reschedule tasks.

To show you how to use templates, let's open a project template and begin entering information.

Opening A Project Template

1. **Open** Microsoft Project
2. **Click** File
3. **Click** OpenBrowse to the directory and file folder of the template file.
4. **Double Click** Project Template File upon saving a template file, a prompt will request a new file name so the template file is not changed (Figure 4-8).

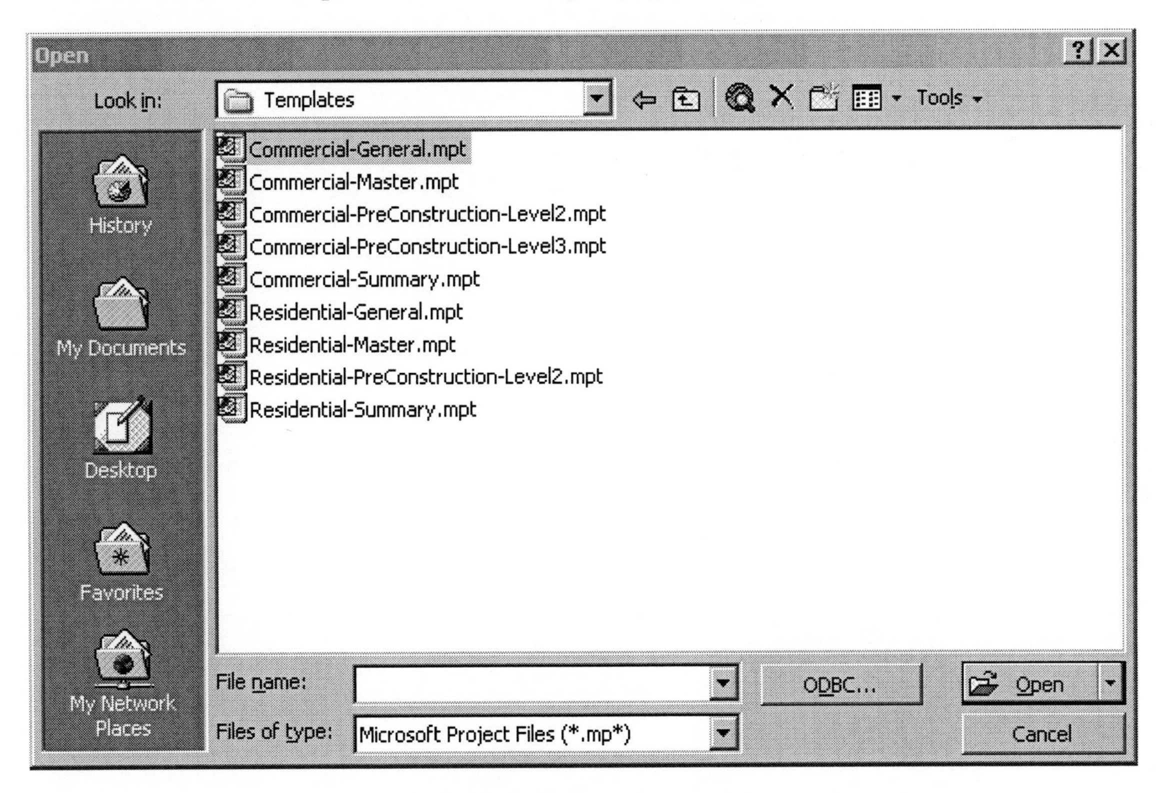

4-8: Open a Project Template

Entering Project Information

1. **Click** Project

2. **Click** Project Information

3. **Enter** Start Date for the Project, The start date should be the anticipated start date of the project (Figure 4-9). The Start Date can always be changed later.

Enter the date in a format xx/xx/xx or use the Calendar in the dropdown menu.

4. **Make Sure** Schedule from: Project Start Date is selected and Calendar is set at Construction. You should always schedule from the Project Start Date. Scheduling from the Finish date will throw off the calculations and your schedule.

5. **Leave** Current date as shown, Status date as NA, and Priority at 500

You can change these later if required.

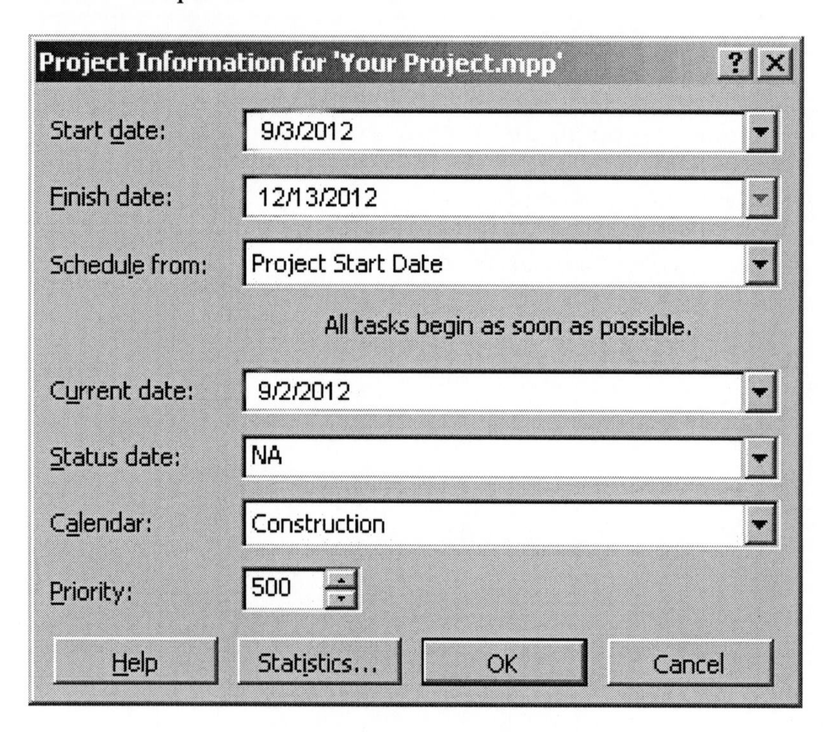

4-9: Enter Project Information

Entering File Properties

1. **Click** File
2. **Click** Properties
3. **Click** Summary Tab
4. **Type** Project File name

   Author

   Manager

   Company

The project file name you enter is the name that is auto filled on inserted projects within a master project and is included as the title of reports. You can enter spaces between names or words (Figure 4-10).

Complete any other text fill-in boxes as desired.

5. **Click** Save preview picture this will allow for a small preview in windows explorer when searching for files.

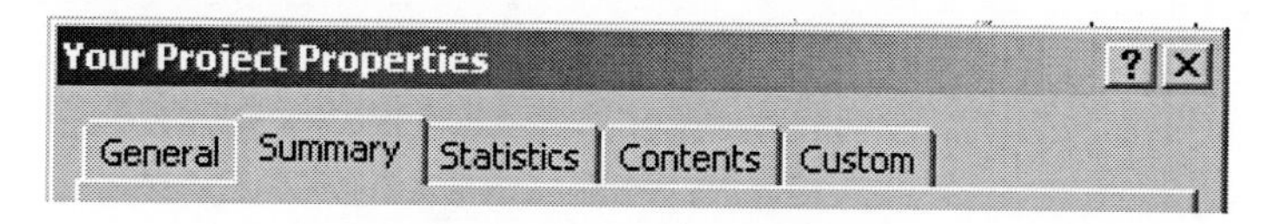

4-10: Property Headings

The other tabs on the properties dialog box are for reference and additional information if you want to use them:

General: Describes the file that stores the document.

Statistics: Provides statistics about the project file including the file name, type, location, size of the file, file creation date, last modified date, and last opened date.

Contents: Displays summary statistics about the project.

Custom: Allows you to enter additional properties to the file to allow searching for the file by the custom properties (Figure 4-11).

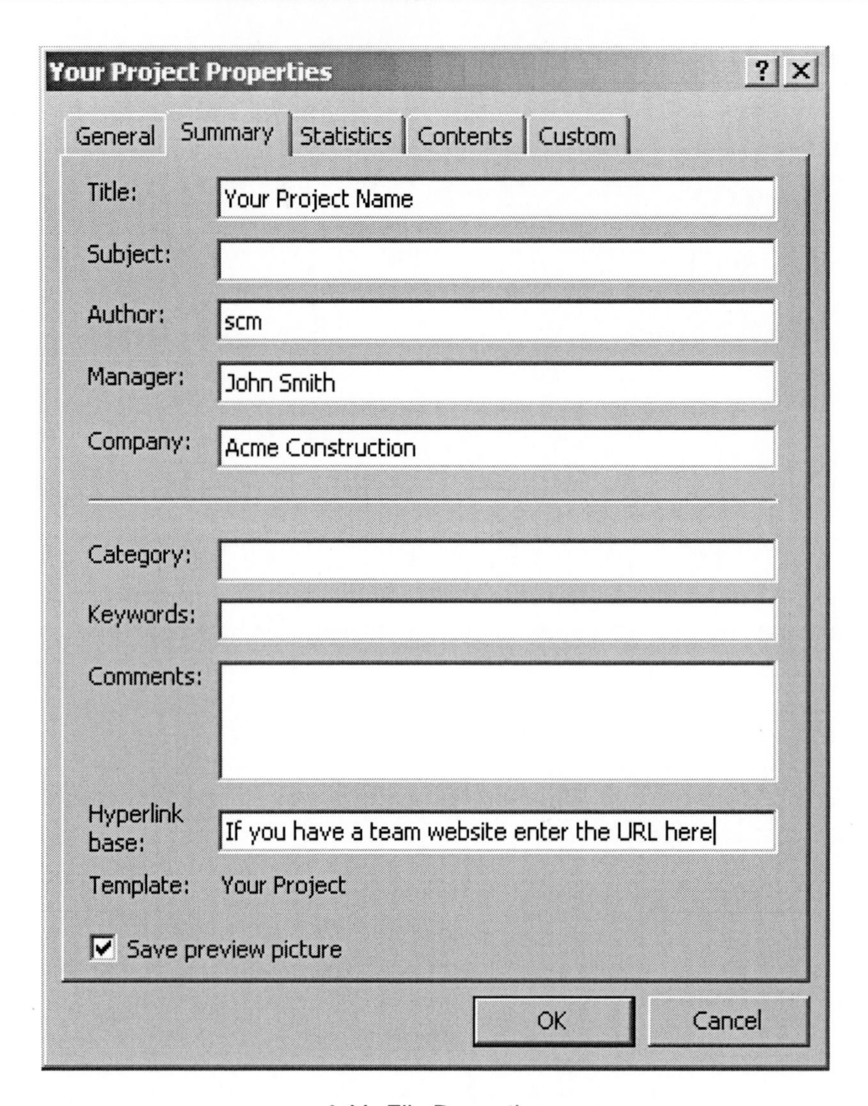

4-11: File Properties

Saving A Project

1. **Click** File

2. **Click** Save As

3. **Click** In Save In dropdown, Browse to the directory and file folder to save in. Save all templates and project files in a subfolder under a main folder named Project Scheduling. This makes navigation easier when working with several projects.

4. **Type** Desired File Name in File name dropdown or choose the file name if you are overwriting. Don't use spaces between words or names so that you have the option to save the project file as a web page. Use "or" as a divider between words or names.

5. **Click** In Save as type dropdown

6. **Click** Project or Template. Save as a Project if you are creating, working with or intending on modifying the file. Save as a template to use the information for future project schedule creation.

7. **Click** Save

Saving project templates and files in one central folder makes future reference easier (Figure 4-12). Keep template files in separate folders from project files to make sure they are not over-written.

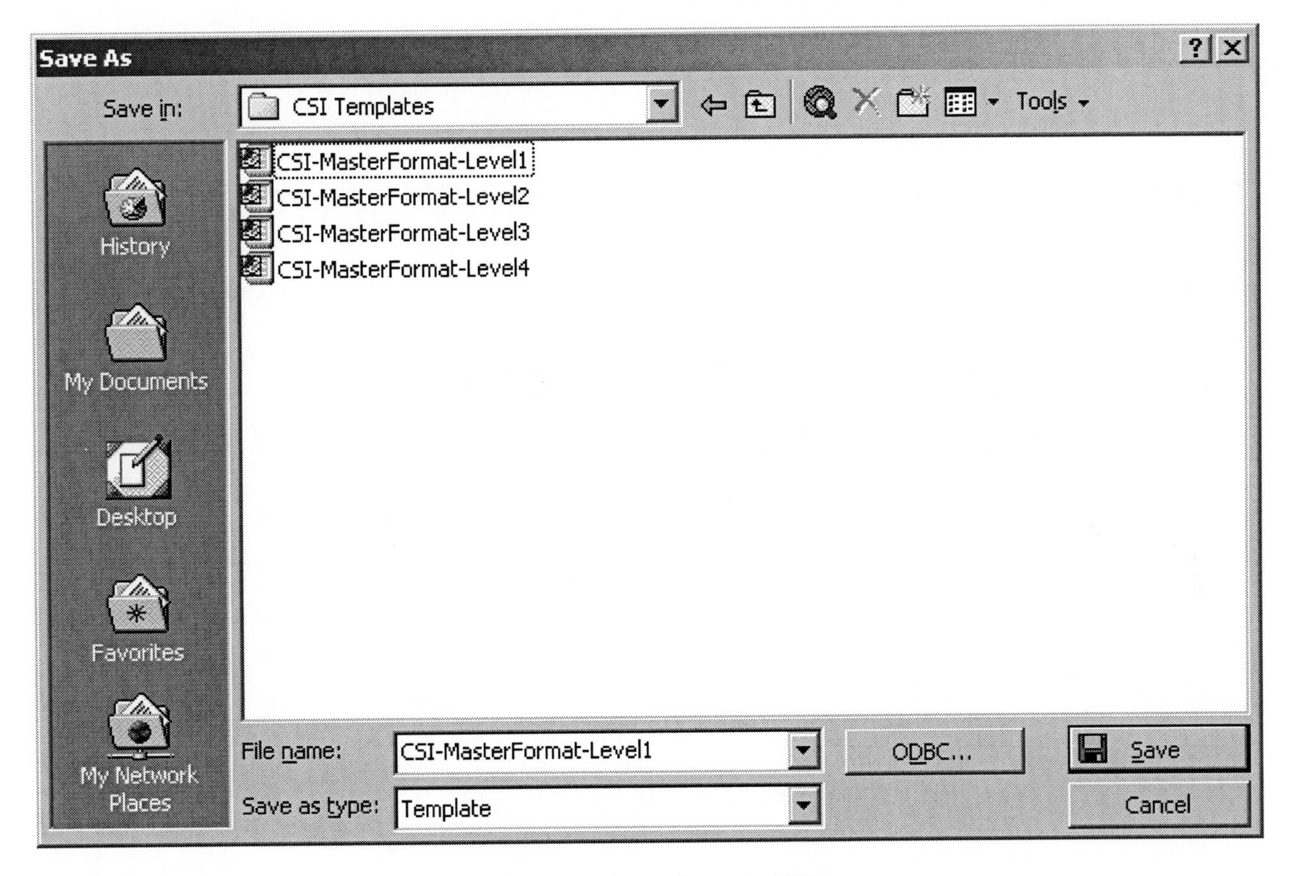

4-12: Save Templates and Files

Changing the Project Calendar

1. **Click** Tools

2. **Click** Change Working Time. Make sure that Construction (Project Calendar) is selected from the dropdown (Figure 4-13).

3. **Click** Days or periods of time in Select Date(s) box. Control click or click and drag to select multiple dates.

4. **Click** Nonworking Time under Set selected date(s) to: Select Use Default to change a nonworking day or time to a working day.

5. **Click** OK

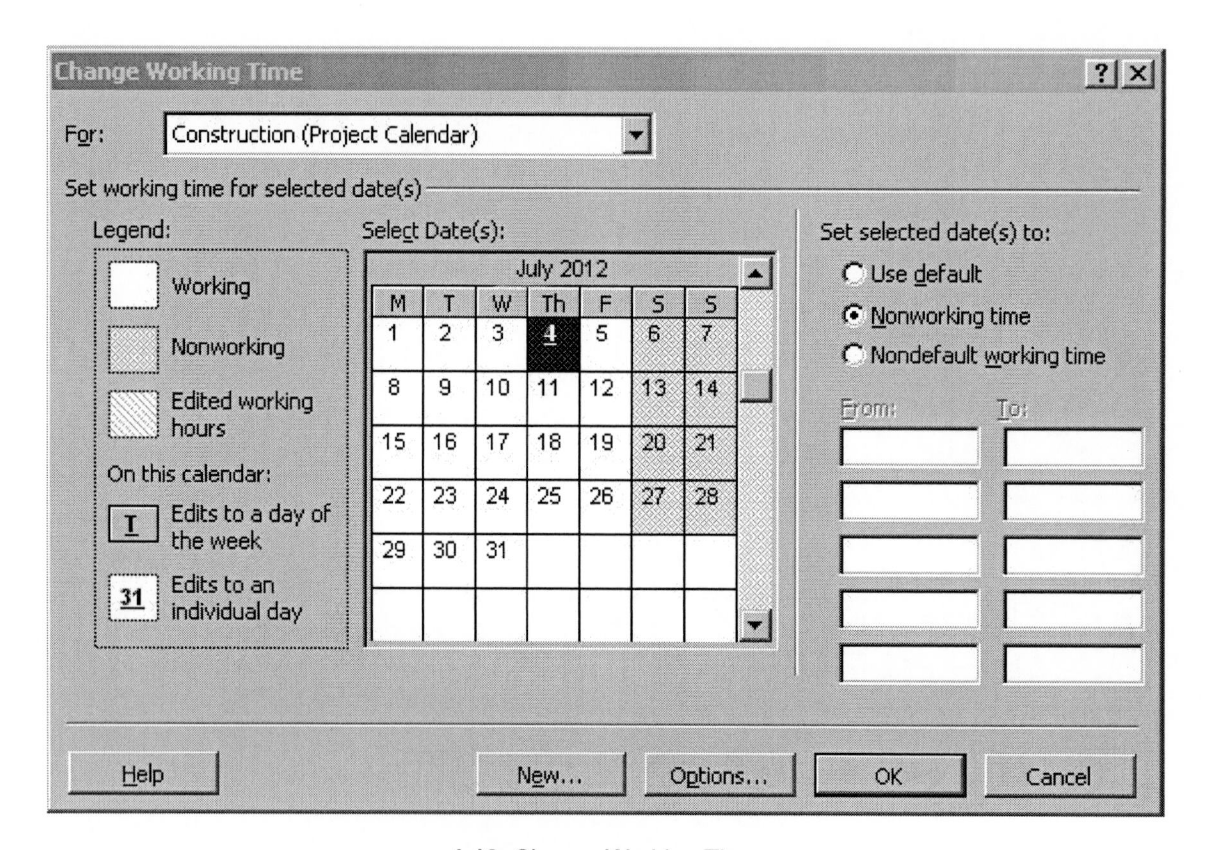

4-13: Change Working Time

Changing the Default Calendar

1. **Click** Project
2. **Click** Project Information
3. **Click** In Calendar drop down. Select desired Calendar to set as the project default calendar.
4. **Click** Calendar selection for the new default calendar (Figure 4-14). If making this change to a project, the change will be specific to the project. If making the calendar change to a template the change will be reflected in all project files that use the template.
5. **Click** OK

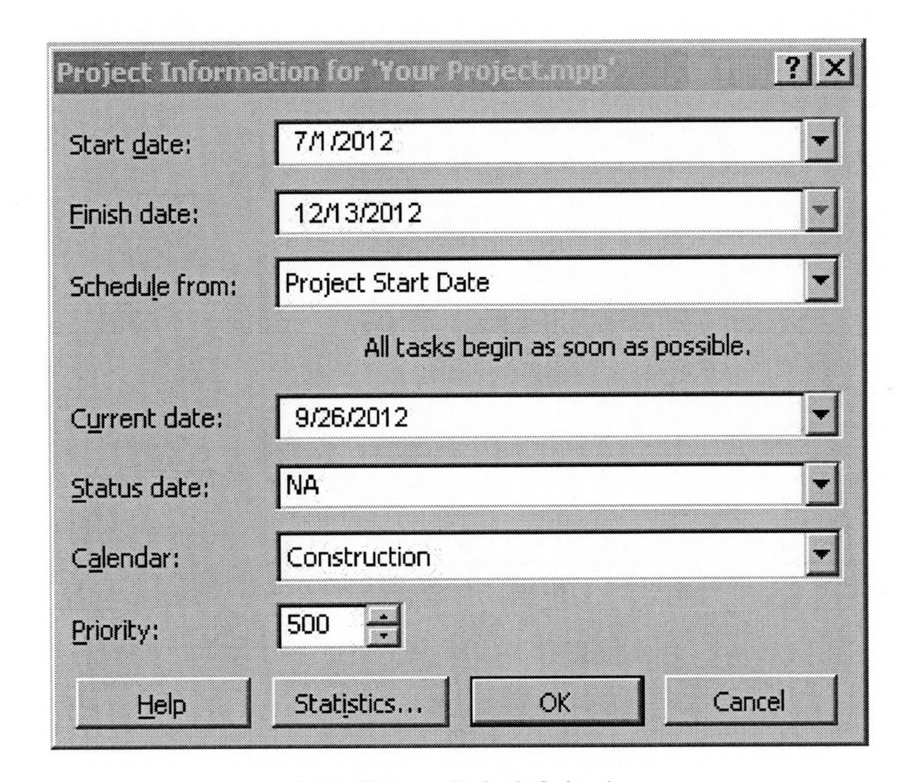

4-14: Change Default Calendar

Adjusting the Timescale

1. **Click** Format
2. **Click** Timescale (Figure 4-15)
3. **Click** In Major Scale Units Dropdowns. Select the major scale calendar options desired.
4. **Click** In Minor Scale Units Dropdown. Select the minor scale calendar options desired.
5. **Click** OK

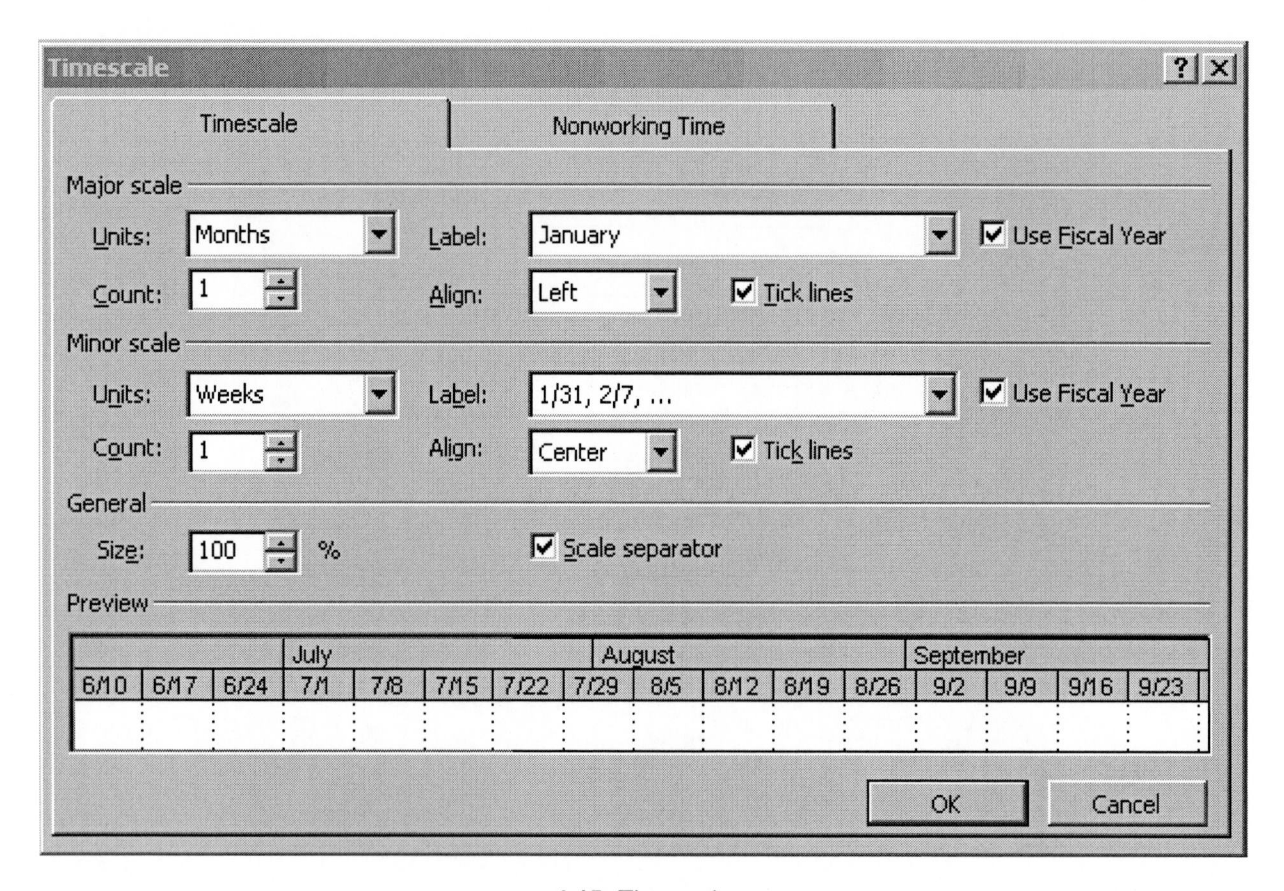

4-15: Timescale

Note: Adjust the timescale as required to capture the attention of the viewers. If using a print copy of the schedule, adjust the timescale to the printer's paper size and layout (portrait or landscape). Schedules that are too large or have too many pages are less effective. Alternately, use the General Size percentage box to view the schedule at a reduced and scaled size. Use the Print Preview to make sure that the timescales show clearly.

Master Entry Table

The Master Entry table and allows you to enter all specific project information in one place. This table includes all of the custom fields for the project to enter, edit or delete information quickly and easily.

1. **Click** View
2. **Click** Table:
3. **Select** Master Entry from dropdown list.
4. **Click** On header of Task Name column. The inserted column will insert to the left of the column clicked on.
5. **Click** Insert
6. **Click** Column (Figure 4-16)

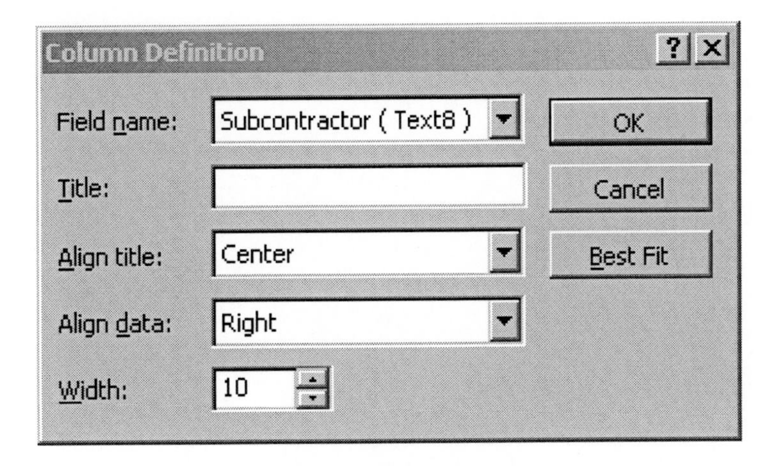

4-16: Column Definition

7. **Click** In drop-down of field name
8. **Select** Task field from dropdown list to insert. This is the task field for specific information to be viewed or tracked for this project.
9. **Continue** Inserting task fields. Insert all of the task fields for this project.
10. **Type or Click** Type name in text field or click on option flag fields for option items.

This is the process of entering known information in the customized fields for viewing, reporting, and filtering information. A quick way to add repetitive information is to enter or click the option desired, click and drag down, and then click on Edit and then Fill Down.

Master Entry Table

11. **Click** Save. The Master Entry Table (Figure 4-17) will save the column format.

12. **Exit** Master Entry Table

Click View→Table→Entry

Note: As needed, return to the Master Entry Table to add, edit, or delete information.

Subcontractor	Subcontract Work	Crew	Task Name
Sites R Us	Yes		SITE CONSTRUCTION
	No	Jones	DEMOLITION
	No	Falco	EXCAVATION AND BACKFILL
	No	Smith	SITE UTILITIES
AAA Paving	Yes		PAVING AND SURFACING
Landscapes	Yes		LANDSCAPING
	No	Adams	CONCRETE FOUNDATION
Concrete Plus	Yes		CONCRETE SLABS
	No	Smith	EXTERIOR CONCRETE
RJ Pre-Cast	Yes		PRECAST CONCRETE
A-1 Masonry	Yes		MASONRY

4-17: Master Entry Table

Setting A Milestone

1. **Click** Task name duration column

2. **Type** 0 In duration column

3. **Press** Enter Or click or move to another cell

In the following example, steel delivery is marked as a milestone (Figure 4-18) although it actually takes time; it is useful to use the symbol to capture the viewer's attention.

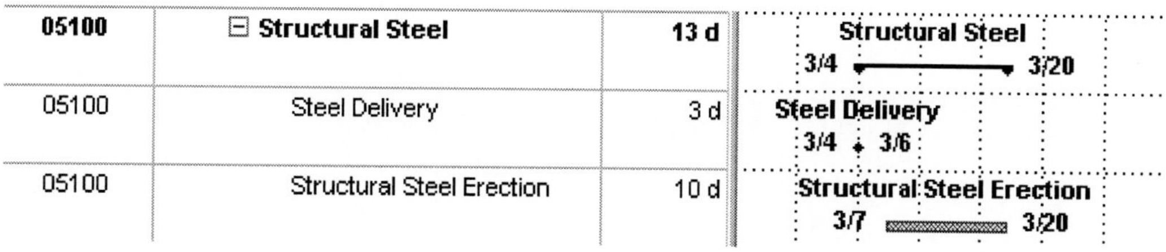

05100	**⊟ Structural Steel**	**13 d**	**Structural Steel** 3/4 — 3/20
05100	Steel Delivery	3 d	**Steel Delivery** 3/4 ◆ 3/6
05100	Structural Steel Erection	10 d	**Structural Steel Erection** 3/7 — 3/20

4-18: Set a Milestone

Marking A Task as A Milestone

1. **Enter** Duration of the task that actually has time to complete

2. **Double Click** The task (i.e., Steel Delivery in the following example). This will bring up the Task Information dialogue box.

3. **Click** The Advanced Tab

4. **Check** Mark task as milestone (Figure 4-19)

5. **Click** OK

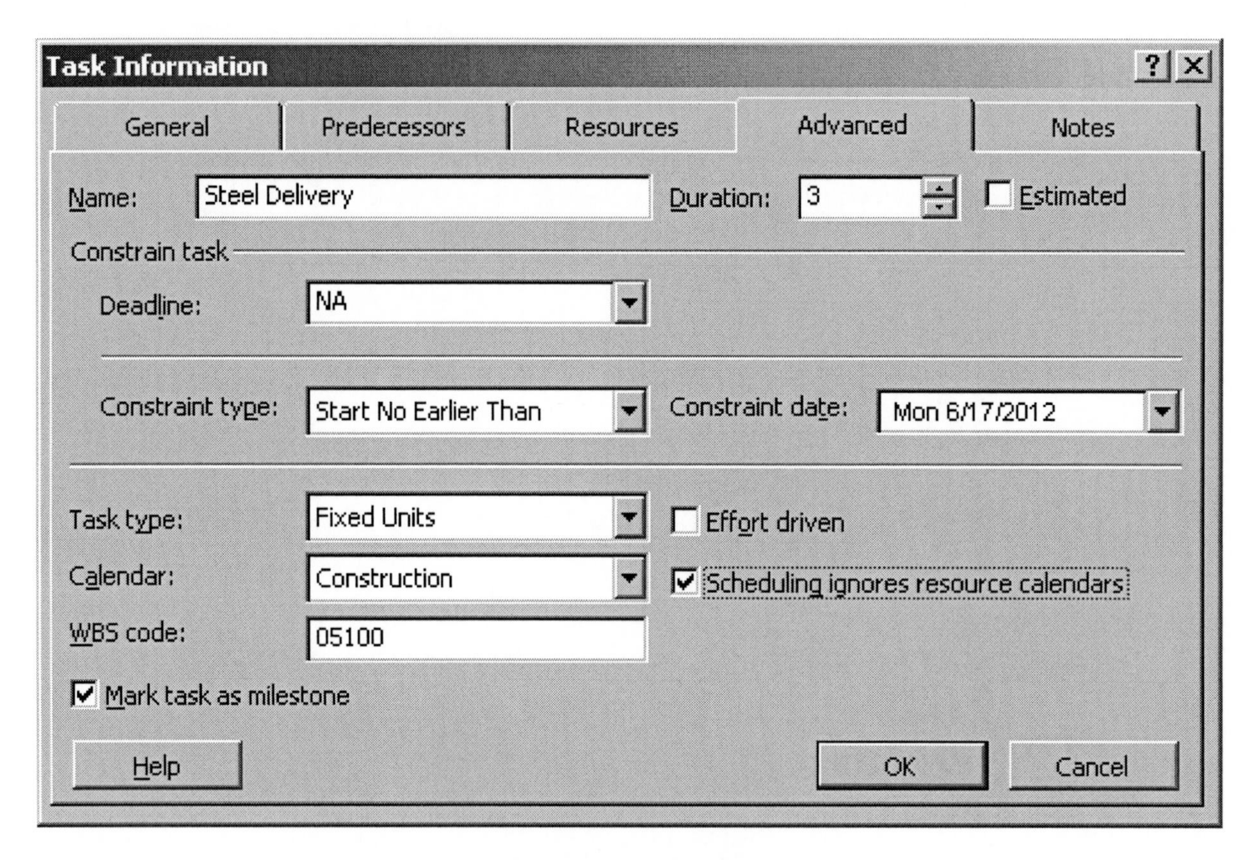

4-19: Mark as a Milestone

Customizing Your Task List

1. **Open** Project Template File

2. **Click** In top summary task cell. This is usually Pre-Construction Requirements.

3. **Select** Tasks or groups of tasks you want to delete. Select individual tasks by pressing the delete key or groups of tasks by control click or click and drag.

4. **Delete** Tasks for the entire task list.

Note: Deleting summary tasks will delete the summary task and subtasks below it.

5. **Go** Back To Top Task in Schedule. Pressing Control Home will move the cursor to the top task.

6. **Add** Tasks. Add desired tasks Press the insert key to insert above the cell selected and type the new task name.

To add a summary task, insert the CSI Division number and the Priority number for top level tasks. These numbers are needed for filtering and sorting purposes. Insert a column to the left of the task and select the division number from the dropdown menu in each cell.

7. **Enter** CSI Codes At Inserted Tasks in the CSI Code number column. Use the same CSI Code number as a similar task or enter a new CSI Code number in the next sequence of numbers to a similar task. Use the same CSI Code number to keep consistency for filtering and reporting. Use any of the template schedules or CSI task listings in the appendices of this book as reference for entering CSI codes.

Estimating Task Durations

1. **Open** Project File If not already open.

2. **Click** In duration column of top subtask under top summary task. This is usually under Pre-Construction Requirements or Project Duration.

Note: You cannot enter durations for summary tasks, as they are representative of the subtasks below them.

3. **Enter** Estimated task durations. Enter durations for individual tasks by entering the number of days, weeks, or months. The default method is days. To enter weeks, enter w after the duration. To enter months, enter mo after the duration.

Note: The task durations entered are windows of time that include normal mobilization, downtime and completion time frames. This allows for a realistic slack in the schedule.

4. **Continue** Entering estimating task durations for all tasks. If the estimated duration for a task is unknown or uncertain, leave the default that shows a question mark as a reminder to go back and enter the task duration when you have more information.

5. **Verify** Questionable or unknown task durations. This process involves contacting the related resource or team member for input on the approximate time frames they require for certain types of work, such as checking with the building department for their review time.

6. **Click** Save

Tip: An effective way to get realistic task durations is to filter for responsibilities, codes, or divisions and send them to the appropriate team member to fill out and send back for entering.

Creating Task Links, Relationships, Lag & Lead Time

1. **Click** Top sub task under top summary task. Do not link summary tasks unless necessary as linking them can cause circular relationships within the schedule. One exception is that the Project Duration summary task is always linked to Substantial Completion.

2. **Click** Window

3. **Click** Split. This will create the Gantt chart on the top of the screen and the Task Form on the bottom. Select the split screen by double clicking on or dragging up the down arrow at the bottom right of the view window.

4. **Verify** The task form displayed has predecessors on left and successors on right If the task form does not have predecessors on the left and successors on the right, then right-click the mouse and choose predecessors and successors to activate this window in the task form

5. **Click** In ID column At left of predecessor entry screen

6. **Type** Predecessor ID number of predecessor tasks

7. **Type** or **Click** Type (relationship) Enter or click dropdown for FF, FS, SF or SS.

8. **Enter** Lag or Lead Time Enter Lag or Lead time to reflect a more accurate relationship between the tasks (Figure 4-20).

9. **Continue** Entering immediate predecessors for the selected task Enter the relationship and lead or lag time for all added predecessors.

10. **Click** Next to go to the next task

11. **Type** Predecessor ID number of predecessor tasks

Follow steps 6 - 9.

12. **Continue** Entering predecessors for all tasks in your schedule Don't worry about entering successors for tasks at this time. The successors will be automatically filled in as task relationships are created.

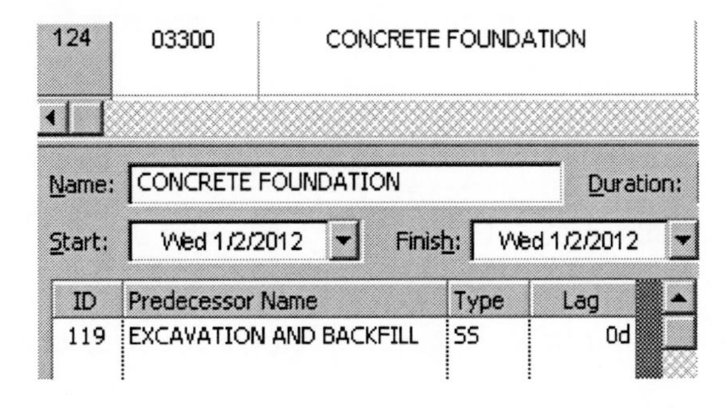

4-20: Link Tasks

Setting Task Constraints

1. **Double Click** Selected task. This is the task on which a date constraint is placed. It is a critical and fixed date in the schedule due to availability and lead time.
2. **Click** Advanced Tab In Task Information Dialogue Box
3. **Click** Constraint Type drop-down
4. **Select** Constraint Type
5. **Click** Constraint Date dropdown
6. **Click** Constraint Date from dropdown
7. **Click** OK

Creating a Resource Pool

1. **Open** Master or Consolidated Project File
2. **Click** View
3. **Click** Resource Sheet (Figure 4-21)
4. **Enter** Resources. The resources entered are those that will be used across multiple or all projects so assignments can be viewed.
   - Resources can be people, trade types, equipment, materials, etc.
   - If you are tracking individuals, use the name of the resource.
   - If you are tracking trade types, use the name of the trade.
   - If you are tracking equipment, materials, etc., enter the name of the associated resource.
   - The Max. Units column is the total number of available resources for each particular type of resource.
5. **Enter** Resource Data. Enter applicable resource data.
   - For individuals enter max. units: 1
   - For trade types enter max. units for the total resource type available (i.e., carpenters: 5 or concrete workers: 8).
   - For equipment, materials, etc. enter max. units to the total amount of the resource available.
   - Enter costs in the associated fields if you intend on tracking costs.
   - Under Calendar -- make sure that Construction is set as the base resource calendar for all resources that follow the base default construction calendar.
6. **Click** View > Gantt Chart This will bring the main Gantt Chart back into view.

	Resource Name	Initials	Max. Units	Group	Email Address	Code	Notes
2	Andrea Wilkins	AJ	1	Sales, Est	ajones@anon.com	094,092	
3							
4	Steve Martin	SM	1	Projd, Projm	smartin@anon.com	094,092,082,105	
5	Tom Jones	TJ	1	Projd, Projm	tjones@anon.com	094,092,082,105	
6							
7	Bill Bones	BB	1	Est	bbones@anon.com	092	
8	Jake Elder	JE	1	Est	jelder@anon.com	092	
9	Tim Drake	TD	1	Est	tdrake@anon.com	092	
10	Tom Golden	TG	1	Est	tgolden@anon.com	092	
11							
12	Kurt Sanders	KS	1	Cad	ksanders@anon.com	082	
13	Jeff Pincher	JP	1	Cad	jpincher@anon.com	082	
14	Bruno Brinkman	BB	1	Cad	bbrinkman@anon.com	082	
15							
16	Ed Easton	EE	1	Site Design	eeaston@anon.com	082	
17							
18	Theresa Billows	TB	1	Support	tbillows@anon.com	085	
19	Mary Michaels	MM	1	Support	mmichaels@anon.com	085	
20	Rachel Linden	RL	1	Support	rlinden@anon.com	085	
21							
22	Jim Tasker	JT	1	Projm	jtasker@anon.com	105	
23	Scott Kintz	SK	1	Projm	skintz@anon.com	105	
24	Wayne Hill	WH	1	Projm	whill@anon.com	105	
25	Michael Stone	MS	1	Projm	mstone@anon.com	105	
26							
27	Mike King	MK	1	Concrete	mking@anon.com	210	
28							
29	Ken Masters	KM	1	Site Work	kmasters@anon.com	140	
30							
31	Superintendents	S	3	Superintendents		110	
32	Operators	O	4	Operators		140	
33	Concrete Workers	C	15	Concrete		210	
34	Masons	M	2	Masons		230	
35	Steel Workers	S	8	Steel		260	
36	Welders	W	2	Steel		262	
37	Carpenters	C	6	Carpenters		270	
38	Painters	P	2	Painters		370	
39	Plumbers	P	1	Plumbers		430	
40	Laborers	L	1	Labor		500	
41							
42	550 dozer	5	2	Equipment		141	
43	850 Doxer	8	1	Equipment		142	
44	Cat Loader	C	2	Equipment		143	
45	JCB Forklift	J	3	Equipment		144	
46	JD Backhoe	J	2	Equipment		145	
47	Excavator	E	3	Equipment		146	

4-21: Resource Sheet

Sharing Resources

1. **Open** Project Files that will share the resources and the Master file that contains the resource pool. If you are using a common resource pool across all projects, open the Master Project or Consolidated Project that contains all of the resource information listed in the resource sheet and the files that will share the resources.

2. **Click** Window. Choose the file or files that will share the resources with the Master or Sharer project file. This will activate the open file to share resources.

3. **Click** Tools

4. **Click** Resources

5. **Click** Share Resources (Figure 4-22)

6. **Click** Use Resources from Dropdown

7. **Select** Master or Sharer File from Dropdown. The file chosen will be the Master or Sharer file that contains the resources and information the files will share.

8. **Click** OK

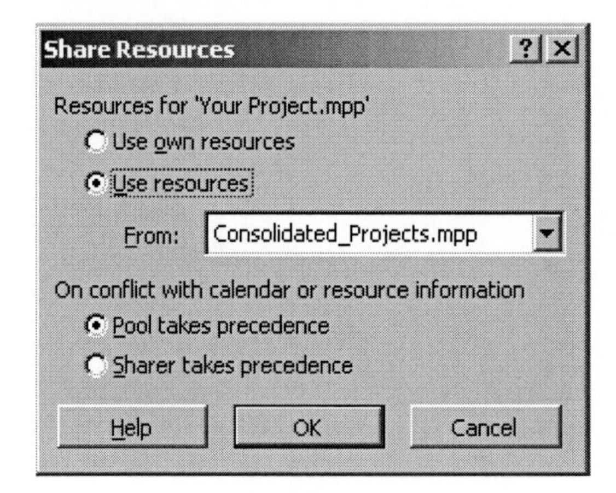

4-22: Share Resources

Entering Resources

1. **Click** View
2. **Click** Resource Sheet
3. **Enter** Resources. Resources can be people, trade types, equipment, materials, etc.
   - To track individuals, use the name of the resource
   - To track trade types, use the name of the trade
   - To track equipment or materials, enter the name of the associated resource
4. **Enter** Resource Data. Enter applicable resource data
   - For individuals enter max. units: 1
   - For trade types enter max. units for the total resource type available i.e. carpenters: 5 or concrete workers: 8
   - For equipment, materials, etc. enter max. units to the total amount of the resource available.

Enter costs in the associated fields to track costs. Under Calendar, set Construction as the base resource calendar for all resources that follow the base default construction calendar

5. **Click** View
6. **Click** Gantt Chart. This will bring the main Gantt Chart back into view.

Resource Name	Initials	Max. Units	Group	Email Address	Code
John Smith	JS	1	Sales, Est	jsmith@anon.com	094,092
Andrea Jones	AJ	1	Sales, Est	ajones@anon.com	094,092
Steve Matzen	SM	1	Projd, Projm	smatzen@anon.com	094,092,082,105
Tom Jones	TJ	1	Projd, Projm	tjones@anon.com	094,092,082,105
Bill Bones	BB	1	Est	bbones@anon.com	092
Jake Elder	JE	1	Est	jelder@anon.com	092
Tim Drake	TD	1	Est	tdrake@anon.com	092

Sample sheet; 4-23: Resource Sheet

Assigning Resources

1. **Click** View
2. **Click** Gantt Chart
3. **Click** Task Name. Select the task to which resources are assigned.
4. **Click** Assign Resources. This will bring up the Assign Resources dialogue box as shown in the following example (Figure 4-24).
5. **Enter** Resources and number of resources. Enter the resources and the total number of resources required to complete the task.
6. **Click** Close to exit the Assign Resources dialogue box.

Assign resources; 4-24: Assign Resources

Note: Resources can be assigned to a summary task. This is useful if the resource is responsible for a group of tasks. For example, if a dozer or backhoe will be on site for the entire period of site work or excavation and backfill, assign the resource to the associated summary task. If the resource's time on the group of tasks increases as the total duration of the subtasks increases, then this is an efficient way to assign the resource to responsibility for all the tasks. However, if the resource's time usage on the group of tasks changes, the resource should be assigned to the individual subtasks and not the summary task.

If assigning a resource full-time to a summary task, don't assign the resource full-time to subtasks under that summary task, or unnecessary over-allocations may occur.

Establishing a Schedule of Values

1. **Click** View

2. **Click** Table

3. **Click** Schedule of Values. This will bring up the Schedule of Values Table.

4. **Enter** Contract Selling Costs in the Total Cost column for each task.

Note: Costs for a summary item can't be entered in the Total Cost field, as these amounts are calculated fields that represent the subtasks below the summary task. However, costs can be entered for a summary item (Pre-Construction Requirements, for example) by inserting the Fixed Cost column and entering the total cost for the summary task. Hide the Fixed Cost column when done to prevent mistakes in entering other costs in this field.

5. **Continue** Entering Selling Costs for all task items. Enter the resources and the total number of resources required to complete the task

6. **Click** View→Table→Entry to exit the Schedule of Values Table

Gather and enter all costs available for the project to complete the Schedule of Values table.

After all costs are input, apply the Schedule of Values filter, which will only show tasks with costs and their related summary tasks. This filter can be used for viewing and printing a schedule of values for an owner, bank, etc. Applying this filter in combination with the Gantt or Tracking view allows you to view the costs in relation to the timescale.

Also, use a Monthly or Weekly cash flow report for the project or for all other projects where the costs have been input into the Schedule of Values table.

Updating a Schedule of Values

1. **Click** View

2. **Click** Table

3. **Click** Schedule of Values. This will bring up the Schedule of Values Table

4. **Enter** The Percentage Complete (% Comp.) for tasks that have been started, are in progress or have been completed. The % of the task completion equals the % of cost complete. Make sure to include any upfront costs (material deliveries, for example) that need to be reflected in the total % complete.

Note: By entering the % complete for a task in the Schedule of Values Table the % complete is automatically updated accordingly in the Tracking Table. If the % complete is entered for a task that does not have an Actual Start date, Microsoft Project assumes that it started according to the current date and establishes the Actual Start date automatically. Later, the Tracking Table can be used to adjust the Actual Start date if necessary.

Note: You should avoid entering the % complete for a summary item. Only enter the % complete for a summary item like Pre-Construction Requirements. Entering the % complete for summary tasks for the actual construction phase will not accurately reflect the % of work and cost complete.

5. **Continue** Entering the Percent Complete (% Comp.) for all current tasks in progress.

The Cost to Date and Remaining Cost field will automatically be updated.

Note: The total cost for the project will be the sum of the highest level summary tasks. In the example, it would normally be the sum of Pre-Construction Requirements and Project Duration.

6. **Click** View>Table>Entry to exit the Schedule of Values Table.

A Schedule of Values has many benefits beyond its use as a billing tool and a reflection of cash flow. If contractors know that the schedule is being used for billing and payments, and the percent equals the cost complete, they may be more manageable.

Applying A Filter

1. **Click** Project
2. **Click** Filtered for:
3. **Select** Desired Filter. Select Filter shown (Figure 4-25) or More Filters for filters not shown.

<u>Note:</u> Task filters can't be applied to resource views or resource filters to task views.

<u>Tip</u>: To turn off a filter, point to Filtered for on the Project menu, click More Filters, click All Tasks if a task filter is applied, or click All Resources if a resource filter is applied, and then click Apply.

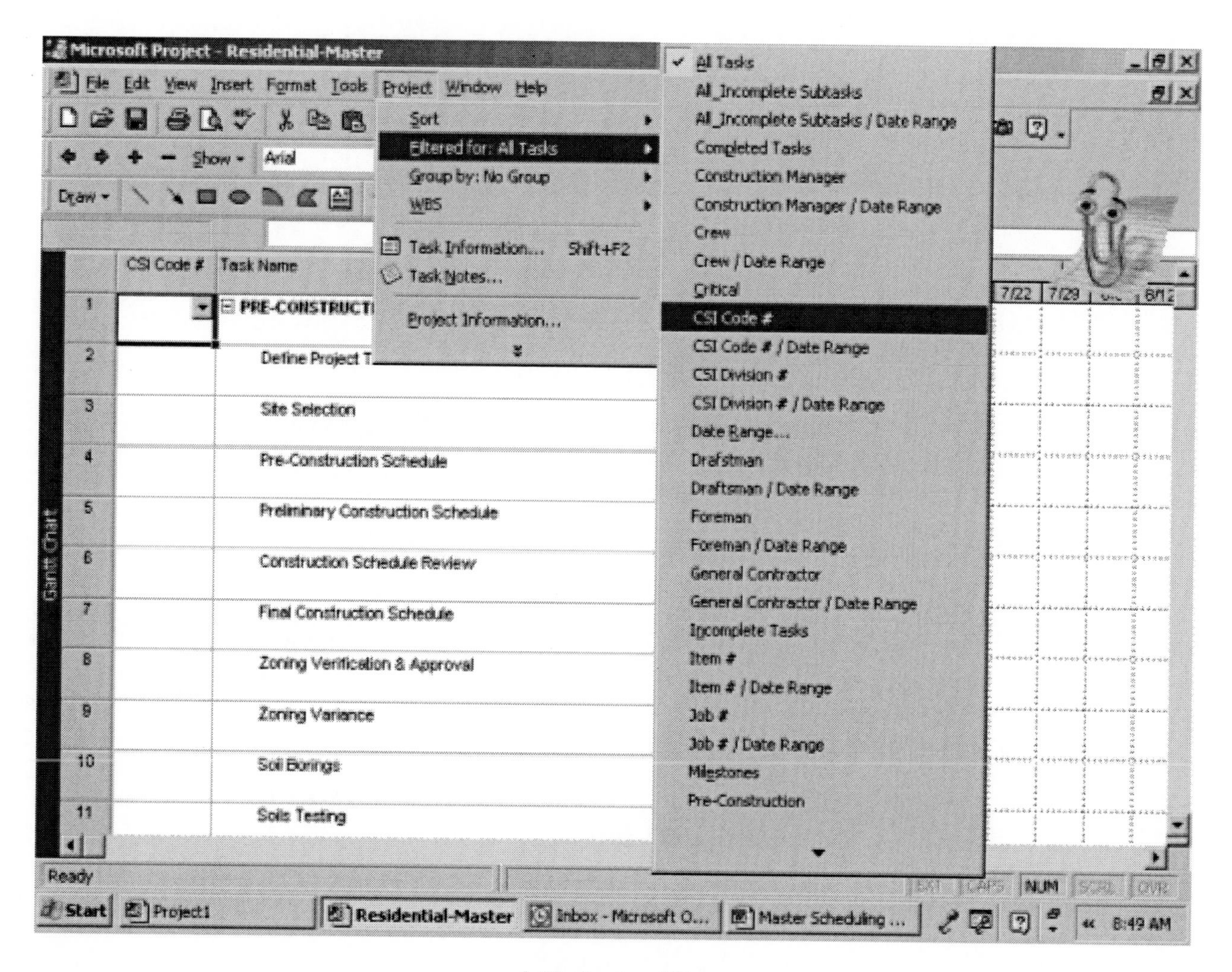

4-25: Apply a Filter

Creating Or Modifying A Filter

1. **Click** Project

2. **Click** Filtered for:

3. **Click** More Filters? (Figure 4-26) at bottom of Filter menu.

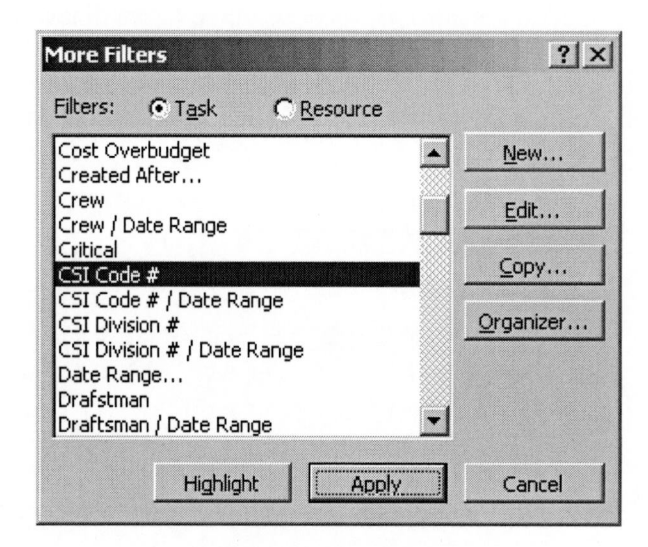

4-26: More Filters?

4. **Select** New (for a new filter with no existing formatting) or an existing filter on which to base your new filter. It is a good idea to view the existing filters to determine which criteria the filters use to help define your new or modified filter.

Selecting New will open the Filter Definition dialogue box (Figure 4-27).

5. **Select** Edit or Copy. This selection is only if a new filter will not be built from scratch. Selecting Edit will modify the existing filter. Selecting Copy will make a copy of the existing filter and then add or modify any changes needed. Create a new name for the filter. (Copying a filter is the recommended method to save the existing filter formatting)

6. **Enter Or Modify** New criteria or Existing criteria. The new criteria set will include Field Names, Tests, and Values.

Refer to Microsoft Project Help or other resource for detailed descriptions of each.

7. **Enter** Name for new filter. Enter a new name for the filter or use the Copy of name to use the existing filter for reference.

8. **Click** OK

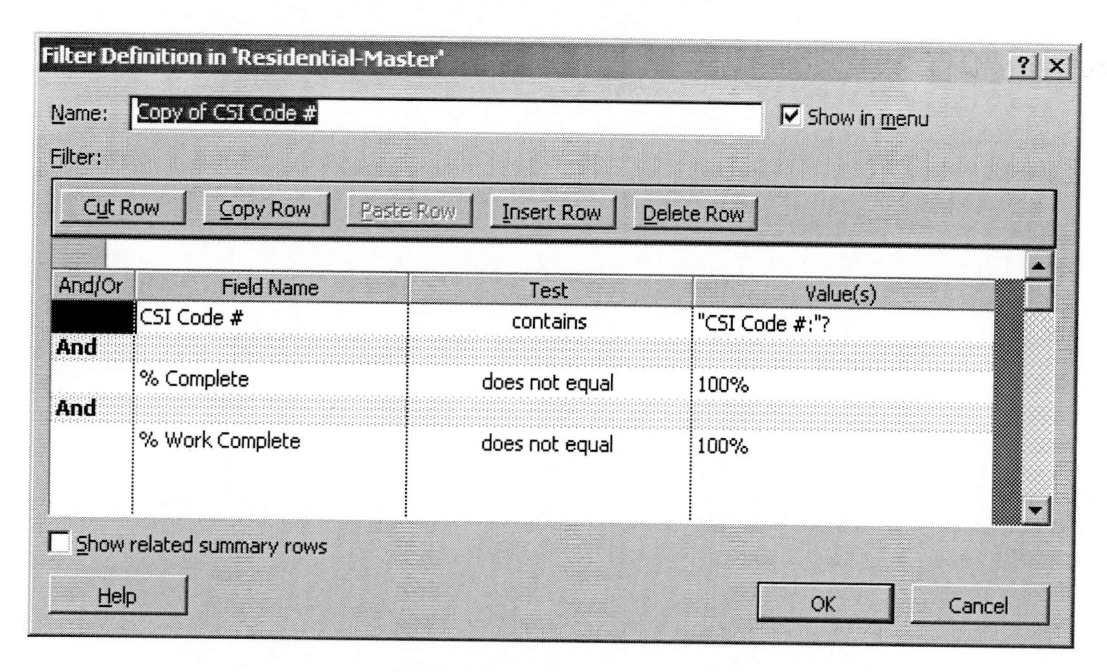

4-27: Filter Definition

Notes: You cannot apply task filters to resource views or apply resource filters to task views.

If entering three or more criteria within one group of expressions, note that the And statements are evaluated before the Or statements. This differs from complex filter expressions in earlier versions of Microsoft Project and may produce different results. Across multiple groups of conditions, however, the And conditions are evaluated in the order that they appear.

Tip: To group criteria so they are evaluated together and not with other criteria, leave a blank line between the sets of criteria, and then select an operator in the And/Or field of the blank row.

Tip: When typing a value with the Equals or Does Not Equal test for a filter, use *(any character) or ? (any character) as a wildcard. The field specified for Field Name must contain a text value, such as Resource Names, rather than a numeric value, such as Duration.

Applying a Sort

1. **Click** Project

2. **Click** Sort

3. **Select** Desired Sort Criteria. Select Sort shown or Sort by to define a criteria. If the active filter has summary tasks, check Permanently renumber tasks and Keep outline structure (Figure 4-28). If your active filter is for incomplete subtasks or you do not want the outline structure to govern the sort order, make sure that keep outline structure is unchecked. (Figure 4-29).

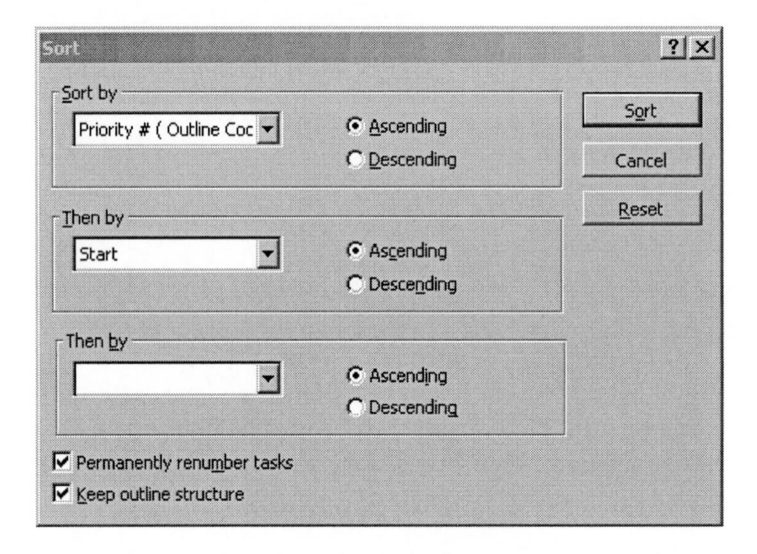

4-28: Sort By - **Option A**

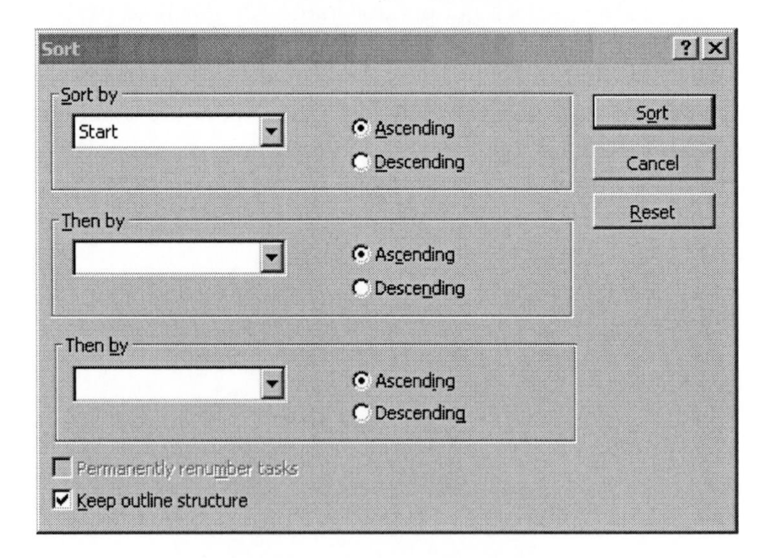

4:29: Sort By - **Option B**

Working With the Critical Path

1. **Click** Project

2. **Click** Filtered for:

3. **Select** Critical

4. **View** Critical Tasks. The Critical tasks are shown in red and are the current tasks that are controlling the project finish date.

5. **Click** Window

6. **Click** Split. This will create the Gantt chart on the top of the screen and the Task Form on the bottom). Select the split screen by double clicking on or dragging up the down arrow at the bottom right of the view window. This should display the task form showing the Predecessors and Successors for the selected task. If this form is not displayed, right click to display it.

7. **Review** Critical Tasks. Review the Critical tasks shown and determine which tasks can be expedited, shortened or delayed depending on whether the schedule is to be shortened or lengthened.

8. **Adjust** Duration. Adjust the Project Duration by:

   - Shortening or lengthening the duration of a task on the critical path.
   - Changing a task constraint.
   - Revise dependencies and relationships.
   - Set or revise lead and lag time between dependent tasks.
   - Communication with team members on expedited schedules and/or deliveries for items that are affecting the critical path.

9. **Continue** Adjusting the Critical Path. Continuously apply the critical path filter; when making changes to tasks on the critical path, new tasks may become critical and affect the project finish date.

10. **Finalize** The Project Duration

11. **Click** Project

12. **Click** Filtered for:

13. **Select** All Tasks or All_Incomplete Subtasks-Depending on the desired filter to apply.

14. **Click** Project

15. **Click** Sort by. This will resort the schedule according to the changes made. Then apply the Summary View or the Cascading View depending on the view desired.

Reports

Microsoft Project 2000 includes 29 predefined task, resource, and cross tab reports but no predefined monthly calendar reports; they will need to be created. Later versions of Microsoft Project have additional reports available. If none of the predefined reports meet project information needs you can create a custom report.

Templates developed by ConstructionScheduling.com include a Weekly and Monthly cash flow report.

Creating a custom report and editing an existing report are very similar; the only difference is the templates used. Editing an existing report requires changing a report that has been predefined to provide certain information, such as critical tasks, cash flow, or resource usage. Each Microsoft Project report, and any custom report created, is based on one of four templates: Task, Resource, Cross tab, and Monthly Calendar.

A custom task report using the Task report template can combine task tables, task filters, and task details (notes, objects, predecessors, and successors) with appearance and sort options to create the report needed. Once created, the custom task report can be edited just as if it were a predefined report. All custom reports are listed in the Reports list in the Custom Reports dialog box and are saved in the project file.

Like all reports, a custom task report uses information from the tables applied in various views. For example, the Task report template uses the task Entry table as its default, which is the default table of the Gantt Chart view. When task names and durations are entered in the Gantt Chart view, the information is used as content in the task report.

Entering Notes

1. **Double Click** Task Name. In the standard method of logging schedule information, use the Project Duration note field (Figure 4-30) to enter project scheduling information.

2. **Click** Notes Tab

3. **Enter** Notes. For the purpose of this example and the recommended documentation of the schedule, enter at least the following information:

1. Date of update

2. A brief description of the update

3. The name or method for whom or how you received the update information

4. The initials or name of who completed the update

5. Any specifics of how the schedule was restructured or substantially changed

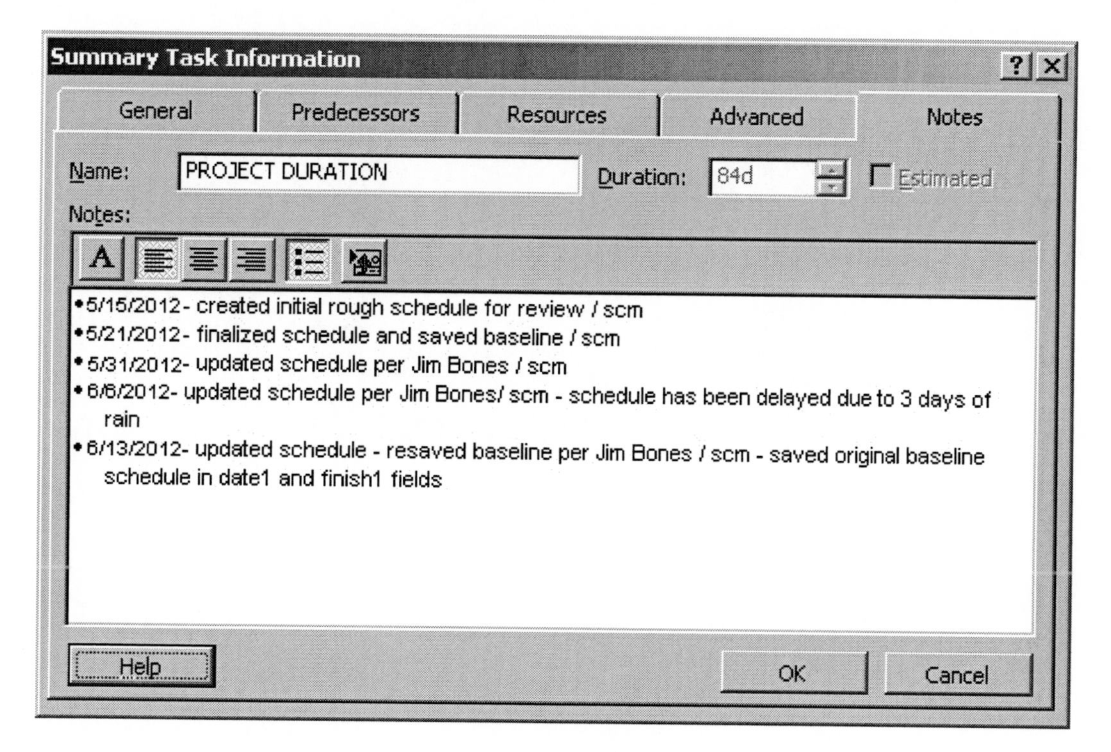

4-30: Project Duration

Saving the Baseline

1. **Click** Tools
2. **Click** Tracking
3. **Click** Baseline
4. **Click** Save baseline and For: Entire project. This method saves the baseline for the entire project. There are other times when the interim plan or the baseline should be saved for selected tasks.

<u>**Note:**</u> If the structure of the schedule changes and a new baseline needs to be saved, save the interim plan into the Start1/Finish1 fields as the original schedule for future reference.

5. **Click** OK The baseline schedule is now saved (Figure 4-31) and can be viewed against the current schedule by displaying the Tracking Gantt View.
6. **Enter** Notes in Project Duration Notes Field. Enter the date the baseline is saved with any related notes to establish a record of what you did and when.
7. **Enter** Date of final schedule In Page Setup>Header> center field

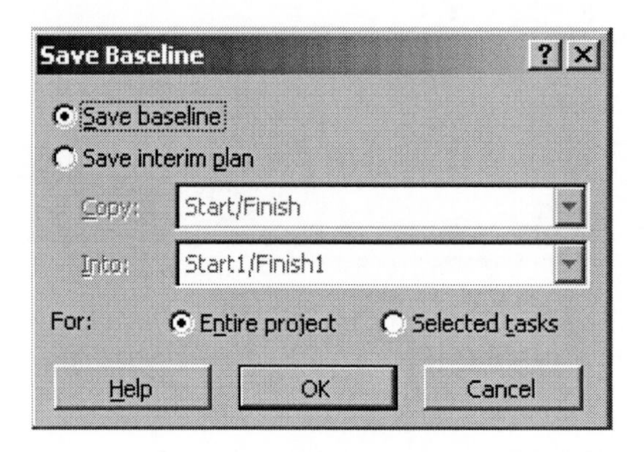

4-31: Save Baseline

Updating the Schedule

1. Click View

2. Click Tracking Gantt. This will bring up the tracking Gantt where all update information will be saved.

3. Click Project

4. Click Filtered for:

5. Click All_Incomplete Subtasks. This filter displays the Cascading View.

6. Click Project

7. Click Sort by?

8. Click Start. Sort by Start should be the only selection in the dropdown boxes.

9. Check Keep outline structure

(See Option B)

10. Click Sort. This sorts the project tasks by start date with no structural reference to the related summary tasks.

11. View Incomplete items since last update

12. Enter Actual Start dates for all tasks that have started

If the schedule is on track, only enter the activity information that falls within the date range of the last update. Notes specific to the tasks also can be entered in the notes field.

13. Enter Actual Finish dates for all tasks that have finished.

14. Enter Percent Complete / % Compl. The percent complete entered should be the percentage of the total cost completed. By entering the % complete in the Tracking Table, the % complete will automatically be updated in the Schedule of Values Table.

Updating the Schedule

When using a Schedule of Values, switch to the Schedule of Values Table to enter the % complete so that the Cost to Date and Remaining Cost can be viewed as the percentage is updated.

15. Enter Any rescheduled dates or task information that has been changed or affects the structure of the schedule. When updating the scheduled start and finish dates, review the new dates and how the schedule has changed relative to the current schedule.

16. Enter Schedule update notes. Enter the schedule update notes in the notes field of Project Duration.

17. Click File>Page Setup to enter the updated date on the schedule. The main date entered in the center of the schedule under the title is the original schedule date of the final schedule going to construction (the date that the baseline is saved).

18. Click Header Tab and Alignment Right Tab. The updated date will be entered on the right side of the header.

19. Enter Updated: (date of update)

20. Click Save. Switch back to any view and the information that has been entered will carry to all views (except any header information such as in step 19). As needed, enter the header information into each view that you are using, typically the Gantt View and Tracking Gantt View.

Inserting A Project

1. **Open** Consolidated or Master project file
2. **Select or Insert** Cell or row in the Task Name field
3. **Click** Insert
4. **Click** Project (Figure 4-32)
5. **Select** Project File to insert. Browse to the directory, file, or folder of the file to be inserted.
6. **Click** Insert. The project is now inserted into the Consolidated Project.

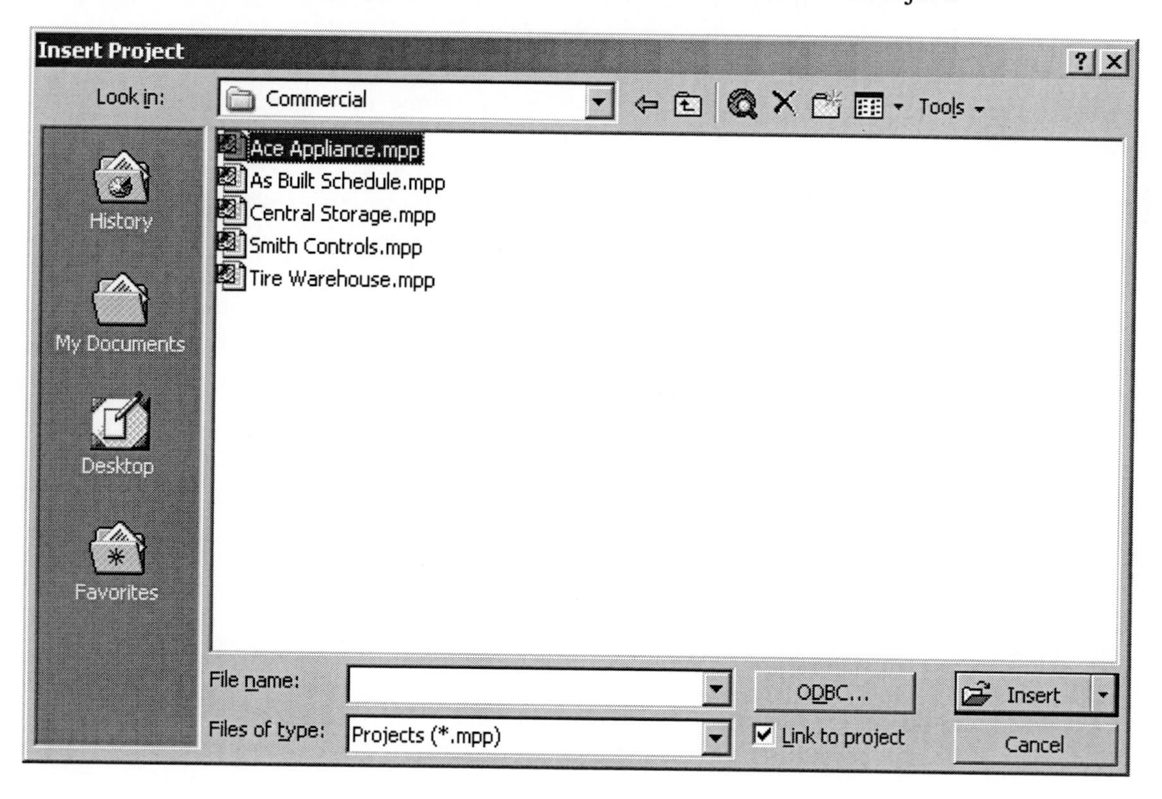

4-32: Insert Project

Using the Organizer

The Organizer contains all of the settings and features that a project uses. The Organizer is a great way to quickly transfer file settings between projects. Microsoft Project stores the default settings in the Global.mpt file. These settings are available in all projects unless altered. File settings can be copied, renamed, and deleted. The Organizer is divided into two lists. The list on the left shows the settings for the Global.mpt by default. The right side of the Organizer shows the current project file. To copy an item from the project file to the Organizer, you can highlight any or all of the options to copy and click

Copy to transfer the settings to the Global file. Deleting a file from the Global.mpt file makes it unavailable to all files unless the file has custom settings.

By default, the custom tables, filters, views, reports, and other items are available in the active project that is open. All of the custom features that are a part of the ConstructionScheduling.com templates are available for the template and output project files that you create. To use any of these custom features for creating project files that do not use the templates, copy any of these customizations using the Organizer.

Click the dropdown arrow in the available in: dropdown box and transfer settings between any open projects. Because template files are used to create all of the construction schedules, leave the Global.mpt settings alone and do not change or delete them. The original default settings will then be available as settings for other project scheduling. The Organizer is a valuable and efficient tool when customizing and copying file attributes from one file to another. It saves a tremendous amount of time and is a great way to keep files consistent.

Using the Organizer

1. Open Project Files. Open the project files to be changed.

2. Click Tools. Bring up the Organizer from any active file.

3. Click Organizer (Figure 4-33)

4. Click 'Drop-down' menus at bottom left and right of the Organizer. Select active project files to be copied, renamed, edited, or deleted.

5. Click Tab of settings to copy or change

6. Highlight Item settings from either file

7. Click Copy, Rename or Delete Continue to make all the changes between the files by clicking each tab.

8. Click Cancel when done

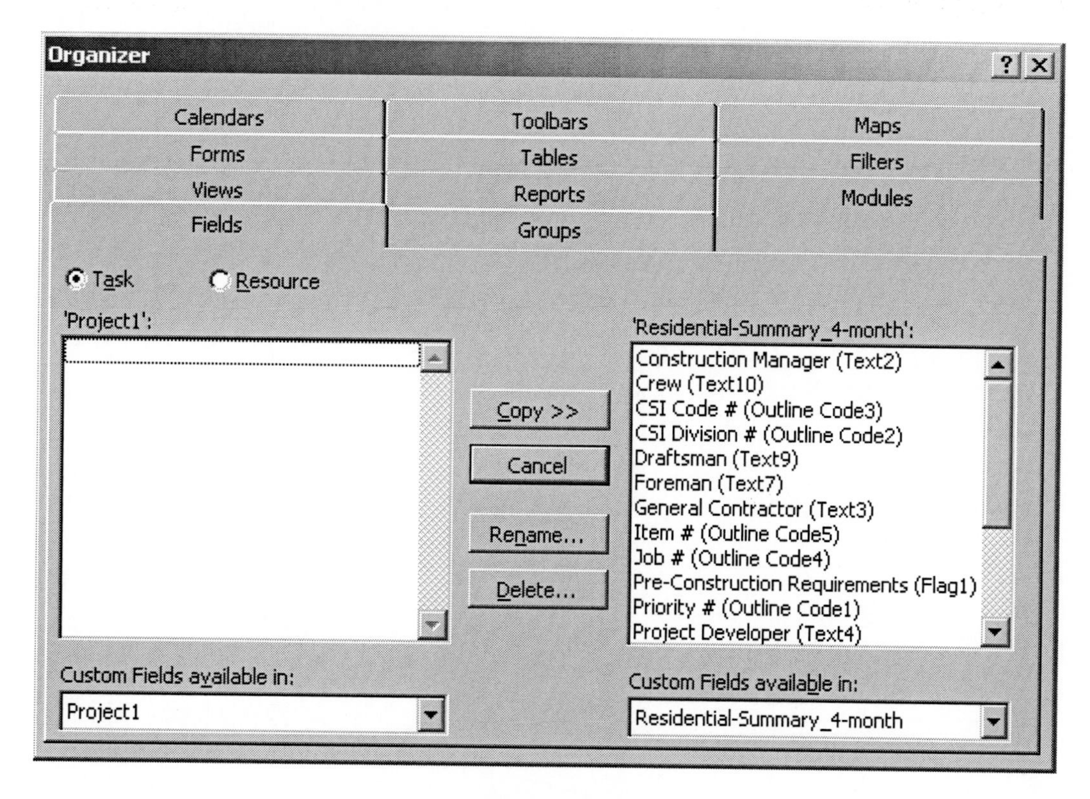

4-33: Organizer

Now that you've seen how Microsoft Project works and, specifically, how to put it to work in construction scheduling, the next chapter gets you started on your first project: pre-construction scheduling.

.

Chapter Five

Pre-Construction Scheduling

Pre-Construction Scheduling

Pre-Construction scheduling is the foundation of a successful project. It is the focus and attention to detail during this phase that determines the success of a project. Pre-Construction scheduling establishes the plan, sets the expectations for the project, and keeps everyone accountable. This phase is scheduled separately but linked directly to the actual construction schedule.

A project begins when you make a commitment to proceed on the project. A schedule needs to be initiated and structured from this point forward to determine all of the steps, milestones, and activities required to get the project ready for construction.

This process is scheduled as a separate phase and is usually managed by people other than those actually managing the construction process. This phase includes the development of a working construction schedule that all team members will eventually follow. It is also in this phase that the project time frame is set and the project goals are conveyed to all project team members.

Typically, this phase is not managed at the level required (if at all) to include the tasks needed to coordinate activities, phases and time frames for project success. Many times, schedules are prepared just prior to the actual construction phase. If the Pre-Construction process (Figure 5-1) is not managed properly, the project may be set up for failure, and you will not know it until you experience all of the problems and delays that can be caused by a lack of organization and planning.

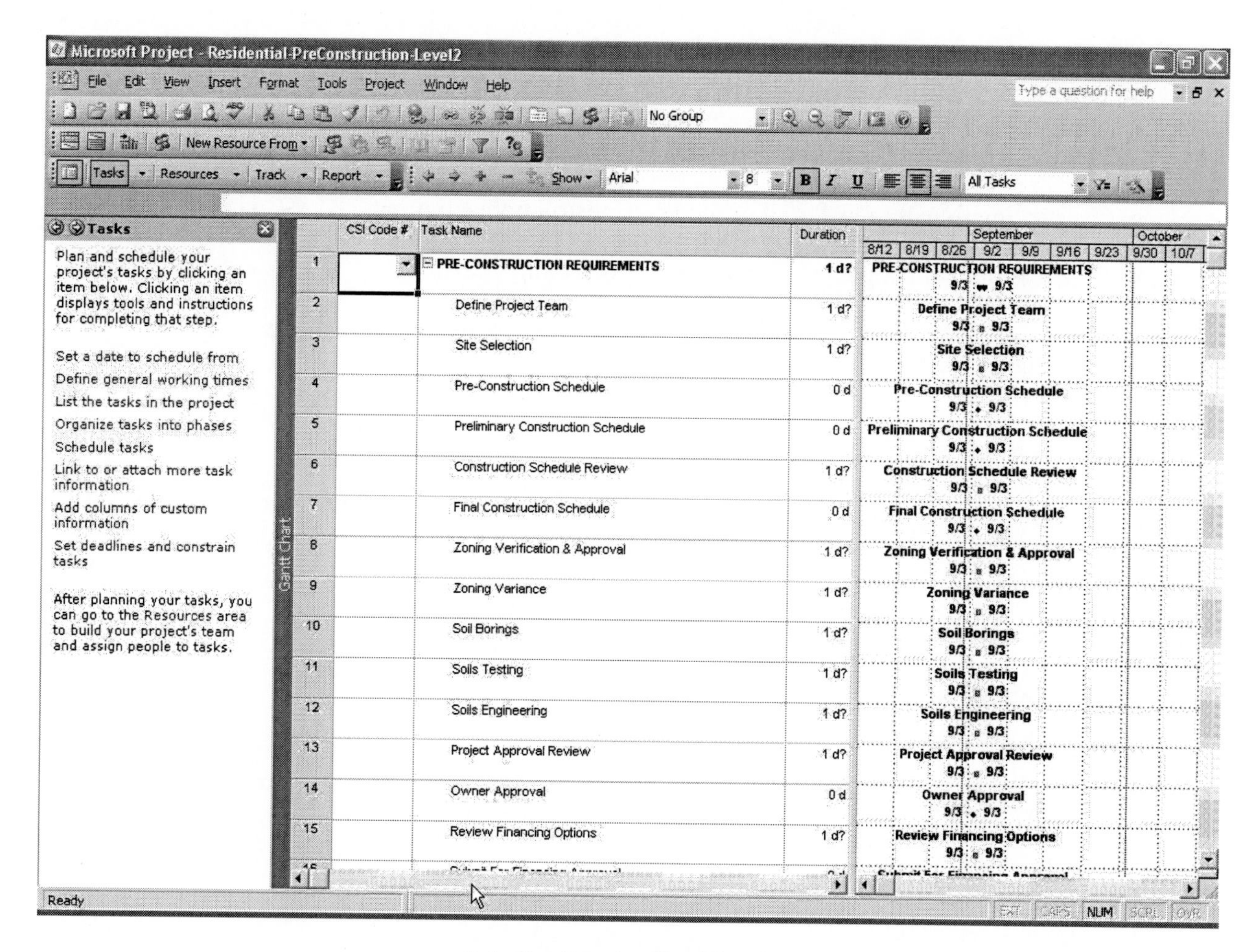

5-1: Pre-Construction Planning

A typical Pre-Construction Requirements summary phase with common Pre-Construction task items for major projects is included below. You can modify it to meet each project's requirements. This phase requires the most task editing, as each project is different in nature. The management of the Pre-Construction process sets the project up for success. The project cannot be completed effectively without a working schedule that provides interim goals, time frames, and milestones that everyone can use and rely upon.

Pre-Construction Scheduling

Pre-Construction scheduling consists of:

- Establishing all of the activities, relationships, and milestones required to get the project ready for construction.
- Estimating time frames for all of the Pre-Construction activities.
- Management of people, milestones, and project approvals.
- Establishing and management of lead and delivery items in accordance with the requirements of the project.
- Communication between everyone who is a part of the Pre-Construction process.

Keeping all of the people who are a part of the Pre-Construction process accountable including:

- Owners
- Architects
- Engineers
- Salesmen
- Draftspeople
- Project Managers
- Support Personnel
- Anyone else who is a part of the Pre-Construction process

Many projects are destined for problems when the project goes to construction. This is because there is insufficient planning, scheduling, and communication during the Pre-Construction phase. An environment must be created that keeps everyone and that allow you to make the necessary adjustments or changes to meet the goals of the project and its customers. This can only be done with effective management at the Pre-Construction phase and coordination by the use of Pre-Construction scheduling. The Pre-Construction schedule is linked directly to the actual construction schedule that contains the physical work items required to complete the project.

Pre-Construction Scheduling

In many projects, the managers ask for schedules from each contractor. This is also the process in many municipal projects. The combination of various individual schedules into a master schedule does not work for several reasons:

1. Most people do not know how to schedule
2. People schedule in an infinite number of ways
3. People can be poor communicators
4. Contractors provide schedules in many different formats and task breakdowns
5. Their schedules will be developed from their agenda and not yours
6. Many contractors are unorganized and do not work with schedules, and if they do, most are not working with them as a tool to help manage, forecast, and maintain a schedule and communication. They let the schedule run them instead of running the schedule themselves.
7. Team members don't know how to use in Microsoft Project or other scheduling software.

To take full advantage of the scheduling and communication features, schedules need to incorporate all phases of a project. This will provide accurate planning, cash flow projections, and management of the people and resources required to successfully complete projects. While specific information is needed from each contractor, the schedule should be prepared by the leaders of the project at this early stage. It is then the management's responsibility to develop a working copy of the schedule that all team members will follow. Hopefully the final schedule can be worked in as an attachment to the contracts so that worker performance is linked financially to the schedule.

The Pre-Construction scheduling phase can be controlled for a single project or across all current projects. Using views, filters, and reports can help the managers communicate at whatever level is required to keep everyone on track and accountable to the schedule. The whole goal of Pre-Construction scheduling is to develop a successful project plan and to keep everyone focused and accountable to that plan. The management of this process is invaluable in ensuring the success of your project and business.

Pre-Construction Requirements Table

- Preliminary Construction Schedule
- Construction Schedule Review
- Final Construction Schedule
- Zoning Verification & Approval
- Zoning Variance
- Soils Testing
- Owner Approval
- Review Financing Options
- Submit for Financing Approval
- Financing Review
- Financing Approval
- Design & Drafting
- Review Progress Plans
- Plan Review Meeting
- Finalize Construction Drawings
- 'Sign-Off' On Final Construction Drawings
- Landscape Design
- Civil Engineering
- Structural Engineering
- Mechanical Engineering
- Electrical Engineering
- Value Engineering
- Prepare Specifications
- Review Specifications
- Finalize Specifications
- Prepare Preliminary Estimate
- Estimate Review
- Prepare Final Estimate
- Notice to Proceed
- Letter of Intent to Begin Drafting
- Letter of Intent to Begin Sitework
- Letter of Intent to Order Steel
- Verify Existing Utility Locations
- Order Rebar
- Rebar Delivery
- Order Trusses
- Order Lumber
- Order Doors and Windows
- Demolition Permit
- Grading Permit
- Foundation Permit
- Shell Permit
- Locate Underground Utilities
- Site Surveying
- Property Corner Stakeout
- Building Stakeout
- Verify Existing Topographic Map
- Project Layout And Preparation
- Submit Payment Requisition for Approval
- Payment Requisition Approval
- Submit for Planning Board Approval
- Planning Department Review
- Concept Planning Board Approval
- Final Planning Board Approval

Pre-Construction Requirements Table

- Submit for Health Department Approval
- Health Department Review
- Final Health Department Approval
- Submit for D.O.T. Approval
- D.O.T. Review
- D.O.T Approval
- Submit for Highway Department Approval
- Highway Department Review
- Highway Department Approval
- Submit for Water Department Approval
- Water Department Review
- Water Department Approval
- Submit for Pure Water Department Approval
- Pure Water Department Review
- Pure Water Department Approval
- Initiate Gas Service Order
- Initiate Electric Service Order
- Electric / Telephone Planning Meeting
- Electric / Telephone Planning
- Building Department Submission
- Building Department Review
- Building Permit Acquired
- Prepare Bidding & Proposal Documents
- Credit Approval
- Contract Negotiations
- Contract Legal Review
- Execute Contracts
- Mechanical / Electrical Subcontractor Review Meeting
- Pre-Construction Meeting
- Turnover Project Communication Flow Chart
- Meeting with Owner
- Interior Design
- Review Design with Owner
- Selection of Finishes
- Finalize Finishes
- Submittal Submission
- Submittal Review
- Submittal Approval
- Shop Drawing Submission
- Shop Drawing Review
- Shop Drawing Approval
- Lead Time Order Items

The steps to preparing and maintaining a Pre-Construction schedule are outlined on the Scheduling Checklist below and developed in the following chapters on Organizing (6), Planning (7), Controlling (8), and Consolidating Projects (9).

Scheduling Checklist

The following checklist is a quick reference guide to preparing your Project File from a Project Template. This is offered as a sequential list and a quick reminder of the steps required in preparing a schedule. Depending on the state of the project and the information known, the steps may alter in sequence, although most steps will need to be complete to finalize the schedule. You can print this out and keep it with your project file as a reminder of the status of the items.

Project Name: ______________________ Date: ___________________

Prepared By: ______________________

*** indicates optional steps**

- ☐ Open a Project Template
- ☐ Enter Project Start Date
- ☐ Enter File Properties
- ☐ Save the Project File
- ☐ Customize Your Task List
- *☐ Define Master Entry Table
- *☐ Share Resources
- *☐ Enter Resources
- *☐ Enter Costs
- ☐ Estimate Task Durations
- ☐ Create Task Links
- ☐ Set Milestones
- ☐ Set Constraints
- ☐ Set Lead & Lag Time
- ☐ Evaluate the Schedule
- ☐ Adjust Project Duration
- *☐ Assign Resources
- ☐ Communicate the Plan
- *☐ Prepare a Schedule of Values
- ☐ Finalize The Schedule
- ☐ Save The Baseline

Chapter Six

Organizing Schedules

Organizing Schedules

Open A Project Template

The first step in building a schedule is to identify the type of project and the level of detail desired. To make this a simpler process. All customizations are detailed in the appendices.

Three Project templates are available, two of which are included with this book:

1. Residential Master (Appendix E) AVAILABLE AT CONSTRUCTIONSCHEDULING.COM
2. Residential General (Appendix F)
3. Residential Summary (Appendix G)

To begin, open a project template:

Opening A Project Template

1. **Open** Microsoft Project
2. **Click** File
3. **Click** Open, Browse to the directory and file folder of the template file (Figure 6-1).
4. **Double Click** Project Template File. Upon saving a template file, a prompt will request a new file name so the template file is not changed.

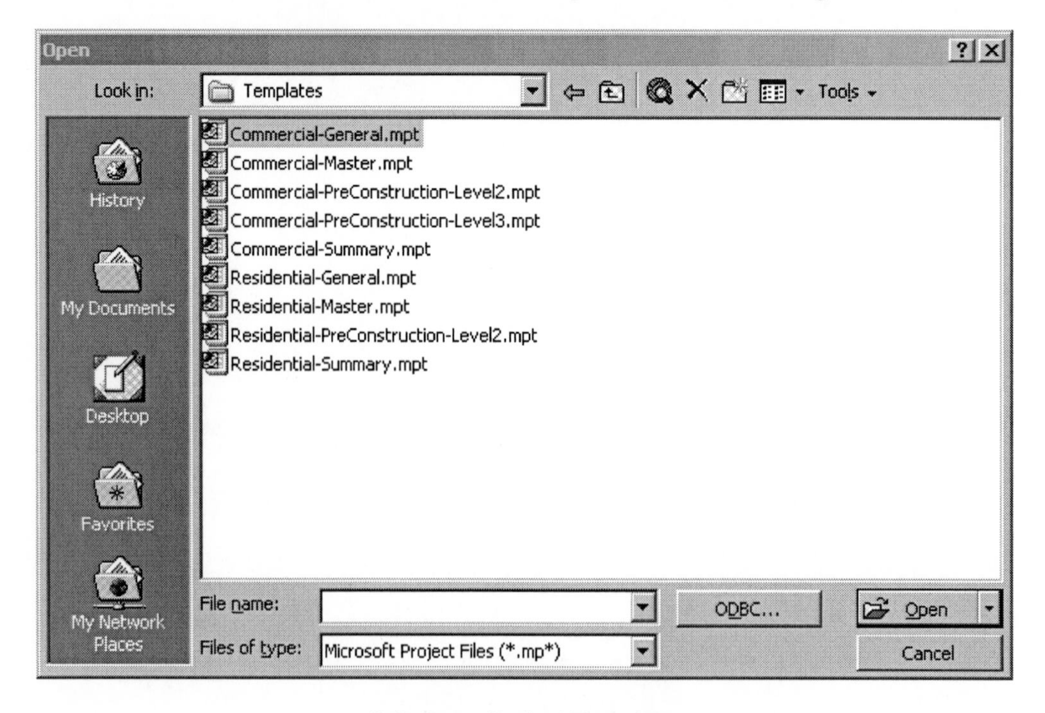

6-1: Open Project Template

Enter Project Start Date

The first step after opening a project template is to enter the Project Start Date in the Project Information dialogue box. The start date entered should be the anticipated start date of the project. If scheduling Pre-Construction activities, the start date should be the date that the Pre-Construction activities are scheduled to start.

Entering Project Information

1. **Click** Project
2. **Click** Project Information
3. **Enter** Start Date for the Project. The start date should be the anticipated start date of the project (Figure 6-2). The Start Date can always be changed later.

Enter the date in a format xx/xx/xx or use the Calendar in the dropdown.

4. **Make Sure** Schedule from: Project Start Date is selected and Calendar is set at Construction. Always schedule from the Project Start Date. Scheduling from the Finish date will throw off the calculations and the schedule.
5. **Leave** Current date as shown, Status date as NA, and Priority at 500. These can be changed later if required.

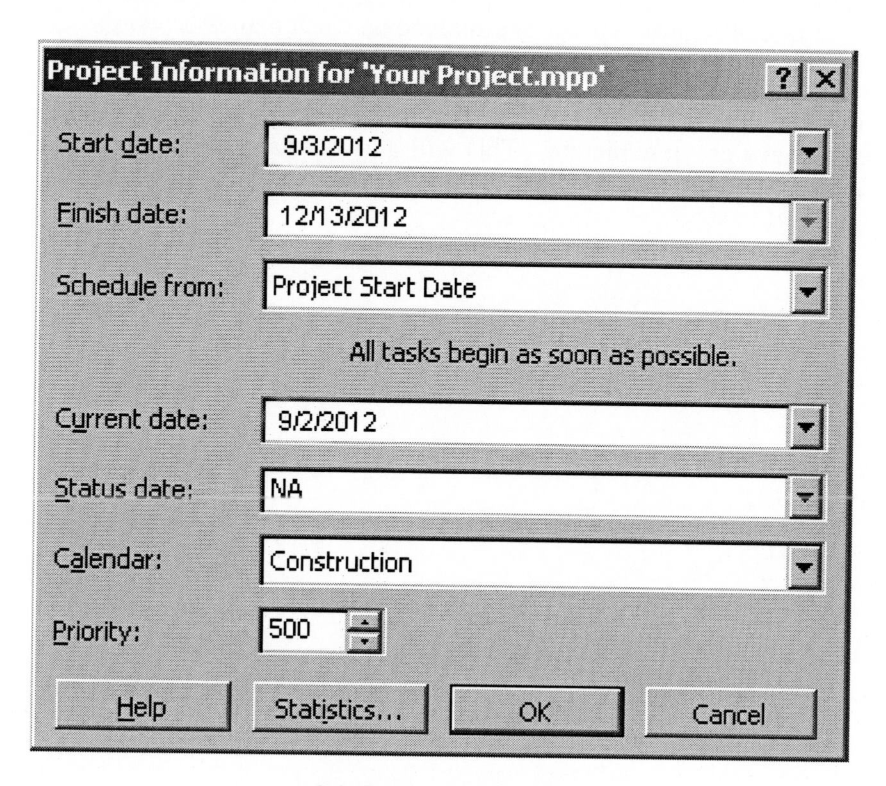

6-2: Project Information

Enter File Properties

After the Project Start date is entered, enter the file properties specific to the project. The file properties will give specific project information and provide automatic fill-in fields for reports and views, as well as file details for searching (Figure 6-3).

6-3: Property Headings

1. **Click** File
2. **Click** Properties
3. **Click** Summary Tab
4. **Type** Project File name
   Author
   Manager
   Company

The project file name entered is the name that is auto-filled on inserted projects within a master project and is included as the title of reports. Spaces are allowed between names or words.

Complete any other text fill-in boxes as desired.

5. **Click** Save preview picture. This will allow for a small preview in Windows explorer when searching for files.

The other tabs on the properties dialog box are for reference and additional information if you want to use them:

General: Describes the file that stores the document.

Statistics: Provides statistics about the project file including the file name, type, location, size of the file, file creation date, last modified date, and last opened date.

Contents: Displays summary statistics about the project.

Custom: Enter additional properties to the file to be able to for the file by the custom properties (Figure 6-4).

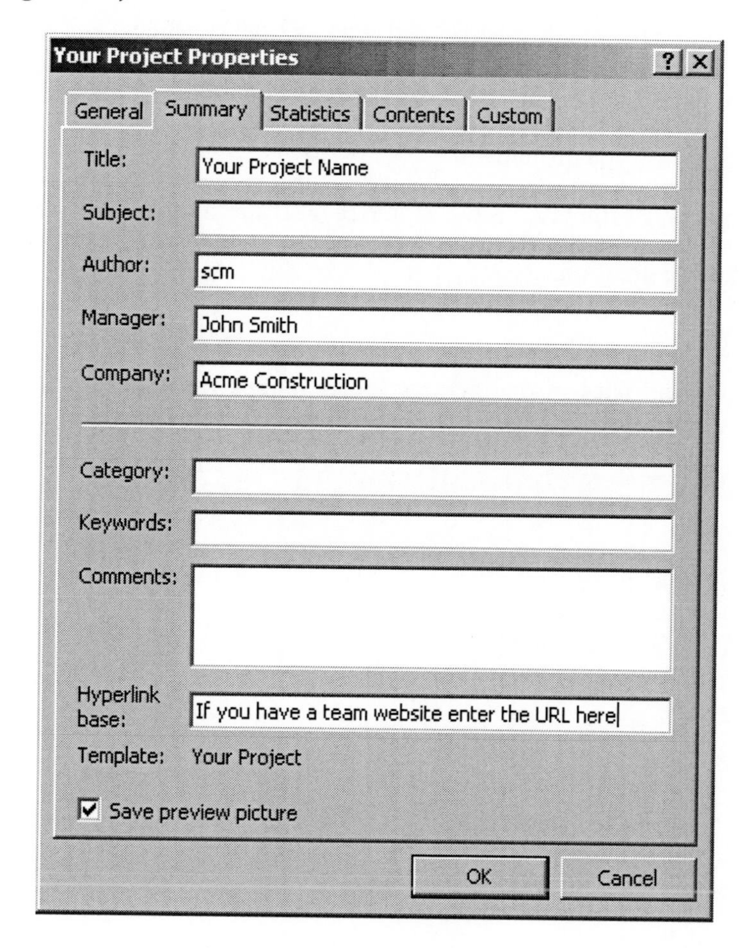

6-4: File Properties

Save the Project File

Save the project file after the file properties are entered. The name chosen will be the name that is auto-filled into headers, reports, etc. so it should be a name that everyone can relate to. If the file will be published on the Internet, use characters (- or _) between names and words, as spaces cannot be used.

1. **Click** File
2. **Click** Save As
3. **Click** In Save In dropdown. Browse to the directory and file folder to save in. Save all templates and project files in a subfolder under a main folder named Project Scheduling. This makes navigation easier when working with several projects.
4. **Type** Desired File Name in File name dropdown or choose the file name if overwritten. Don't use spaces between words or names, so that the project file can be saved as a web page. Use a - or _ as a divider between words or names.
5. **Click** In Save as type dropdown
6. **Click** Project or Template. Save as a Project if the intent is to create, work, or modify the file (Figure 6-5). Save as a template to use the information for future project schedule creation.
7. **Click** Save

Saving project templates and files in one central folder makes future reference easier. Keep template files in separate folders from project files to make sure they are not over-written.

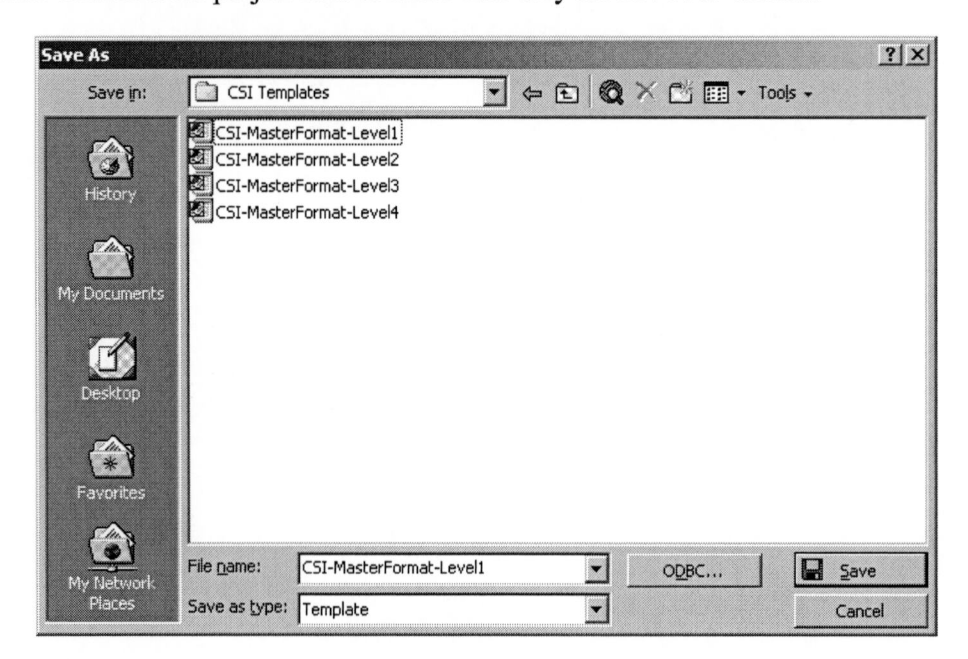

6-5: Save As

Note: When organizing schedules, refer to Chapter 4 for instructions on performing specific tasks. Also refer to Chapter 10 for advanced techniques.

Customize the Task List

The next step is to create a meaningful task list specific to the project. This process will first involve deleting unneeded task items listed in the template. The initial task list will be a rough draft that should be shared with all of the available team members. Communication with everyone who is part of a project is the key to building a successful schedule.

The Greater detail included in the task list will provide greater accountability and control. If the project specifications are available, use the specification listing to match the task list as much as possible. Doing so will keep 'project-specific' information in the schedule. This will create more attention and accountability to the schedule. It will also be another reminder of the respective specification items and will make the schedule more relevant to the entire scope of the project. You will also be able to take advantage of the Schedule of Values feature (if costs are entered) that is included with the templates.

If using a CSI specification, consider changing the code for each of the respective task items so that it will directly correspond to the specification listing.

Note: If adding tasks, add a CSI code number for the task. It is best to use the same CSI code number as one that the task relates to. This will help when filtering for certain code items. Use any of the templates or the CSI coding task lists for entering CSI codes.

If using another coding structure, add a custom outline code field to enter these numbers. Don't overwrite the CSI codes in the Outline Code 3 field, as this numbering structure will be common to all projects. The same base coding structure can then be used for all projects so they can be filtered for items across multiple projects.

The task list section under Pre-Construction Requirements requires the most modifications. The detail must be specific and tailored for the particular project being worked on. Filter for Pre-Construction under the filter menu and distribute the Pre-Construction task list to the appropriate team members for review, comments, and suggestions (Chapter 4). Remember that the management of the pre-construction process will directly affect the outcome of the entire project.

Note: If adding Pre-Construction tasks, insert the column field Pre-Construction Requirements and mark the added task Yes (Chapter 10). This allows you to filter for all Pre-Construction tasks when the Pre-Construction filter is applied. Hide the Pre-Construction Requirements column when done.

To customize the task list:

1. **Open** Project Template File
2. **Click** In top summary task cell. This is usually Pre-Construction Requirements.
3. **Select** Tasks or groups of tasks to delete. Select individual tasks by pressing the delete key or groups of tasks by control click or click and drag.
4. **Delete** Tasks not required for the entire task list.

Note: Deleting summary tasks will delete the summary task and subtasks below it.

5. **Go** Back To Top Task in Schedule. Pressing Control Home will move the cursor to the top task.
6. **Add Tasks** Add desired tasks Press the insert key to insert above the cell select and type the new task name.

To add a summary task, insert the CSI Division number and the Priority number for top level tasks. This is needed for filtering and sorting purposes. Insert a column to the left of the task and selecting the division number from the dropdown menu in each cell.

7. **Enter CSI Codes** At Inserted Tasks in the CSI Code number column Use the same CSI Code number as a similar task or enter a new CSI Code number in the next sequence of numbers to a similar task. Use the same CSI Code number to keep consistency for filtering and reporting. Use any of the template schedules or CSI task listings in the appendices of this book as reference for entering CSI codes.

Define Master Entry Table

After the task list has been edited, define the responsible parties and information to be tracked and communicated relative to the scheduled tasks. The Master Entry Table is built to enter this information all in one place and have one place to edit the information.

Defining the Master Entry Table is the process of filling in known information in the customized fields for viewing, reporting, and filtering information (Chapter 4). This information consists of fields for different resources working on the project and flag fields to identify any items to be filtered (Chapter 10). It is important to use the same fields for specific information and resources across all projects so that they can be viewed, filtered, and reported on globally.

It is possible that not all information will be available at the initial scheduling phase, but enter whatever information is known. It can be updated any time. By defining these items in their respective fields, they can be viewed, filtered, and reported. For example:

- The Responsibility field can be used to identify the various subcontractors and team members responsible for each respective task item. The Subcontractor field can be used, although all of the task items may not be completed by subcontractors. If the Responsibility field is used, the architect, engineer, subcontractor, or other in the same field can be listed. If the files are shared with others using the same scheduling structure, the Subcontractor field can be available and used for their subcontractors. Add whatever fields are most practical for the project.

- With the Flag 3 field (Subcontract Work), each subcontract item can be marked Yes (Chapter 4). You can then filter for intended or contracted subcontract work. This can aid in buying out a project or viewing the amount of subcontract work or potential subcontract work within a single project or across several projects in a Consolidated Master project. By viewing the filtered results and evaluating the workload you can decide whether to subcontract more work or to 'self –perform' work if the workload is light.

Note: The custom fields used should be the same in all projects. This allows filtering for relative information across all projects in a Master Project or Consolidated project file. The custom fields that are a part of each project file are the same as the ones used in the Consolidated Master project file.

Share the Resources

To take advantage of managing resources across many or all projects, create a resource pool in a Master or Consolidated project for resources that all projects share (Chapter 9). This is not required, but it offers all of the available features to report, view, and filter information across all of the projects and includes business planning information beyond a specific project file.

If a resource pool is used, the next step is to share resources with the file that contains the resource pool (Chapter 4).

Enter Resources

Resources are the people, equipment, or other entities responsible for actually completing the tasks in the project. There are two types of resources: work resources and material resources. Scheduling does not require that resources are defined and scheduled; however, if resources are defined it will make viewing, filtering, and tracking information relative to them easier.

Within this system of scheduling, resources are defined within the Microsoft Project resource sheet as the people, equipment or other entity in YOUR ORGANIZATION to be tracked and or communicated with to provide management of these resources. There are two reasons that only resources specific to your organization are entered:

1. To provide a method of managing all people and/or equipment that are directly responsible or accountable to a project throughout all phases of construction, i.e. sales, marketing, drafting, project management, support personnel, as well as the actual people and equipment that are part of completing the physical construction on a project. This provides for a general management tool that will use all people or other resources within your organization for completing a project (Chapter 10).

2. The other reason only the resources only are scheduled is to create a common resource pool that can be managed across all projects. In a consolidated project or master project including all active projects, a common resource pool can be used to view, filter, and report on the resources at a global level. If different subcontractors and other entities are entered into the resource sheet for different projects, the resource pool would be very large and unmanageable.

Combining projects into a consolidated project and viewing information across all projects is one of the greatest features within Microsoft Project. Schedules should not all be different and stand alone. There should be a commonality across all projects.

There are other resources that are a part of the individual projects scheduled, such as subcontractors, architects, engineers, and other outside entities. You are not responsible for managing or coordinating their individual resources. You are, however, responsible for managing these various groups and creating coordination and accountability to create accurate and effective scheduling. The scheduling system will assist in entering, tracking, filtering, and viewing information for all outside resources in the Master Entry Table (Chapter 4) with the customized fields (Chapter 10) that are programmed into Microsoft Project. This is an effective way of providing a list format for all outside resources. These additional fields allow for quick viewing of information in a column listing format across the different layers of responsibility for the various tasks of the project. These additional resources are entered and displayed in tables and views.

Included with this book is a project file named Consolidated Projects.mpp. This is an empty file pre-programmed with all of the features and customizations that are a part of each of the template files but has additional formatting features to enhance viewing. This is the file where shared resources should be entered.

Resources as entered into the resource sheet in this system would be similar to the following:

- People in your organization that are directly involved in and accountable for the completion of tasks. This would include: all individuals within the various departments or across your organization i.e.

  Sales

  Drafting

  Support Personnel

  Project Managers

  Superintendents

- Types of trade people within the organization that have a specialized skill or complete certain types of work i.e.

  Concrete Workers

  Steel Workers

  Carpenters

  Painters

  Operators

  Laborers

- A listing of available major equipment typically used and tracked, such as:

  550 Dozer

  850 Dozer

  JCB Forklift

  Loader

  JD Backhoe

Schedulers should understand the purpose of entering and using resources as well as what this information should provide. Microsoft Project uses effort-driven scheduling as the default method for determining the calculation for the duration of the tasks, meaning that the task duration is dependent upon the amount of resource work or the number of resources assigned to the tasks.

The default scheduling method can be modified to fixed duration. With this method, Microsoft Project leaves the project duration unchanged by the amount of work or units assigned to a task. The reason that fixed duration is used as the default is because there are many different trades that need to manage their individual resources to maintain the project schedule. The resources must be managed to meet a schedule; the resource productivity cannot determine the schedule. Typically, with enough time, planning, communication, and accountability, anything can be completed. There is always another way of getting the job done. If you or a subcontractor cannot complete the task schedule for one or more items, either add resources, add overtime to the existing resources, or subcontract a partial or all the tasks work to maintain the project schedule. The reason that most schedules are delayed is because they do not have the vision and the time to make the above adjustments, and consequently they must deal with the resources available at the time to complete the work. This potentially drives the costs higher, delays the task item, and potentially the schedule finish date itself.

Resources can be filtered for your organization, subcontractors, or other entities to isolate and view associated tasks and communicate them accordingly.

Specifically, use resources in projects when the goal is to:

- View, filter, report, and communicate the task items associated with people and equipment in the organization.
- Ensure high accountability for and understanding of the project. When responsibilities are clear, there is a decreased risk of tasks being overlooked.
- Monitor resources.
- Provide for resource accountability.

The resources assigned to the individual tasks in the schedule are the resources that are directly responsible for getting the work completed. Schedulers mainly work with people and equipment. Using these main resources helps create accountability, evaluate availability, and use the features within Microsoft Project to include with the views and reports.

Resources can be entered in one of three ways:

1. Create a resource list within the project plan.

Create a resource list within the project plan when resources are to be used solely by the project. Enter the resource information all at once in the Resource Sheet view. Resource information in the templates includes the resource name, type, initials, group, and max. units. Resource information also can be entered while making assignments.

Note: If you enter resources in a separate project file, you will not be able to take advantage of managing the resources across multiple projects using a resource pool as described in item #3.

2. Share the resource list created in another project plan.

Share the resource list created in another project plan when the resources needed for the current project are listed in the other plan. When another project's resources are used, that resource list is shown in the project plan.

Note: If you enter resources in a limited number of project files, you will not be able to take advantage of managing the resources across multiple projects using a resource pool as described in item #3.

3. Share resources with other projects, a Consolidated Project, or Master Project from a resource pool. (Recommended)

Share resources with other projects from a resource pool when several projects repeatedly use the same resources. A resource pool is a project file that contains only resource information to be shared by several projects.

*This method is most desirable because it allows you to manage and view resources across all active projects.

To use this method, enter all project resources in the Master Project or Consolidated_Projects.mpp file accompanying this book.

The method used depends on your project requirements, resource use in a specific department, and resource use throughout the company. Now that resources are outlined, the next step is to determine the method of scheduling resources and to enter them (Chapter 4).

Enter Costs

In Microsoft Project, costs for materials, equipment, and resources can be entered. However, many contractors have found that in the current evolution of scheduling software, tracking costs for the above items is ineffective. This is true due to the fact that this method of scheduling uses Fixed-Duration scheduling rather than Effort-Driven scheduling. Schedules are used to manage time and not cost. In many ways they do not work directly together. If using scheduling to track costs, use 'Effort-Driven' scheduling. In this method, the resource productivity determines the schedule, and adjustment to the schedule will be required if delays occur. It becomes a cumbersome effort to try to maintain and expedite a schedule. Using Fixed Duration scheduling, normal windows of time are established allowing for normal production and downtime. In construction, delays and setbacks occur and schedulers need to adjust accordingly. Using Fixed Duration scheduling allows for managing and using slack time to help compensate for various delays. Using costs for tasks, billing and cash flow is effective and provides a great tool to use in coordination with billing by using the custom Schedule Of Values table. By entering selling costs for tasks or levels of tasks, a schedule can be used as a billing mechanism and produce effective cash flow reports as a byproduct. In addition, using selling costs which are usually public and available to most team members on a project, schedules can be shared electronically without the worry of revealing actual costs.

Using a Schedule of Values

A schedule of values is a listing of costs used for billing purposes that is provided to banks, owners, or other payers. The billing cost is presented as the percent of cost complete for the task or summary item.

Using the Schedule of Values, the following example uses the percent of cost complete to equal the percent of work complete for the task and summary items so that the schedule can be used as a billing tool. Using the custom Schedule of Values table within schedules for billing purposes will force more attention and accountability to the schedules. In addition, it will force the people who are managing the schedule to make sure that the schedule is 'up-to-date' and accurate.

Using the schedule with the custom Schedule of Values table provides a single source of information that displays the percent of cost and work complete for a task or summary item. Ideally these items should coordinate, but they seldom do. By entering the percent complete for a task work item, the current cost and remaining cost are automatically updated.

Use this view of the Schedule of Values table (Chapter 10) to serve as an attachment or backup for billing purposes. People understand pictures. Using the schedule with this table will also serve as a foundation for cash flow filters, views, and reports.

View Sample Schedule of Values

Using the Schedule of Values to mirror the percent of work complete will result in much greater attention to the schedule. Team members will look forward to getting the schedule and will be more motivated to complete their work rather than being focused on how much they can bill for.

Cash Flow

Cash flow is critical to your business. Access to an accurate picture of cash flow is limited by the ability to see costs relative to the actual work in process and to be completed.

Using a schedule of values in the schedules gives a clear and accurate picture of the cash flow. Cash flow and billing are directly linked to the progress and completion of work. By using the schedule to track the percent complete and the cost complete you will have the ability to filter, view, report, and forecast the financial picture so that intelligent financial decisions can be made.

There is a customized monthly and weekly cash flow report in all of the templates and schedule files. These reports automatically output information if a Schedule of Values is used to reflect the selling costs. The Schedule of Values table provides the ability to input and update all costs in one place. Cash flow can be viewed for a particular project or across all projects.

Using a cash flow report with the schedules provides the ability to:

- Report to banks or lenders to increase the company's financial strength
- Report to owners so that they are aware of their financial responsibility
- Control purchasing and deliveries relative to available funds
- Provide a foundation for construction draws
- Eliminate guessing or inaccurate estimating of cash flow
- Instill confidence in lenders, banks, and owners that you are in control and have accurate figures and reporting
- Adjust the workload in accordance with available funds and outstanding loan commitments
- See cash flow relative to current obligations
- Adjust profit margins on future work

Cash flow and profits are the bottom line. Even if projects are profitable, not having an accurate picture of cash flow and of financial requirements can reduce the power needed for managing your business.

The next step is to Establish a Schedule of Values and Enter Costs (Chapter 4).

Using Filters

A filter is an action applied to the schedule to view information desired for tasks and/or resources. Filters limit the view of entries displayed in the task or resource sheet for the criteria defined in the filter.

There are many pre-defined filters within Microsoft Project. Custom filters also can be created to filter out unneeded information. Simplicity is the key to communication. Filters can be applied to task, or resource, and the resulting information can be passed on by Internet, e-mail, or physical distribution. Applying and effectively using filters is a very powerful tool in communication and planning.

Applying a filter (Chapter 4)

Within the standard templates are included custom filters for tasks and resources. These filters are common filters used in the construction industry to isolate the information for particular tasks, groups, people, etc. Use the filters as shown or create new filters based on from the ones included in the templates.

Create or modify a filter (Chapters 4 and 10)

To use undefined filters, enter and define the information in the fields associated with the filter. This can be done in any table view or in the Master Entry Table as described earlier in this chapter.

The filters included with the templates filter for the criteria described and also filter out all completed tasks. The purpose for this is to allow you to view current and future information relative to the project or group of projects.

Date Range filters have been provided to view all information for particular criteria with a specified date range. Use this filter to specify the dates for information to be viewed.

Filters are one of the most powerful and useful features in Microsoft Project. Use filters to eliminate unrelated tasks, view relevant information for the responsible party, and distribute this information accordingly. People tend to pay attention when information is condensed and specific to their responsibilities. If a detailed schedule is delivered, specific information tends to get lost in the vast amount of tasks, and they seem less important. Filtering for specific tasks, resources, and responsible parties increases accountability and action.

Using Notes

Notes are very useful to document information relative to each task as it relates to the schedule (Chapter 4). If delays or specific information are important, enter this information in the note field attached to the particular task. Memories are short and nobody remembers the same thing. Documenting important information will create a record of what actually happened and why. No one can dispute a written record of events. Liability will be greatly reduced by maintaining notes of information that affect, the schedule. For example, if weather delays extend the time of a project, note this in the Notes field for the affected tasks. Make sure that whoever is communicating the information is aware that they should report any critical comments that should be attached to the schedule.

For the purpose of this system of scheduling and reporting, this chapter only focuses on the information as it relates to the dates used to update the schedule, who provided the information for the update, and any critical information as it relates to the schedule as a whole. All other note information is left to discretion, but use the notes field as much as required to create an accurate accounting of the schedule and its history.

To enter updating information as it relates to the schedule use the Project Duration note field to log and input the following information:

1. Date of update
2. A brief description of the update
3. The name or method from whom the update information was received
4. The initials or name of who completed the update
5. Any specifics of how the schedule was restructured or substantially changed.

If this information is not entered, there is no way to look back and see how well the schedule was maintained and how often it was updated. In addition, it is a good place to document items such as when the baseline was saved and what was done with the information.

When people know that things are being documented and that they are being held accountable for communicating information, there will be increased attention, action, and enthusiasm.

Viewing the Schedule

A view is the display of information shown on the output schedule with an associated timescale. View information can include all tasks, filtered tasks, costs associated with tasks, etc. Information can be displayed in many different formats for any particular purpose necessary. The version of the schedule viewed is a combination of the active filter with the sort criteria.

Effective communication lies in the ability to provide clear and understandable information so that anyone, regardless of their experience and knowledge in working with schedules, can view and understand them. Using views in combination with tables, filters, and reports gives everyone the 'picture' of the schedule and their associated responsibilities.

There are two primary timescale views with all tables, filters, and reports:

Summary View

The Summary View (Chapter 4) shows the timescale for all tasks and resources structured by priority under Pre-Construction Requirements, Project Duration, CSI Division number and CSI Code number.

The Summary View is helpful when viewing the overall schedule for phases and for Division responsibilities.

Cascading View

The Cascading View (Chapter 4) shows the timescale for all tasks and resources that flow under Pre-Construction Requirements and Project Duration. All other summary tasks are filtered out and the tasks are shown in sequence by start date.

The Cascading View allows you to see the priority of work items sorted by start date while keeping the tasks under the appropriate Pre-Construction and Project Duration phases.

Using the Timescale

How the timescale is displayed depends on what methods are used to distribute and convey the schedule (Chapter 4).

Basically, the goal is to minimize the timescale as much as possible.

- For most overall schedules it is most useful to use the Major scale as months and the Minor scale as weeks.

- When the Major scale is displayed as weeks and the Minor scale as days, attention is focused on the specific date in relation to the calendar format. This is effective when the timescale displayed is not too large and can fit easily on the screen or printed page.

Adjust the timescale as required to capture the attention of the viewers. If printing a hard copy of the schedule, adjust the timescale to the paper size. Schedules that are too large or have too many pages are less effective. Also use the General Size percentage box to view the schedule at a reduced and scaled size. Verify the size using Print Preview to make sure that the timescale clearly shows the needed information.

Using a filtered version of the schedule with the Major scale as weeks and the Minor scale as days is very effective when the area is not too large.

Sorting

Sorting the task or resource list enables you to view information designated by a certain criteria.

1. Sort tasks or resources by criteria, such as CSI Code, CSI Division, Priority number, Task Name, Deadline, Resource Name, etc. Sorting can be useful to see tasks in a particular sequence.

2. Sorting is not maintained when switching views, but is saved when a project file is closed. However, a custom sort cannot be saved.

3. Using an active filter allows sorting by criteria for that filtered information only.

Applying a Sort (Chapter 4)

Sorting is often done by start date so the tasks can be viewed in sequence. Enter the criteria for the sort so the sort information can be viewed as desired. For example, when sorting by start date with a summary task view, Microsoft Project will sort the Summary Tasks first and then the subtasks.

Keep the CSI Coding structure when sorting so that everyone is familiar with the outline structure. A sort criteria is predefined by using the custom Outline Code 1 field so that the sort numbers can be redefined if necessary.

Using the desired filter with the sort options will allow you to view and distribute specific information in a particular order. Examples of two main filter views are as follows:

1. All_Incomplete subtasks

This filters out all summary tasks excluding Pre-Construction Requirements and Project Duration. When this filter is applied and then sorted by start date, it results in a Cascading view of all tasks separate from their respective summary tasks. This gives a clean flow of task items but also keeps them as subtasks under Pre-Construction or Project Duration so that the two phases of construction are isolated.

This is shown in the following example. Note that the tasks flow by start date and the summary tasks are filtered out. The CSI Code number is unrelated to the sort criteria.

2. All Tasks

The All Task filter maintains the outline CSI Coding structure and then sorts by start date. In this case the summary tasks can be sorted Priority number and then by start date.

The next area of project development is planning schedules.

Chapter Seven

Planning Schedules

Estimate Task Durations

Estimating task durations is a key to building a successful schedule. This process involves experience and knowledge of average time frames required to complete certain types of work.

In our system of scheduling, we are using fixed duration as the default method of scheduling tasks.

With fixed duration scheduling, Microsoft Project leaves the project duration unchanged by the amount of work or units assigned to a task. The reason that we use fixed duration as the default is because we are dealing with many different trades that need to manage their individual resources to maintain the project schedule. They usually need to manage their own resources to meet a schedule the resource productivity cannot determine the schedule. Generally, with enough time, planning, communication, and accountability anything can be completed. There is always another way of getting the job done. If a contractor or subcontractor cannot complete the task schedule for one or more items, the options are to add resources, add overtime to existing resources, or subcontract a partial or all the tasks work to maintain the project schedule. The reason that most schedules are delayed is because they do not have the vision and the time to make the above adjustments and consequently they must deal with the resources available at the time. This potentially drives the costs higher, delays the task item, and potentially the schedule finish date itself.

Estimating task durations requires you to enter average time frames that are reasonable and that include normal mobilization and downtime (Chapter 4).

After creating the rough draft and creating the Task Links discussed (later in this chapter), the next step is to distribute the schedule to all related team members for their input and comments. The schedule will be more realistic and team members will feel like a part of the schedule.

Set Milestones

A milestone is a task with no duration (zero days) used to identify significant events in the schedule, such as an important task or completion of a major phase. Milestones are displayed as a ◊ and are meant to capture attention.

It is a good idea to use the milestone symbol for various tasks or items in the schedule such as:

◊ Submit For Building Permit

◊ Receive Building Permit

◊ Deadlines For Customers

◊ Major Delivery Dates

◊ Project Completion

Setting A Milestone

1. **Click** Task name duration column
2. **Type** 0 In duration column
3. **Press** Enter or click or move to another cell

In the following example, steel delivery is marked as a milestone. Although the delivery actually takes time it is useful to use the symbol to capture the viewer's attention (Figure 7-1).

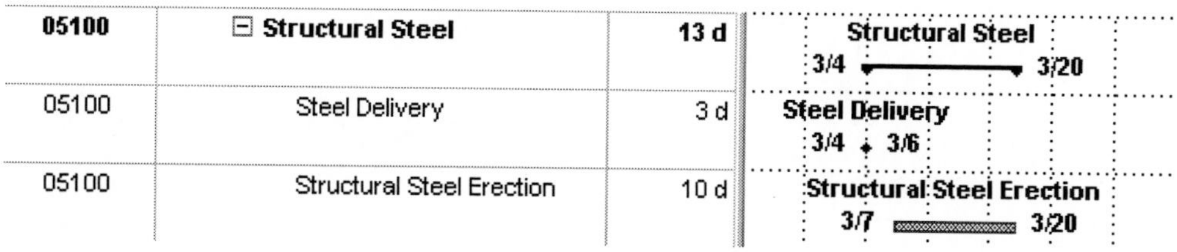

05100	⊟ **Structural Steel**	**13 d**	**Structural Steel** 3/4 – 3/20
05100	Steel Delivery	3 d	**Steel Delivery** 3/4 ◆ 3/6
05100	Structural Steel Erection	10 d	**Structural Steel Erection** 3/7 – 3/20

7-1: Set a Milestone

Marking A Task as A Milestone

1. **Enter** Duration of the task
2. **Double Click** The task (i.e. Steel Delivery in the following example; Figure 7-2) This will bring up the Task Information dialogue box
3. **Click** The Advanced Tab
4. **Check** Mark task as milestone
5. **Click** OK

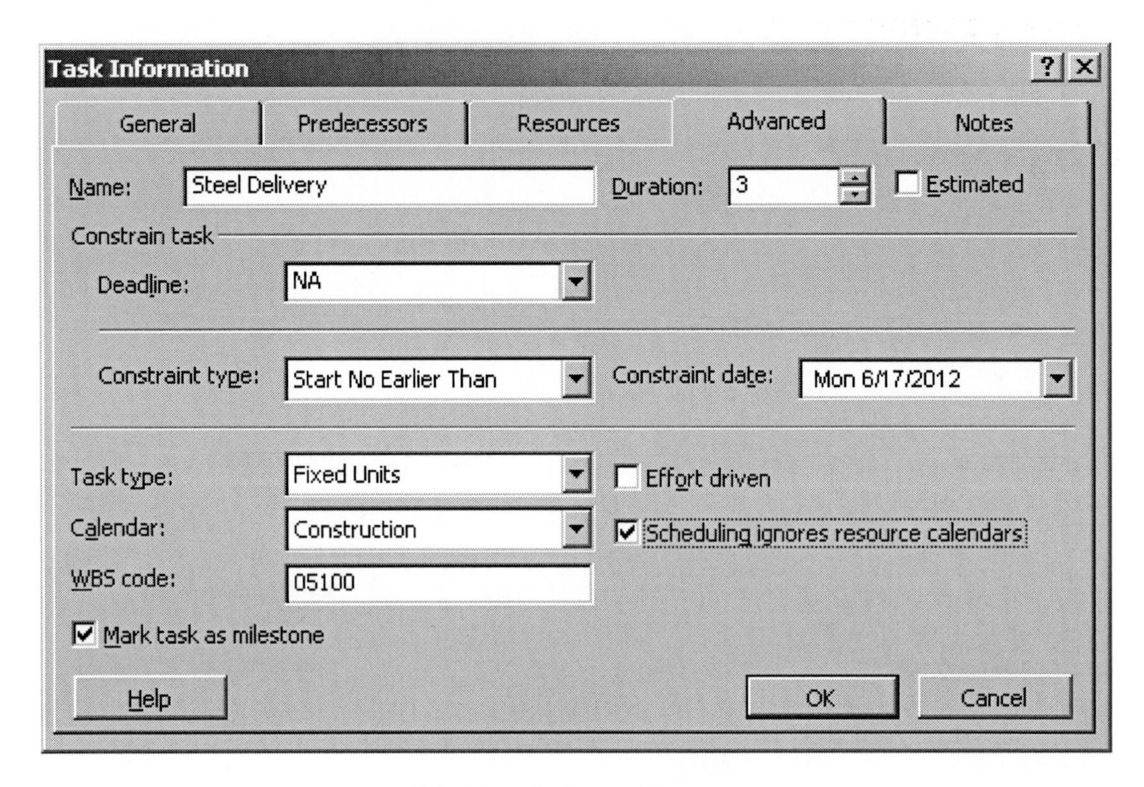

7-2: Mark Task as a Milestone

Create Task Links

If you are unfamiliar with Task Relationships, please refer to Understanding Task Relationships in Chapter 4.

When creating a project, create task links for all tasks in the project so that it can be quickly updated for any delays or adjustments to the schedule. Essentially, all tasks are required to complete the project, and it is the coordination of all tasks that will complete the project in an organized manner with the least amount of effort and input.

Even if a task is not directly dependent on another task, enter a relationship to the task with the nearest degree of dependency. It is not a good idea to link summary tasks in the project. The summary tasks are representative of the sub tasks below them. Linking summary tasks can throw off a task dependency of a predecessor or successor task and create circular relationships. This can lead to Microsoft Project being unable to calculate the project relationships and dates.

After the task list is finalized, task durations are entered and milestones are set, create all links for all tasks throughout the project (Chapter 4)

Note: Establishing lead time and lag time are included in this step to save you from having to go back to enter them later.

Enter Task Constraints

If you are unfamiliar with constraints, please refer to Understanding Constraints in Chapter 4.

Allow Microsoft Project to calculate the start and finish dates initially for all tasks. By entering durations for linked tasks and allowing Microsoft Project to calculate the project start and finish, the possibility that artificial restraints will affect the schedule is eliminated. Enter task constraints for major task items and milestones that will drive the schedule, such as Steel Delivery, Building Permit Acquired, Owner Move In, and the dates of other long lead time items.

To set task constraints:

1. **Double Click** Selected task. This is the task on which a date constraint is placed. It is a critical and fixed date in the schedule due to availability and lead time.
2. **Click** Advanced Tab In Task Information Dialogue Box
3. **Click** Constraint Type drop down
4. **Select** Constraint Type
5. **Click** Constraint Date dropdown
6. **Click** Constraint Date from dropdown
7. **Click** OK

All projects deal with certain parameters, delivery schedules, and expectations that you try to build the schedule around. It is these items that have specific dates relative to goals and availability. After constraints have been entered, related tasks can be scheduled accordingly.

Viewing and Changing Task Constraints

Periodically review the existing constraints to make sure they are valid. Constraints are often mistaken for deadlines and are therefore erroneously applied. This restricts Microsoft Project's flexibility in scheduling.

Change a task constraint if the constraint is not needed or should be changed to another type or another date. Must Start On and Must Finish On are the most inflexible constraints. Start No Earlier Than and Finish No Earlier Than are moderately flexible constraints. As Soon As Possible and As Late As Possible are the most flexible constraints.

When specifying a start constraint for a task or milestone, use Start No Earlier Than. This constraint schedules the task or milestone to be dependent on the availability or deadline for the task. If the task is moved by updating the schedule, evaluate the dependencies and reschedule the task or predecessor tasks as needed.

Use Lead Time and Lag Time

If you are unfamiliar with Lead Time and Lag Time, please refer to Understanding Lead Time and Lag Time in Chapter 4.

Use lead and lag time to shorten or lengthen a schedule. Many tasks in a schedule will overlap, and the necessary lead time and lag time may need to be entered to accurately reflect the relationships between tasks. In the initial schedule, enter normal Lead and Lag time that would be typical for the project. As the project progresses, work with the Lead and Lag time between tasks to reflect current scheduling events and also to adjust the project duration.

Evaluate the Schedule

After the schedule has been roughed in with tasks, links, relationships and constraints, evaluate the schedule to see the overall project duration as it relates to the project goals and to review the overall time frame for the project.

Many projects are typical, and the approximate time frames it takes to complete certain types of projects are known. By viewing the overall project duration against the history of similar projects you can gain insight into the practicality of the time frame.

Evaluating the schedule is critical and involves review of:

- The owner's expectation of the scheduled time frame for the project
- The practicality of the schedule with present resources and availability
- The overhead cost to complete the project (the length of the project directly affects the overhead of running the project)
- The practical or expedited relationships between tasks
- Realistic durations for tasks
- Realistic delivery dates
- The feasibility of how the people and team members will work with and maintain the project schedule
- The relationship between this and other projects
- The liability of completing the project in the time frame
- The risk of not completing the project on time

Schedules change, so there must be a system and process in place for maintaining the schedule and conveying project schedule information. Review of the schedule should be performed by the management of the team required to complete the project and by the scheduler.

Adjust the Project Duration

After the schedule is evaluated, adjust the Project Duration accordingly. To adjust the Project Duration, refer to the following sections in this chapter: Use Lead and Lag Time, Work with the Critical Path, and Adjust Task Durations.

Remember to communicate with critical subcontractors and other team members with regard to realistic or expedited delivery dates, expedited work schedules, and alternate methods to expedite the schedule (if required). Once the schedule is communicated to an owner and other outside team members, it becomes a concrete document. It is important to create a realistic schedule (no matter how aggressive) prior to publishing the schedule to outside team members and stakeholders.

Work with the Critical Path

The critical path is the series of tasks (or even a single task) that dictates the calculated finish date of the project. A task that is in the critical path has a direct impact on the finish date of the project. It is important for a project to finish on schedule, and close attention must be given to the tasks on the critical path and the resources assigned to them. These elements determine whether the project will be finished on time. You can also use the critical path to shorten or lengthen the project finish date depending on the time available. The series of tasks are generally interrelated by task dependencies. Although there are likely to be many such networks of tasks throughout the project plan, the network finishing the latest is the project's critical path.

Note that the critical path can change from one series of tasks to another as the schedule progresses. The critical path can change as critical tasks are completed or as tasks in another series are delayed. There is always one overall critical path for any project schedule. The new critical path becomes the series of tasks tracked more closely to ensure the desired finish date. By knowing and tracking the critical path for the project, as well as the resources assigned to critical tasks, it is easier to determine which tasks can affect the project's finish date and whether the project will finish on time. The Gantt Chart and Tracking Gantt chart view have been built to show the critical tasks with red bar styles. By viewing and working with these highlighted tasks, along with the Critical filter, the project duration can be adjusted.

To shorten the project finish date, shorten the dates of the critical path tasks. This is also known as "crashing." To do this:

1. **Shorten** the duration or work on a task on the critical path.
2. **Change** a task constraint to allow for more scheduling flexibility.
3. **Break** a critical task into smaller tasks that can be worked on simultaneously by different resources.
4. **Revise** task dependencies to allow more scheduling flexibility.
5. **Set** lead time between dependent tasks where applicable.
6. **Schedule** overtime.
7. **Assign** additional resources to work on critical path tasks.

Be aware that if the dates of the critical path tasks are shortened, a different series of tasks could become the new critical path. However, if the task durations on the critical path are shortened, and another series of tasks does not overtake it, then the finish date of the project can be brought in.

When trying to control the finish date of the project, use the Critical Filter to identify what tasks are affecting the finish date. Apply the critical filter as many times as necessary to shorten the schedule and to view the new critical tasks.

To work with the Critical Path:

1. **Click** Project
2. **Click** Filtered for:
3. **Select** Critical
4. **View** Critical Tasks. The Critical tasks are shown in red and are the current tasks that are affecting the project finish date.
5. **Click** Window
6. **Click** Split. This will create the Gantt chart on the top of the screen and the Task Form on the bottom). Select the split screen by double clicking on or dragging up the down arrow at the bottom right of the view window. This should display the task form showing the Predecessors and Successors for the selected task. If this form is not displayed, right click to display it.
7. **Review** Critical Tasks. Review the Critical tasks shown and determine which tasks can be expedited, shortened, or delayed depending on whether the schedule is to be shortened or lengthened.
8. **Adjust** Duration. Adjust the Project Duration by:
   - Shortening or lengthening the duration of a task on the critical path.
   - Changing a task constraint.
   - Revising dependencies and relationships.
   - Setting or revising lead and lag time between dependent tasks.
   - Communicating with team members on expedited schedules and or deliveries for items that are affecting the critical path.

9. **Continue** Adjusting the Critical Path. Continuously apply the critical path filter; when making changes to tasks on the critical path, new tasks may become critical and affect the project finish date.

10. **Finalize** The Project Duration

11. **Click** Project

12. **Click** Filtered for:

13. **Select** All Tasks or All Incomplete Subtasks depending on the desired filter to apply.

14. **Click** Project

15. **Click** Sort byspace. This will resort the schedule according to the changes made. Then apply the Summary View or the Cascading View depending on the view desired.

Assign Resources

If scheduling resources, first complete the draft of the schedule and then assign resources. The steps to assigning resources are:

For a single project:

- Enter resources (Chapter 6)

For multiple or Master Consolidated Projects

- Set up a resource pool (Chapter 9)
- Enter resources (Chapter 6)
- Share resources (Chapter 5)

Assigning resources can be done in an individual project or in a Consolidated Master project file.

Using a Consolidated Master project file allows you to view current workloads, allocations, and other information across all projects in order to balance your workload.

Assigning Resources

1. **Click** View
2. **Click** Gantt Chart
3. **Click** Task Name. Select the task to which resources are assigned.
4. **Click** Assign Resources. This will bring up the Assign Resources dialogue box as shown in the following example (Figure 7-3).
5. **Enter** Resources and number of resources. Enter the resources and the total number of resources required to complete the task.
6. **Click** Close to exit the Assign Resources dialogue box.

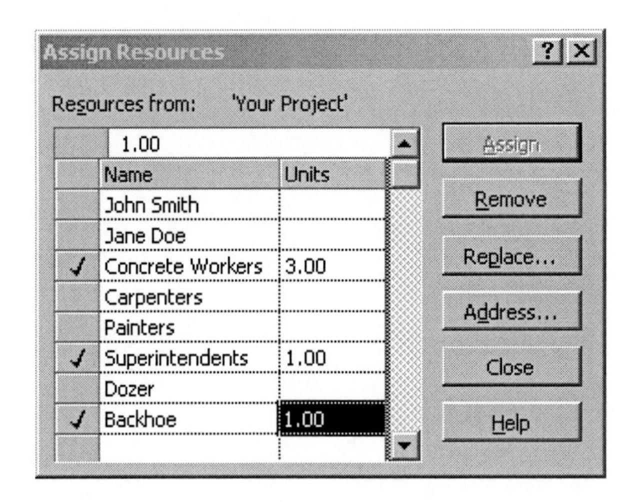

7-3: Assign Resources

Communicating the Plan

Communicating the schedule is critical to ensure that everyone takes part in and understands the final schedule going to construction. Too often, the schedule is not conveyed to all of the team members and their input is not a part of the final schedule. This is when the schedule loses power and effectiveness.

A schedule is a combined plan of all team members that are a part of a project and it needs to reflect all of the tasks and related time frames associated with each team member and contractor. Whenever possible, the final schedule should be incorporated as an attachment to contracts to increase attention and accountability.

The schedule is still preliminary until final feedback and information is received from all team members to incorporate into the final construction schedule. It is at this phase of review from all of the team members that a Schedule of Values can be prepared.

View Sample Schedule of Values

Use this view of the Schedule of Values table to serve as an attachment or backup for billing purposes. People understand pictures. Using the schedule with this table will also serve as a foundation for managing cash flow, filters, views, and reports. You want the schedule to control the project and not vice versa. Using the Schedule of Values to mirror the percent of work complete will force much greater attention to the schedule. Team members will look forward to getting the schedule and will be more motivated to complete their work, rather than focusing just on how much they can bill for.

Establishing a Schedule of Values

1. **Click** View
2. **Click** Table
3. **Click** Schedule of Values. This will bring up the Schedule of Values Table
4. **Enter** Contract Selling Costs in the Total Cost column for each task

Note: Costs for a summary item can't be entered in the Total Cost field, as these amounts are calculated fields that represent the subtasks below the summary task. However, costs for a summary item (Pre-Construction Requirements, for example) can be entered by inserting the Fixed Cost column and entering the total cost for the summary task. Hide the Fixed Cost column when done to prevent mistakes in entering other costs in this field.

5. **Continue** Entering Selling Costs for all task items. Enter the resources and the total number of resources required to complete the task
6. **Click** View→Table→Entry to exit the Schedule of Values Table

To update a Schedule of Values:

1. **Click** View
2. **Click** Table
3. **Click** Schedule of Values. This will bring up the Schedule of Values Table
4. **Enter** The Percentage Complete (% Comp.) for tasks that have been started, are in progress, or have been completed. The % of the task completion equals the % of cost complete. Make sure to include any upfront costs (material deliveries, for example) that need to be reflected in the total % complete.
5. **Continue** Entering the Percent Complete (% Comp.) for all current tasks in progress. The Cost To Date and Remaining Cost field will automatically be updated.

Note: The total cost for the project will be the sum of the highest level summary tasks. In the previous example, it would normally be the sum of Pre-Construction Requirements and Project Duration.

6. **Click** View→Table→Entry to exit the Schedule of Values Table.

Refer to Chapters 4 and 10 for additional information on Schedule of Values.

Finalize the Schedule

Once the process of building the schedule is complete, request any final comments, conflicts with or structural changes for the schedule from all team members. This is the final schedule that will go to construction. Plan also to review the schedule to reflect any changes or adjustments that have been made since the last preliminary schedule update.

Save the Baseline

Saving the baseline of the schedule is critical to maintaining a gauge and comparison of current events against the original scheduled plan. The Tracking Gantt view shows the original scheduled bar directly against the current schedule bar. This information is invaluable in keeping team members accountable to the project schedule. It will also serve as a baseline for updating the schedule and an agenda for meetings with the associated team members.

To keep control over and accurate accounting of the scheduling progress, it is critical that schedules be updated regularly (preferably weekly). This is necessary to keep everyone focused and accountable and to help documenting and controlling progress.

The schedule baseline in combination with Schedule Updating serve to:

- Control the scheduling process
- See where the current schedule stands as compared to the original baseline schedule
- Keep everyone accountable
- Keep everyone's attention
- Decrease liability
- Maintain liability for performing contractors
- Forecast conflicts
- Help prioritize actions
- Review delivery status
- Pinpoint any team member who is lagging behind
- Adjust team members' schedules in response to accelerations and delays
- Maintain current report information
- Set current expectations for all team members
- Identify 'over-allocations'
- Balance the workload
- Prevent crisis situations
- Keep owners and other stakeholders 'up to-date'

To save the baseline:

1. **Click** Tools
2. **Click** Tracking
3. **Click** Baseline
4. **Click** Save baseline & For: Entire project. This method saves the baseline for the entire project (Figure 7-4). There are other times when the interim plan or the baseline should be saved for selected tasks.
5. **Click** OK. The baseline schedule is now saved and can be viewed against the current schedule by displaying the Tracking Gantt View.
6. **Enter** Notes in Project Duration Notes Field. Enter the date the baseline was saved with any related notes to establish a record of what was done and when.
7. **Enter** Date of final schedule In Page Setup→Headercenter→ field

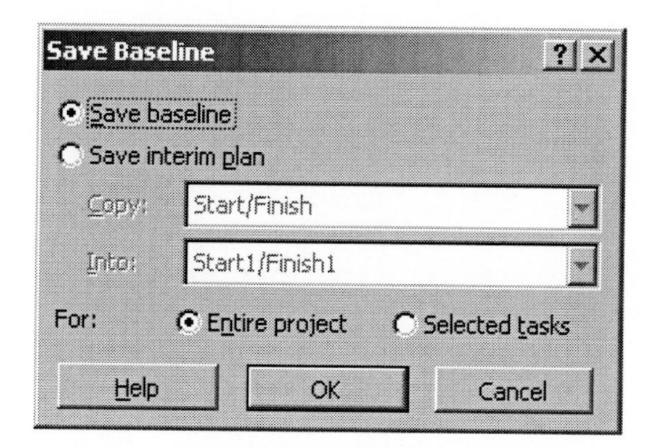

7-4: Save the Baseline

Saving the baseline is one of the most important aspects in maintaining control in a project.

The project schedules have been organized and planned. The next chapter covers controlling schedules.

Chapter Eight

Controlling Schedules

Updating the Schedule

To keep control over and accurate accounting of the scheduling progress, it is critical that schedules be updated regularly, preferably weekly. This is necessary to keep everyone focused and accountable and to help in documenting and controlling progress and adjusting actions.

Schedule updates serve to:

- Control the scheduling process
- See where the current schedule stands as compared to the original baseline schedule
- Keep everyone accountable
- Keep everyone's attention
- Decrease liability
- Maintain liability for performing contractors
- Forecast conflicts
- Help prioritize actions
- Review buyout status
- Review delivery status
- Pinpoint any team member who is lagging behind
- Adjust team member's schedules in response to accelerations and delays
- Maintain current report information
- Set expectations for all team members
- Identify 'over-allocations'
- Balance the workload
- Prevent crisis situations
- Keep owners and other stakeholders 'up-to-date'

Updating schedules is critical to controlling the scheduling process; this is where many projects fail. Often, people do not take the time to monitor and communicate changes. It is at this point when time controls the project instead of the contractor controlling the time. Updating the schedule does not take long if it is current and realistic. If a schedule is properly linked and has all of the information available, an average detailed schedule can be updated in less than an hour, and a simple schedule in minutes. Taking this time is invaluable.

To update a schedule:

1. **Click** View
2. **Click** Tracking Gantt. This will bring up the tracking Gantt where all update information will be saved.
3. **Click** Project
4. **Click** Filtered for:
5. **Click** All Incomplete Subtasks. This filter displays the Cascading View.
6. **Click** Project
7. **Click** Sort by?
8. **Click** Start. Sort by Start should be the only selection in the dropdown boxes.
9. **Check** Keep outline structure
10. **Click** Sort. This sorts the project tasks by start date with no structural reference to the related summary tasks.
11. **View** Incomplete items since last update
12. **Enter** Actual Start dates for all tasks that have started
13. **Enter** Actual Finish dates for all tasks that have finished.
14. **Enter** Percent Complete / % Compl. The percent complete entered should be the percentage of the total cost completed. By entering the % complete in the Tracking Table, the % complete will automatically be updated in the Schedule of Values Table.
15. **Enter** Any rescheduled dates or task information that have been changed or that affect the structure of the schedule. Upon updating the scheduled start and finish dates, review the new dates to see how the schedule has changed relative to the previous schedule.

16. **Enter**	Schedule update notes. Enter the schedule update notes in the notes field of Project Duration.
17. **Click**	File→Page Setup. To enter the updated date on the schedule. The main date entered in the center of the schedule under the title is the original schedule date of the final schedule going to construction (the date that the baseline was saved)
18. **Click**	Header Tab and Alignment Right Tab. The updated date will be entered on the right side of the header.
19. **Enter**	Updated: (date of update)
20. **Click**	Save. Switch back to any view and the information that has been entered will carry to all views (except any header information saved in step 19). As needed, enter the header information into each view that is being used, typically the Gantt View and Tracking Gantt View.

Note: A useful tool to create an action or a response to the schedule updates can be found at the end of Chapter 10.

Reviewing the Baseline

After the schedule is updated, review the current schedule against the baseline.

The Tracking Gantt view is configured to show the current schedule bar below the baseline bar for the task bar. By viewing the relationship of the current scheduled task against the original baseline, the schedule can be quickly evaluated for which tasks have been accelerated or delayed. The revised scheduled dates will adjust the current summary bar for the subtasks below it and will show the status of the summary item respectively.

The main bar to pay attention to is the Project Duration summary bar. The baseline summary bars remain constant and are shown as white with a black outline. The current scheduled summary bar is shown as a solid black line that is hidden behind the white summary bar. When the project is delayed, the current Project Duration will be shown as a solid black line extending past the Project Duration summary bar. The project has been delayed in Figure 8-1:

8-1: Delayed Summary Bar

When the bottom right black arrow is shown to the left of the right baseline summary bar down arrow the schedule has been accelerated, as shown in Figure 8-2:

8-2: Accelerated Summary Bar

When updating the schedule, constantly check the status of the current scheduled finish date in relation to the baseline scheduled finish date.

1. If the project has been extended as shown in the first example, work with the critical path to adjust the Project Duration (Chapter 4).

2. If the project has been accelerated as shown in the second example, make sure that all related team members can adjust to the accelerations and that there will be no conflicts in material deliveries, labor availability, etc.

In both cases, it is critical that the final schedule update be conveyed to all associated team members so they can adjust their schedules accordingly. Using the Schedule Update form will be very helpful in promoting action, communication, and accountability.

Communicating Project Information

Communication is everything! To maximize communications, develop a system that works automatically. This system should be one that is effective and useful to all team members. Scheduling, if effectively used, has the potential to control the projects and aid in planning business operations. But if it cannot deliver accurate information to the people who need it, in a way that they can understand and use, the potential will be lost.

Each organization and project needs a system to communicate information. Establish the quickest and most effective method of delivering the schedule. Convey a complete picture of the schedule, or use filters and views to narrow down information specific to date ranges or responsibilities. Keeping information concise and specific will result in greater attention and accountability. Many people do not pay close attention to individual tasks when they are included in a large schedule of crossed responsibilities.

Depending on the nature of the project and the technological 'know-how' of team members, options to convey the schedule and information include:

Internet

Using the internet is probably the best way to communicate a schedule and related information; however, the effectiveness of using this method is dependent on the use of the internet by the project team members. The Share Point Team Website solution (provided through our site) is a great way to post schedules, related information and documents that other team members can view and respond to right from their browser. This method has the least overhead for managing, viewing, and conveying information.

With a Team Website, there is one communication source which everyone can access, use, and comment on. Using a Project Intranet over the Internet is the way of the future. The technology is available now, and it is your challenge to make sure that your team members are using it.

Several versions of the schedule can be posted on a Project Intranet:

- An overall copy of the schedule
- A current 'two-week' look ahead
- A picture of the current critical path
- Individual schedules
- Specifications
- Meeting minutes

E-Mail Distribution

Email is a very powerful means to distribute schedules and related communication with low overhead. Deliver the schedule as an attachment or a copy displayed in the body of the email.

- Use PDF files so that anyone can open and read the files
- Use an attached 'picture' GIF file of the schedule
- Attach an actual copy of the Microsoft Project file
- Include a Schedule Update form and a narrative message of the current events and needs of the team members.

Physical Distribution

This option is the least desirable, although depending on the technological capabilities use of the team members it may be the most effective. In either event, try to get everyone accustomed to using email and the Internet, as this will improve communication, lower overhead, and streamline the process.

Hard-Copies of Schedules can be delivered in various ways:

- Distribute copies at project meetings
- Use U.S. Mail, UPS, or other shipping means
- Facsimile

All schedules need to have a Schedule Update form to generate prompt attention and reaction. If the schedules are printed, using 17" wide by 11" high format will allow team members to view information with as much of the future timescale as possible. In all of the above methods, the goal is to get the schedule into the hands of all team members to keep them accountable and updated.

Working with The Team

It is essential to maintain communication with all team members to make sure that they are adhering and paying attention to the schedule.

Note that most people are not 'schedule-oriented' and need constant communication to make sure that they are maintaining their part of the project schedule.

- Always remember the following:
- Most people do not know how to schedule effectively
- People schedule in an infinite number of ways
- People can be poor communicators
- Contractors provide schedules in many different formats and task breakdowns
- Schedules and agendas typically are not coordinated
- Many contractors are unorganized and do not work with schedules. If they do, most are not working with them as a tool to help manage, forecast, and maintain a schedule and communication. They let time and the schedule run them instead of the other way around.
- Adjust schedules to keep in mind both, the project and the team members. Each project is particular in nature and it is important to determine the necessary methods and means required to maximize the effectiveness of the schedules and communication.

Ensure that everyone is working within the parameters and scope of the schedule. Through communication, conflicts can be approached creatively and effectively.

Updating the Schedule

1. **Click** View
2. **Click** Tracking Gantt. This will bring up the tracking Gantt where all update information will be saved.
3. **Click** Project
4. **Click** Filtered for:
5. **Click** All Incomplete Subtasks. This filter displays the cascading-view.
6. **Click** Project
7. **Click** Sort by
8. **Click** Start. Sort by Start should be the only selection in the dropdown boxes.
9. **Check** Keep outline structure
10. **Click** Sort. This sorts the project tasks by start date with no structural reference to the related summary tasks.
11. **View** Incomplete items since last update
12. **Enter** Actual Start dates for all tasks that have started

If the schedule is on track, only enter the information that falls within the date range of the last update. Notes specific to the tasks also can be entered in the notes field.

13. **Enter** Actual Finish dates for all tasks that have finished.
14. **Enter** Percent Complete / % Compl. The percent complete entered should be the percentage of the total cost completed. By entering the % complete in the Tracking Table, the % complete will automatically be updated in the Schedule of Values Table.

When using a Schedule of Values, switch to the Schedule of Values Table to enter the % complete so that the Cost to Date and Remaining Cost can be viewed as the percentage is updated.

15. **Enter** Any rescheduled dates or task information that have been changed or that affect the structure of the schedule. When updating the scheduled start and finish dates, review the new dates to see how the new schedule has changed relative to the previous schedule.
16. **Enter** Schedule update notes. Enter the schedule update notes in the notes field of Project Duration.

Updating the Schedule

17. **Click** File→Page Setup. To enter the updated date on the schedule. The main date entered in the center of the schedule under the title is the original schedule date of the final schedule going to construction (the date that the baseline was saved)

18. **Click** Header Tab and Alignment Right Tab. The updated date will be entered on the right side of the header.

19. **Enter** Updated: (date of update)

20. **Click** Save. Switch back to any view and the information that has been entered will carry to all views (except any header information saved in step 19). As needed, enter the header information into each view used, typically the Gantt View and Tracking Gantt View.

Tracking Progress

Tracking progress is a daily responsibility of the managers of a project. Depending on the organization structure, make sure that whoever is responsible for communicating schedule information is conveying that information to the scheduler and other responsible parties as often as is necessary to maintain the project schedule. This is the only way to successfully manage a project. Management needs to evaluate changes and communicate the revised plan to everyone involved so that they know how to adjust their actions. Many times, a schedule is prepared for a project and the tracking is either ineffective or not done at all. This takes away the effectiveness of the schedule as a tool to manage the project. Managing the schedule and tracking progress is not difficult if it is done as a part of the daily responsibilities of the project manager, superintendent, etc.

Establish a Schedule and Reporting System

Scheduling in itself is very powerful and can provide everything needed to run and control the business, but if it is not integrated into a useful system that everyone uses and follows, the effectiveness will be lost. To achieve maximum results, there needs to be written procedures on how to use the schedules as well as the established methods of communication and reporting. They can be attachments to the contracts or an established procedure within business operations.

A system is defined as:

- A group of interacting, interrelated, or interdependent elements forming a complex whole

- A condition of harmonious, orderly interaction

- An organized and coordinated method: a procedure

A business system is:

- A repeatable process that produces a profit

Controlling Schedules

Establish the best methods to integrate and make use of the scheduling and reporting systems developed. Examples of scheduling and reporting systems include:

- A guide within the organization outlining scheduling and reporting procedures
- An attachment to contracts outlining scheduling and reporting procedures
- Regular meetings to review the scheduling and report information
- Specific reports require to be delivered on specific dates and at regular intervals
- A written meeting minute agenda for all projects to follow
- A communication flow chart for each project and each department within the organization

The whole goal is to develop a flow of communication that operates automatically and that delivers specific and relevant information. We are all busy and we can get so consumed with being busy that we do not do the things that we know we should. Written procedures and requirements build expectations into everyone's personal agendas.

Project Reporting

Reports can provide the exact information required by the people that see them. Reports narrow down information into a format that is useful for a specific purpose. Microsoft Project contains many 'built-in' report formats to make communication with other team members as effortless as possible (Chapter 4). Use these reports 'as is' or use them as a foundation to create new ones.

Reports are used for:

- Reviewing overall project information
- Owners
- Management
- Costs
- Cash Flow
- Assignments
- Upcoming Tasks
- Workload

All that you need to do to generate reports in Microsoft Project is set up the framework for what information needs to be reported. The software automatically generates the reports based on the structure given. In the included Scheduling Templates and Project Files available online at www.ConstructionScheduling.com are two main reports that are typically used in the construction industry.

Monthly Cash Flow

This report is invaluable for running business operations and providing necessary information to owners, banks, etc. to ensure that all financial arrangements, approvals, etc. are complete when payment is ready to be made. By entering selling costs and using the custom Schedule of Values table, you can create a Monthly Cash Flow report that gives the foundation to control the financial aspect of the process.

Weekly Cash Flow

The weekly cash flow report is not used as much as the Monthly Cash Flow report, but it is available for special needs and circumstances. Remember, reports can easily be modified to fit your business. Also remember that the greater the detail provided in the schedule, the greater the accuracy of the forecasts and reports.

Chapter 9 shows how contractors can enhance efficiency and profits by consolidating schedules.

Chapter Nine

Consolidating Schedules

Consolidating Schedules

A Consolidated Project is a Master scheduling file that contains some or all of the active projects. This file offers business planning and insight that takes scheduling beyond the individual project.

Each project file is inserted into the Consolidated Project file. Any changes made in either the individual project file or the Consolidated Master project file is reflected in both files.

The information can be filtered, viewed, and reported on across all of the projects for global project information and insight. This file can be viewed and worked with as individual projects, combinations of projects, or all of the projects.

To use the Consolidated Project file, all projects must be structured the same way (using the same fields). The ConstructionScheduling.com templates offer the framework to structure all of the schedules with similar customizations so that they can be consistently created.

We provide a Consolidated Project file that has all of the same coding and customizations as the rest of the scheduling templates and files available at www.ConstructionScheduling.com. Any schedules created from similarly structured templates can be inserted into this Consolidated Project file to view, filter, and report on information across all of the projects.

Using a Master Consolidated Project file can be used to:

- Adjust profit margins for future work depending on workload.
- View, filter, and report on costs in progress, costs completed, costs remaining across all projects.
- Run a cash flow report across all projects.
- View and report on all outstanding Pre-Construction work.
- View, filter, and report on any task or resource across all projects.
- View, filter and report for work type (concrete, site work, carpentry, etc.) and view workload and 'over-allocations.' With this information, decisions can be made on whether to subcontract work, hire more people or reschedule work.
- Filter for CSI Division number or CSI Code number across all projects and use for purchasing contracts and materials.
- View workloads for individual contractors, subcontractors, etc.
- View, filter, and report on responsibilities across all projects.

- Group project files by Project Manager, Potential Construction, etc. and view, filter, and report on information specific to any of these categories.
- Keep everyone accountable.

Using the Consolidated Project file gives management the information and insight to make intelligent decisions and establish forces accountability across all projects.

Establish the Consolidated Project Layout

A consolidated project is simply a file containing inserted projects that enables you to combine any or all of the inserted projects so that information can be filtered, viewed and reported on.

Notice in the Consolidated Project sample that the main summary items are in blue bold and the inserted projects are in red bold. We have configured these text settings to appear automatically when a summary item is created and inserted in a project. This is intended to draw attention and focus to the different items. These settings can be changed if desired (Chapter 10).

The Consolidated Project can be set up with any combination of tables and views.

The sample shows:

- The summary tasks are Potential Projects and the name of each Project Manager for projects under construction. By viewing the potential projects in the consolidated project, any Pre-Construction schedule can be combined with projects under construction to view overall workloads and other information.
- The Schedule of Values table. Showing this table with the summary names of the Project Manager and their associated projects offers a quick summary view of the project, the cost status for each project, and an overall view for the respective project manager. This is a good quick view of the financial status of projects under construction.

<u>**Note:**</u> Only specific information can be filtered on an expanded summary task or expanded group of summary tasks, such as the Project Manager or Inserted Project(s). This is a good reason to group projects by summary under each Project Manager or other responsible group. Their inserted projects can then be expanded to filter and view information specific to the Project Manager or Group.

By default, the entry table appears with the Updated field inserted (Chapter 10).

Tables and views can be modified to any need. Any table and view combination saved will appear the next time the consolidated project is opened. It is a good idea to establish the standard tables and views commonly used so these settings can be established prior to saving the Consolidated Project file.

Inserting Projects

After the layout and setup of the summary task structure is established, individual projects can be inserted (Chapter 4). Keep in mind that the projects inserted into the Consolidated Master project are the same in both, the individual project and the consolidated Master project. Any changes made to the schedule in either the individual project file or in the Consolidated file will reflect the same changes in each file. Make sure that the projects contained in the Consolidated Project are current and updated. The reports and information created will only be effective if the information is 'up-to-date' and accurate.

Creating a Resource Pool

Resources are the people, equipment, or other entities responsible for actually completing the tasks in the projects. Using a resource pool for shared resources across many or all of the projects will organize valuable information and keep people accountable.

Scheduling does not require that resources be defined and scheduled; however, defining resources will allow for viewing, filtering, and tracking information. Specifically, use resources in the project plan when the goal is to:

- View, filter, report, and communicate the task items associated with people and equipment in the organization
- Ensure high accountability for and understanding of the project. When responsibilities are clear, there is greater production and a decreased risk of tasks being overlooked
- Monitor resources
- Provide for resource accountability

Within this system of scheduling, resources are defined within Microsoft Project as the people, equipment, or other entities in YOUR ORGANIZATION that you want to track and/or communicate with to provide management. There are two reasons that you should only enter resources specific to your organization:

1. This will provide a method of managing all people and/or equipment that are directly responsible for or accountable to a project throughout all phases of construction, i.e. sales, marketing, drafting, project management, and support personnel, as well as the actual people and equipment that are a part of completing the physical construction on a project. This provides a general management tool that will use all people or other resources within the organization.
2. The other reason that you should schedule only your own resources is that you want to create a common resource pool that you can manage across all projects. In a Consolidated Project or Master Project including all active projects, you can use a common resource pool to view, filter, and report on your resources at a global level across all projects. If different subcontractors and other entities are entered into the resource sheet for different projects, the resource pool would be very large and unmanageable.

There are other resources that are part of the individual projects, such as subcontractors, architects, owners, engineers, and other outside entities. You are not responsible for coordinating their individual resources. You are, however, responsible to coordinate their functions within the project for accurate and effective scheduling. Use the Master Entry Table (Chapter 10) to enter, track, filter, and view information for all outside resources using the customized fields (Chapter 10) in the supplied template. This is an effective way to provide a list format for all outside resources. These additional fields allow you to quickly view information in a column listing format across the different layers of responsibility for the various tasks in the projects. These additional resources are entered and displayed in tables and views.

Included with this book is a project file named Consolidated_Projects.mpp. This is an empty file pre-programmed with all of the features and customizations that are a part of each of the template files. This is the file you should use for entering shared resources.

Resources as entered into the resource sheet in this system would include:

People in the organization that are directly involved with and accountable for the completion of tasks. This would include all individuals within the various departments across the organization, i.e.

- Sales
- Drafting
- Support Personnel
- Project Managers
- Superintendents

Types of trade people within the organization that have a specialized skill or complete certain types of work, such as:

- Concrete Workers
- Steel Workers
- Carpenters
- Painters
- Operators
- Laborers

A listing of available major equipment that to be tracked, such as:

- 550 Dozer
- 850 dozer
- JCB Forklift
- Loader
- JD Backhoe

Using both resources within the resource sheet and outside resources listed in the table listing format in the schedules will create higher accountability as well as a system for reporting on individuals, groups, equipment, and trades.

The next step is to Create a Resource Pool:

1. **Open** Master or Consolidated Project File
2. **Click** View
3. **Click** Resource Sheet
4. **Enter** Resources. The resources entered are those that will be used across multiple or all projects so assignments can be viewed (Figure 9-1).

   -Resources can be people, trade types, equipment, materials, etc.

   -For tracking individuals, use the name of the resource.

   -For tracking trade types, use the name of the trade.

   -For tracking equipment, materials, etc., enter the name of the associated resource.

   -The Max. Units column is the total number of available resources for each particular type of resource.

5. **Enter** Resource Data. Enter applicable resource data,

   -For individuals, enter max. units: 1.

   -For trade types, enter max. units for the total resource type available i.e. carpenters: 5 or concrete workers: 8.

   -For equipment, materials, etc., enter max. units to the total amount of the resource available.

   -Enter costs in the associated fields to track costs. Under Calendar, make sure that Construction is set as the base resource calendar for all resources that follow the base default construction calendar.

6. **Click** View > Gantt Chart. This will bring the main Gantt Chart back into view.

	Resource Name	Initials	Max. Units	Group	Email Address	Code	Notes
2	Andrea Wilkins	AJ	1	Sales, Est	ajones@anon.com	094,092	
3							
4	Steve Martin	SM	1	Projd, Projm	smartin@anon.com	094,092,082,105	
5	Tom Jones	TJ	1	Projd, Projm	tjones@anon.com	094,092,082,105	
6							
7	Bill Bones	BB	1	Est	bbones@anon.com	092	
8	Jake Elder	JE	1	Est	jelder@anon.com	092	
9	Tim Drake	TD	1	Est	tdrake@anon.com	092	
10	Tom Golden	TG	1	Est	tgolden@anon.com	092	
11							
12	Kurt Sanders	KS	1	Cad	ksanders@anon.com	082	
13	Jeff Pincher	JP	1	Cad	jpincher@anon.com	082	
14	Bruno Brinkman	BB	1	Cad	bbrinkman@anon.com	082	
15							
16	Ed Easton	EE	1	Site Design	eeaston@anon.com	082	
17							
18	Theresa Billows	TB	1	Support	tbillows@anon.com	085	
19	Mary Michaels	MM	1	Support	mmichaels@anon.com	085	
20	Rachel Linden	RL	1	Support	rlinden@anon.com	085	
21							
22	Jim Tasker	JT	1	Projm	jtasker@anon.com	105	
23	Scott Kintz	SK	1	Projm	skintz@anon.com	105	
24	Wayne Hill	WH	1	Projm	whill@anon.com	105	
25	Michael Stone	MS	1	Projm	mstone@anon.com	105	
26							
27	Mike King	MK	1	Concrete	mking@anon.com	210	
28							
29	Ken Masters	KM	1	Site Work	kmasters@anon.com	140	
30							
31	Superintendents	S	3	Superintendents		110	
32	Operators	O	4	Operators		140	
33	Concrete Workers	C	15	Concrete		210	
34	Masons	M	2	Masons		230	
35	Steel Workers	S	8	Steel		260	
36	Welders	W	2	Steel		262	
37	Carpenters	C	6	Carpenters		270	
38	Painters	P	2	Painters		370	
39	Plumbers	P	1	Plumbers		430	
40	Laborers	L	1	Labor		500	
41							
42	550 dozer	5	2	Equipment		141	
43	850 Doxer	8	1	Equipment		142	
44	Cat Loader	C	2	Equipment		143	
45	JCB Forklift	J	3	Equipment		144	
46	JD Backhoe	J	2	Equipment		145	
47	Excavator	E	3	Equipment		146	

9-1: Resource Sheet

Using Shared Resources

Resources can be entered and assigned in the same way as individual project files.

To enter resources:

1. **Click** View

2. **Click** Resource Sheet

3. **Enter** Resources. Resources can be people, trade types, equipment, materials, etc.

   -To track individuals, use the name of the resource

   -To track trade types, use the name of the trade

   -To track equipment or materials, enter the name of the associated resource

4. **Enter** Resource Data. Enter applicable resource data,

   - For individuals, enter max. units: 1

   - For trade types, enter max. units for the total resource type available i.e. carpenters: 5 or concrete workers: 8

   - For equipment, materials, etc., enter max. units to the total amount of the resource available.

Enter costs in the associated fields to track costs, Under Calendar, set Construction as the base resource calendar for all resources that follow the base default construction calendar

5. **Click** View

6. **Click** Gantt Chart, This will bring the main Gantt Chart back into view.

Resource Name	Initials	Max. Units	Group	Email Address	Code
John Smith	JS	1	Sales, Est	jsmith@anon.com	094,092
Andrea Jones	AJ	1	Sales, Est	ajones@anon.com	094,092
Steve Matzen	SM	1	Projd, Projm	smatzen@anon.com	094,092,082,105
Tom Jones	TJ	1	Projd, Projm	tjones@anon.com	094,092,082,105
Bill Bones	BB	1	Est	bbones@anon.com	092
Jake Elder	JE	1	Est	jelder@anon.com	092
Tim Drake	TD	1	Est	tdrake@anon.com	092

9-1: Sample Sheet

To assign resources:

1. **Click** View
2. **Click** Gantt Chart
3. **Click** Task Name. Select the task to which resources are assigned.
4. **Click** Assign Resources. This will bring up the Assign Resources dialogue box as shown in Figure 9-2.

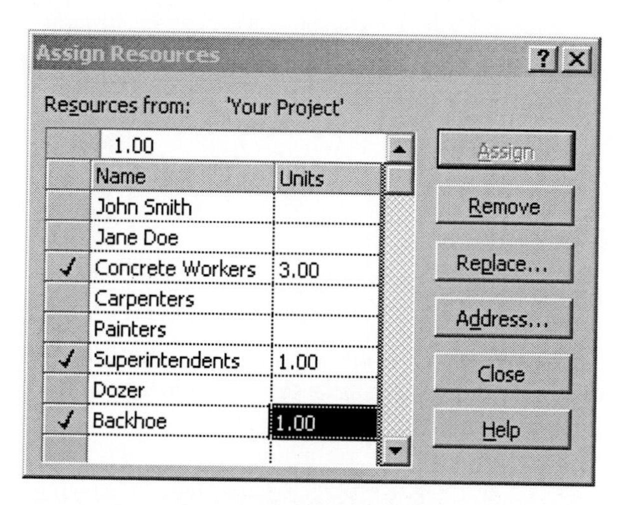

9-2: Assign Resources

5. **Enter** Resources and number of resources. Enter the resources and the total number of resources required to complete the task.
6. **Click** Close to exit the Assign Resources dialogue box.

The benefit of using a Consolidated Master project is that it can combine all active projects to view and work with common resources. Workloads, 'over-allocations,' and responsibilities can be viewed globally.

The following filters can be used in the Consolidated Master project file to view, filter, report and communicate information:

- Resource Group
- Resource Group_Incomplete Subtasks
- Resource Intials
- Resource Name
- Using Resource In Date Range
- Using Resource

Using the above filters in the Consolidated Master project, keeps everyone focused on and accountable to the individual projects and uses the filters as a global business management tool.

Using the Power Of Filters

Filters are among the greatest tools in Microsoft Project. A filter is an action applied to the schedule that allows viewing information desired for tasks and/or resources while filtering out all unneeded information. Filters are a very powerful tool in communication and planning. The view and report information can be defined and limited for just about anything. This filtered information allows users to focus on specific information.

Simplicity is the key to communication. A task or resource filter can be applied and the report distributed by Internet, email, or in person. Microsoft Project has many pre-defined filters that can be used. To create custom filters:

1. **Click** Project
2. **Click** Filtered for:
3. **Click** More Filters at bottom of Filter menu.
4. **Select** New (for a new filter with no existing formatting) or an existing filter. It is a good idea to view the existing filters to determine which criteria the filters use to help define the new or modified filter (Figure 9-3).

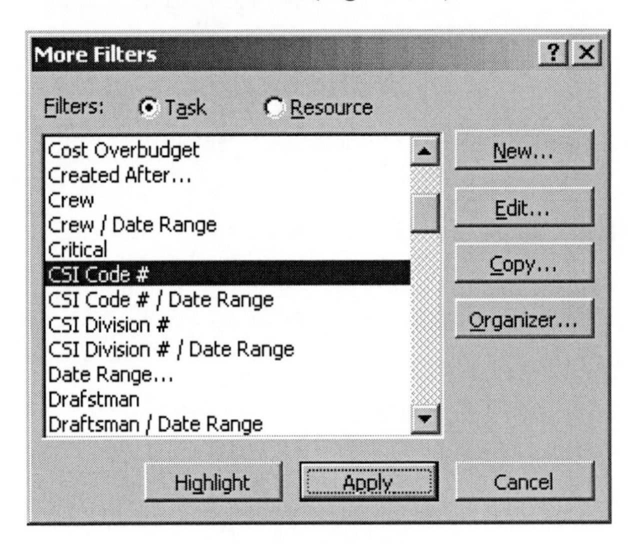

9-3: More Filters?

Selecting New will open the Filter Definition dialogue box.

5. **Select** Edit or Copy. Select this option only if a new filter will not be built from scratch. Selecting Edit will modify the existing filter. Selecting Copy will make a copy of the existing filter, and then add or modify any changes needed. Create a new name for the filter. (Copying a filter is the recommended method to save the existing filter format)

6. **Enter Or Modify** New criteria_Existing criteria. The new criteria set will include Field Names, Tests, and Values.

7. **Enter** Name for new filter. Enter a new name for the filter or use the Copy of name to use the existing filter name.

8. **Click** OK

The Scheduling Templates provided include many custom filters that are useful in the construction industry. The custom filters can be used to filter for:

- CSI Division number
- CSI Code number
- Pre-Construction work
- Subcontract Work
- Any defined trade or type of work (concrete work, carpentry, etc.)
- Job number
- Work Item number
- Custom Priorities
- Responsibility
- Construction Manager
- General Contractor
- Project Developer
- Project Manager
- Superintendent
- Foreman
- Subcontractor
- Draftsman
- Crew

Use filters for a single project or a group of projects (if they are formatted the same way and using the same fields). Also use any of these filters to view all associated tasks or for tasks within a specified date range. Filtering can give you the focus and insight to promote action and responsibility.

Using Views

Using views in a Consolidated Master project is the key to communicating specific information relative to the active projects. Use views in the consolidated Master project the same way as an individual project (Chapter 6).

The important thing is to know what information is needed so that the combination of filters and views can be structured accordingly. In most cases, the Cascading View in the Consolidated project will be used with the summary tasks filtered out. This is due to the number of summary tasks and the size of the combined files. In most views in a Consolidated Master project, insert a column (Project) to the left of the Task Name. Then there will always be a reference to the source project. When the Project column is inserted, the project name is automatically filled in.

When working with the Consolidated Master project file, columns and fields will constantly be added and hidden to meet the specific requirements of the desired view.

Note: Prior to saving and exiting the Consolidated Master file, return to the default main summary view so it will reopen the same way when the project file is reopened.

Using Fields

Using fields is directly related to working with tables and views. When filtering and viewing information, remember that the settings or field content can be modified, offering the ability to correct responsibilities and 'over-allocations' within the Consolidated Master file.

The tables and views in the Consolidated Master file are not static; they may be changed as often as necessary. This is a working file that offers invaluable information.

Note: The view that is saved with the current inserted fields will remain with that selected table. To keep consistency within the Consolidated Master project, return to the default table views so they will be available when reopening the Consolidated Master project.

Updating the Consolidated Project

Updating the Consolidated Master project should be a daily, or at least a weekly, task. This file can offer insight into planning business decisions, coordinating schedules, keeping people accountable, etc. To accomplish these things, the information within each project file that is in the Consolidated Master project must be current and 'up-to-date.' This is also another incentive for management to make sure that each project manager, superintendent or other member forwards the current information for their schedules.

To help track how current the information in the Consolidated Master project file is, use the custom Updated field (Chapter 10). The date entered in this field is the last date the schedule was updated.

Note: Given one project in the Consolidated Master project that is inaccurate or not current will reflect inaccurate information within the combined views and reports. It is critical that each responsible party maintain their schedule updates regularly (Chapter 8).

By maintaining the Consolidated Master project 'up-to-date' with the Updated field adjacent to the project, people directly accountable for providing the information and schedule updates can be kept informed. If individuals or teams do not maintain their individual project schedules, it affects the ability to plan, manage, and forecast global business solutions.

Reporting On Multiple Projects

A Consolidated Master project contains all of the information required to manage information across all of the projects, but it is critical that the information is used and communicated effectively.

Establish standard reports for the people or groups that need to receive them. Without a system of communicating effective information, the potential power of your reports is lost. Reports can be any information that people need to react, to plan, and manage their operations. Determine the specific reports that individuals, groups and management need to see, and supply the reports on a regular schedule.

Once people see the potential of effective scheduling, they will value the process and become dependent on the information. Meet with all respective team members to see and understand what information they need. Accordingly, you can structure the views and reports to meet those specific needs.

Effective information is required for a project to be successful. Most people do not realize the potential power and information available through effective scheduling. Let others know what the schedules can provide and empower them to communicate as a team. Most people struggle without this information. It does take effort and communication, but the benefits are immeasurable.

Chapter Ten

Advanced Scheduling

Reporting On Multiple Projects

Once you have mastered the basic skills of using Microsoft Project for scheduling, advanced tools can be used to work smarter and easier. This chapter contains more techniques for using consolidated projects, resource sheets, customized fields, date and flag fields, outline codes, text fields, tables, task and resource filters, views, cash flow reports, project calendars, option settings, and the schedule update form. These powerful tools will help reduce scheduling time and increase bottom-line profits for your construction or remodeling business.

Consolidated Projects

The Consolidated_Projects.mpp file that accompanies this book has all of the same customizations as the other templates files. The only differences in this file are the use of the Updated field as described in Chapter 9 and the special formatting features as described in the following pages.

The structure of the file is completely up to the scheduler. With proper structuring of this file, it will become the most effective planning tool that can be used. The consistent use of this file with effective communication will require others to keep their schedules 'up-to-date' and accurate. Use the Consolidated 'Master' file as much as possible. It is very easy to use and offers powerful benefits.

Updated Field

When the Consolidated_Projects.mpp file is open, the Updated (Date 1) field is on the left hand side of the screen (Figure 10-1). Use the Updated field by entering the date of the last project update or selecting the date from the dropdown arrow in the cell adjacent to the inserted project.

Updated	Name	Duration
NA	*Team 0 - Steve Matzen*	*298 d*
05/20	AAA Credit Union	136 d
06/05	Cornell Properties	73 d
05/25	NYS Office Building	229 d
06/30	Drake Industries	25 d
NA	*Team 1 - James Tasker*	*296 d*
05/25	Baltimore Fire	192 d
05/25	Appliance Giant	206 d
06/02	St. James Church	189 d
05/28	Arrow Distrubuting	144 d

10-1: Updated Field

The Updated field in the Consolidated Projects file offers a quick look at the status of the individual project updates and shows how current the information is. It is also a good method to keep project managers accountable for keeping the project schedules current.

It is critical that all projects are current and 'up-to-date.' When the combined project information is used to view, report, and forecast, one 'out-of-date' schedule can give the wrong information. Make sure that everyone who is a part of updating and keeping schedules current knows that current updates are required so that consolidated information is available. Require that all schedules be updated and turned in by a specific day of every week.

Note: There should be only one person or one selected group that has access to and is using the consolidated project file. Protect this file with a password so that only authorized people have access.

Text Settings

The following text settings have been programmed into the Consolidated_Projects.mpp file. When a Project file is inserted (Figure 10-2), the file name will show as a red bold font. When a summary task is created (Figure 10-3), the task name will show as a bold blue font.

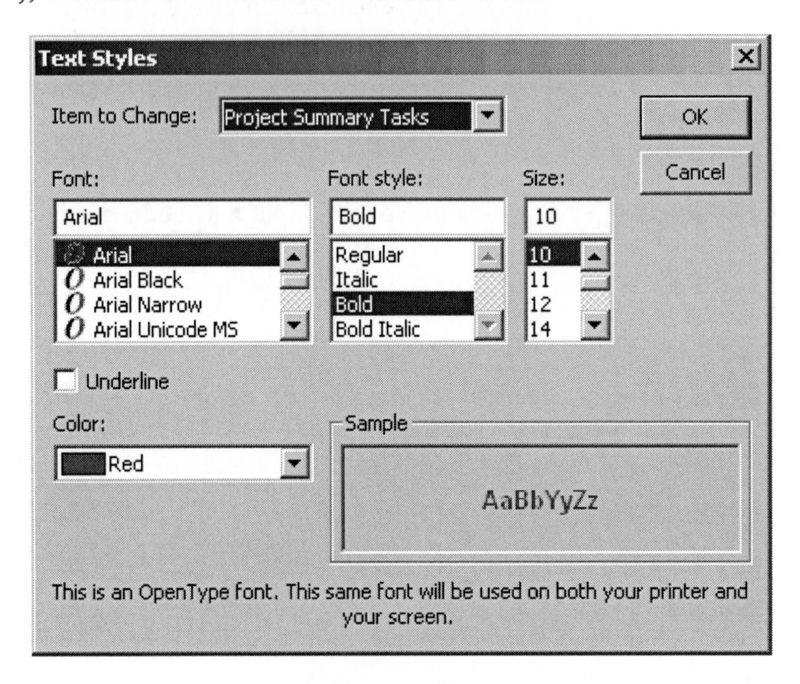

10-2: Project Text

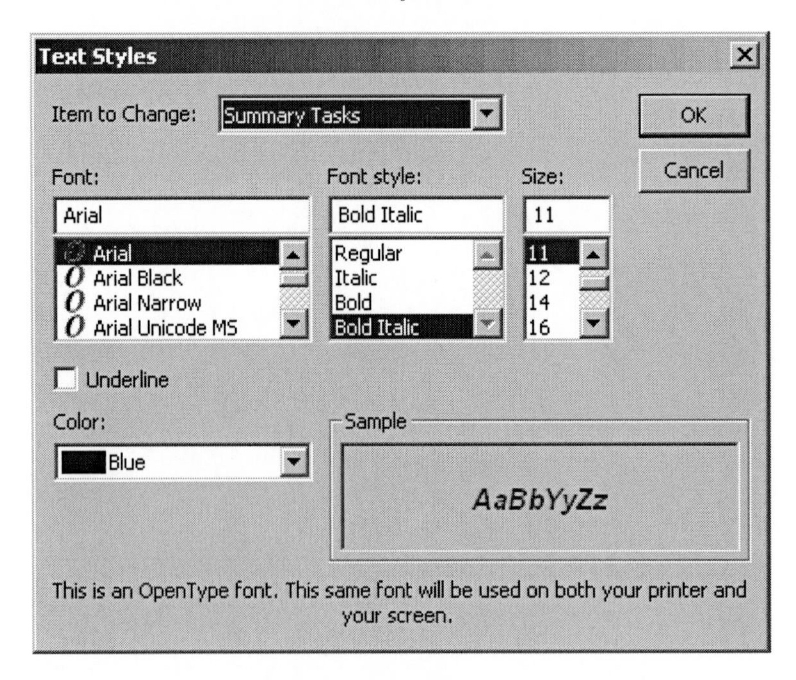

10-3: Summary Text

Resource Sheet

	Resource Name	Initials	Max. Units	Group	Email Address	Code	Notes
2	Andrea Wilkins	AJ	1	Sales, Est	ajones@anon.com	094,092	
3							
4	Steve Martin	SM	1	Projd, Projm	smartin@anon.com	094,092,082,105	
5	Tom Jones	TJ	1	Projd, Projm	tjones@anon.com	094,092,082,105	
6							
7	Bill Bones	BB	1	Est	bbones@anon.com	092	
8	Jake Elder	JE	1	Est	jelder@anon.com	092	
9	Tim Drake	TD	1	Est	tdrake@anon.com	092	
10	Tom Golden	TG	1	Est	tgolden@anon.com	092	
11							
12	Kurt Sanders	KS	1	Cad	ksanders@anon.com	082	
13	Jeff Pincher	JP	1	Cad	jpincher@anon.com	082	
14	Bruno Brinkman	BB	1	Cad	bbrinkman@anon.com	082	
15							
16	Ed Easton	EE	1	Site Design	eeaston@anon.com	082	
17							
18	Theresa Billows	TB	1	Support	tbillows@anon.com	085	
19	Mary Michaels	MM	1	Support	mmichaels@anon.com	085	
20	Rachel Linden	RL	1	Support	rlinden@anon.com	085	
21							
22	Jim Tasker	JT	1	Projm	jtasker@anon.com	105	
23	Scott Kintz	SK	1	Projm	skintz@anon.com	105	
24	Wayne Hill	WH	1	Projm	whill@anon.com	105	
25	Michael Stone	MS	1	Projm	mstone@anon.com	105	
26							
27	Mike King	MK	1	Concrete	mking@anon.com	210	
28							
29	Ken Masters	KM	1	Site Work	kmasters@anon.com	140	
30							
31	Superintendents	S	3	Superintendents		110	
32	Operators	O	4	Operators		140	
33	Concrete Workers	C	15	Concrete		210	
34	Masons	M	2	Masons		230	
35	Steel Workers	S	8	Steel		260	
36	Welders	W	2	Steel		262	
37	Carpenters	C	6	Carpenters		270	
38	Painters	P	2	Painters		370	
39	Plumbers	P	1	Plumbers		430	
40	Laborers	L	1	Labor		500	
41							
42	550 dozer	5	2	Equipment		141	
43	850 Doxer	8	1	Equipment		142	
44	Cat Loader	C	2	Equipment		143	
45	JCB Forklift	J	3	Equipment		144	
46	JD Backhoe	J	2	Equipment		145	
47	Excavator	E	3	Equipment		146	

10-4: Resource Sheet

Customized Fields

The following table shows a list of the customized fields in all of the supplied scheduling templates and files. If more fields are added, use the other available fields so that the files can be shared and integrated easily with others using the same system.

Field	Use
Date 1	Updated
Flag 1	Pre-Construction Requirements
Flag 2	Summary Tasks
Flag 3	Subcontract Work
Outline Code 1	Priority #
Outline Code 2	CSI Division #
Outline Code 3	CSI Code #
Outline Code 4	Job #
Outline Code 5	Item #

Field	Use
Text 1	Responsibility
Text 2	Construction Manager
Text 3	General Contractor
Text 4	Project Developer
Text 5	Project Manager
Text 6	Superintendent
Text 7	Foreman
Text 8	Subcontractor
Text 9	Draftsman
Text 10	Crew

Date Fields

(Date 1) Updated

The Updated (Date 1; Figure 10-5) field is typically used only in a Consolidated 'Master' project file and is used in a column next to the inserted Project name in the Task Name field. This field shows the current update status of the inserted schedule. Manually change this date every time the schedule is updated.

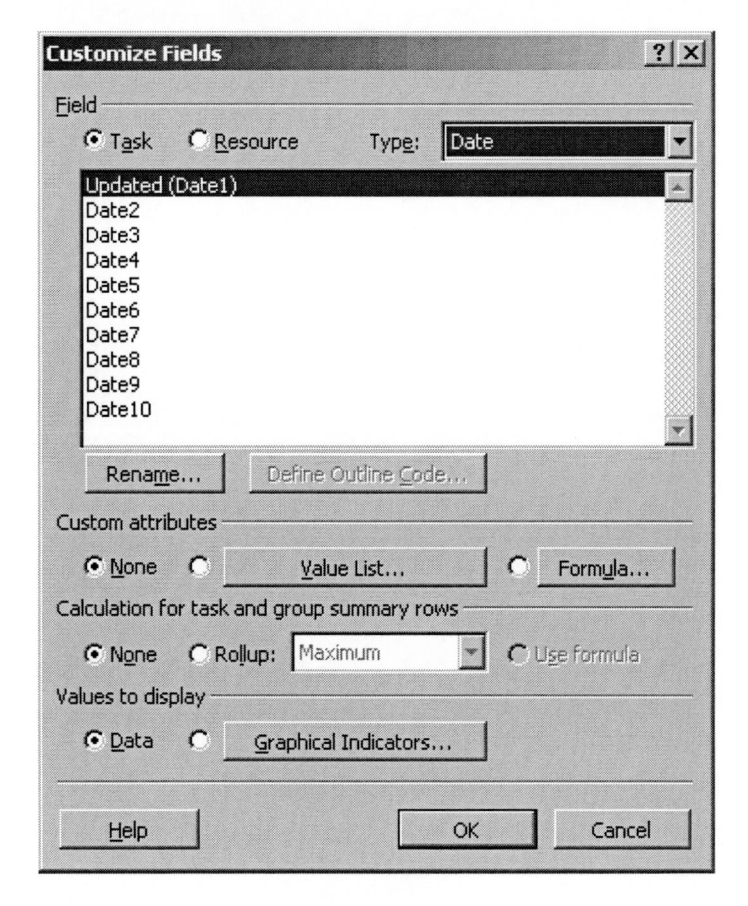

10-5: Date 1

Flag Fields

Flag fields (Figure 10-6) are Yes/No fields that offer options to filter out items that match the filter criteria.

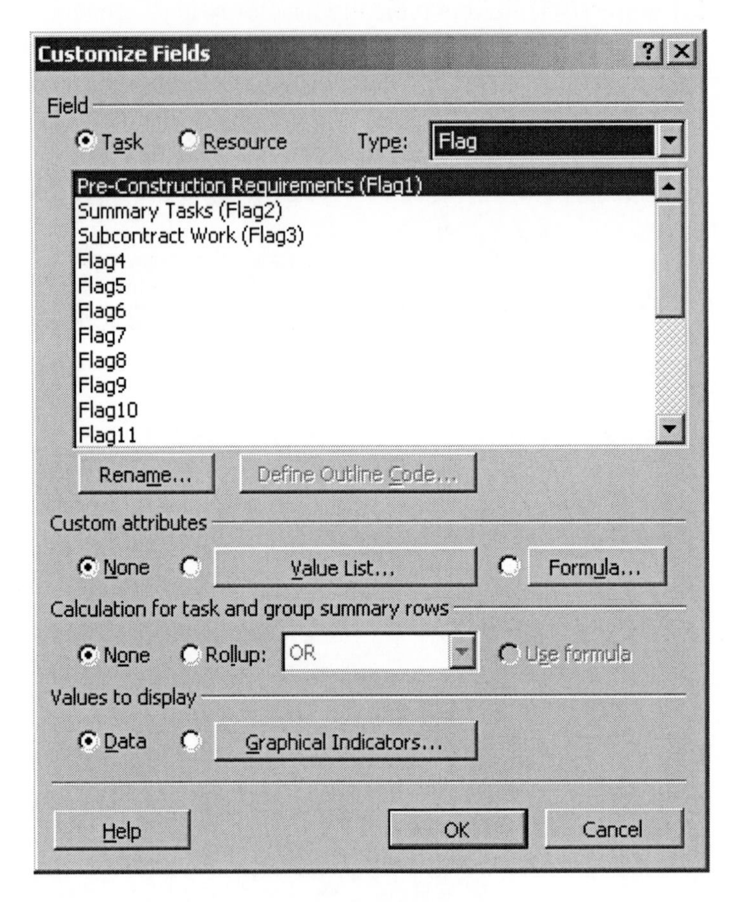

10-6: Flag Fields

(Flag 1) Pre-Construction Requirements

This flag is designed to filter out all summary tasks, and subtasks, excluding any Pre-Construction Tasks. Use this flag by applying the Pre-Construction Requirements filter with Sort by Start Date. The task view will show only the Pre-Construction Tasks.

Pre-Construction Requirements (Flag 1)

PRE-CONSTRUCTION TASKS

No	PRE-CONSTRUCTION REQUIREMENTS
Yes	Pre-Construction Schedule
Yes	Soils Testing
Yes	Code Review
Yes	Design And Drafting
Yes	Plan Review Meeting
Yes	Structural Engineering
Yes	Mechanical Engineering
Yes	Electrical Engineering
Yes	Verify Existing Utility Locations
Yes	Prepare Steel Shop Drawings
Yes	Submit For Building Permit
Yes	Site Surveying
Yes	Planning Department Review
Yes	Pre-Construction Meeting
Yes	Submittal Review
Yes	Interior Design

Note: If any Pre-Construction Tasks are added, also insert the Pre-Construction Requirements column and mark the added task as Yes.

This is extremely useful in planning incomplete Pre-Construction subtasks within a single project or across all current projects in a multiple project file by the consolidating all of the active projects.

(Flag 2) Summary Tasks

This flag is designed to filter out all summary tasks, excluding Pre-Construction Requirements and Project Duration, and is used in the All Incomplete Subtasks filter. Use this flag by applying the All Incomplete Subtasks filter with Sort by Start Date with Keep outline structure unchecked. The task view will show only the Pre-Construction Requirements and Project Duration Summary Tasks with all relative subtasks within each summary task sorted by start date.

Summary Tasks (Flag 2)

SUMMARY TASK

No	PRE-CONSTRUCTION REQUIREMENTS
No	PROJECT DURATION
Yes	GENERAL REQUIREMENTS
Yes	SITE CONSTRUCTION
Yes	CONCRETE
Yes	MASONRY
Yes	METALS
Yes	CARPENTRY
Yes	THERMAL AND MOISTURE PROTECTION
Yes	DOORS AND WINDOWS
Yes	FINISHES
Yes	SPECIALTIES
Yes	EQUIPMENT
Yes	FURNISHINGS
Yes	SPECIAL CONSTRUCTION
Yes	CONVEYING SYSTEMS
Yes	MECHANICAL
Yes	ELECTRICAL
Yes	POST-CONSTRUCTION WORK

Note: All Summary Tasks are marked yes (including all sub summary tasks under each major top level summary tasks) excluding Pre-Construction Requirements and Project Duration.

Check the summary tasks for marked Yes by filtering for Summary Tasks in the filter menu.

(Flag 3) Subcontract Work

This flag is designed to filter out all tasks, excluding tasks, that are usually subcontracted or might be subcontracted due to an increased workload in a particular time frame. Use this flag by applying the Subcontract Work filter with Sort by Start Date. The task view will show only the Subcontract Task Items.

Subcontract Work (Flag 3)

Flag	Task
No	**SITE CONSTRUCTION**
Yes	Cuts And Fills
No	Foundation Excavation & Backfill
Yes	Asphalt Paving
No	CONCRETE
No	Concrete Foundation
Yes	Concrete Slabs
No	Exterior Concrete
Yes	MASONRY
Yes	Brick Veneer
Yes	Block Partitions
Yes	METALS
Yes	Steel Delivery
Yes	Steel Erection
No	CARPENTRY
No	Interior Trim
Yes	Millwork
Yes	THERMAL AND MOISTURE PROTECTION
No	DOORS AND WINDOWS
Yes	Exteiror Windows
No	Interior Doors
Yes	FINISHES
Yes	MECHANICAL
Yes	ELECTRICAL

<u>**Note:**</u> All subcontract tasks are marked Yes. This is a great way to look ahead at potential subcontract work to focus on buying out these trades and making special deals on multiple contracts in a single project or a multiple contract (consolidated project) file.

Outline Codes

Custom outline codes (Figure 10-7) are tags defined for tasks or resources that provide an alternate structure for the project; these codes are different from WBS codes or outline numbers. Up to ten sets of custom outline codes can be created in the project for systems such as accounting cost codes for tasks and job codes for resources. After outline codes are defined and assigned to tasks or resources, they can be used to sort, filter, or group tasks or resources in the project.

Outline codes; 10-7: Outline Codes

Each level of an outline code can consist of uppercase or lowercase letters, numbers, or characters (any combination of uppercase and lowercase letters and numbers) and a symbol to separate the levels of the code. The total length of an outline code can be up to 255 characters.

Templates are pre-coded with the following outline code fields:

Outline Code 1 Priority #

Outline Code 2 CSI Division #

Outline Code 3 CSI Code #

Outline Code 4 Job #

Outline Code 5 Item #

Outline Code 1 – Priority #

This field is pre-defined to reorganize the task list by a pre-defined priority list.

Outline Code 2 – CSI Division #

This field is pre-defined for the CSI Divisions 1-16 with 17 added for 'post-construction' work. This outline code allows you to filter for specific divisions to view and/or distribute to the responsible parties.

Outline Code 3 – CSI Code #

This field is pre-defined for the CSI Code numbers. These code numbers will allow you to filter for specific code numbers or groups of similar code numbers to view and/or distribute to the responsible parties.

Outline Code 4 – Job #

This field is not defined and is provided for the internal job or project numbering.

Outline Code 5 – Item #

This field is not defined and is provided for the internal item or task coding.

(Outline Code 1; Figure 10-8) Priority Number

PRIORITY #	SUMMARY TASK
0	PRE-CONSTRUCTION REQUIREMENTS
5	PROJECT DURATION
10	GENERAL REQUIREMENTS
20	SITE CONSTRUCTION
30	CONCRETE
40	MASONRY
50	METALS
60	CARPENTRY
70	THERMAL AND MOISTURE PROTECTION
80	DOORS AND WINDOWS
90	FINISHES
100	SPECIALTIES
110	EQUIPMENT
120	FURNISHINGS
130	SPECIAL CONSTRUCTION
140	CONVEYING SYSTEMS
150	MECHANICAL
160	ELECTRICAL
170	POST CONSTRUCTION WORK

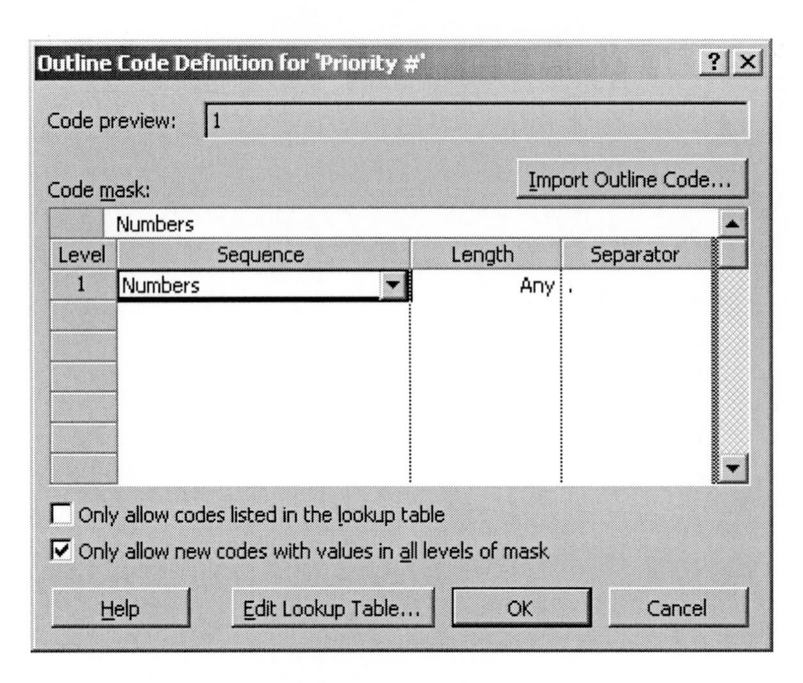

10-8: Outline Code 1

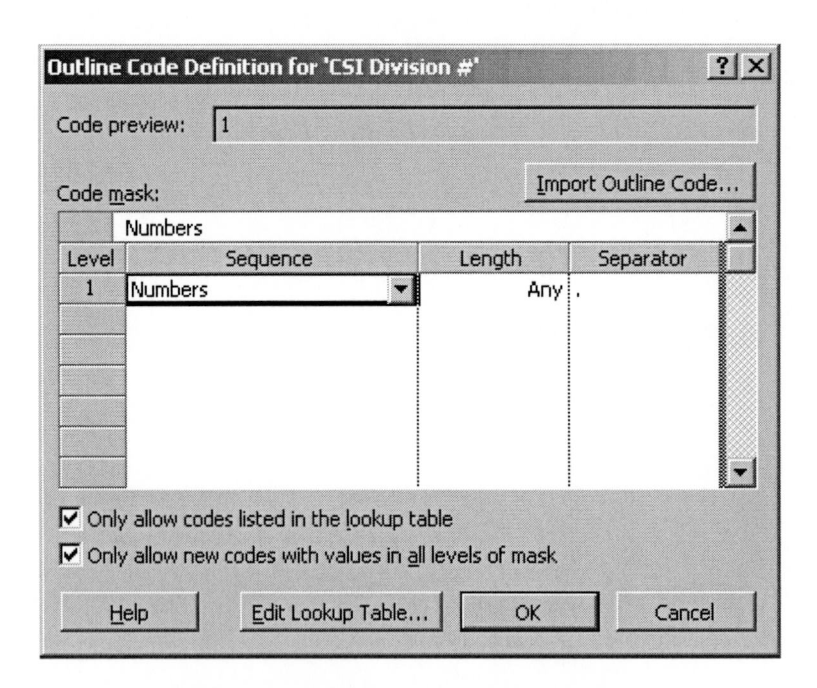

(Outline Code 2; Figure 10-9) CSI Division #

CSI Division # SUMMARY TASK

0. PRE-CONSTRUCTION REQUIREMENTS
1. GENERAL REQUIREMENTS
2. SITE CONSTRUCTION
3. CONCRETE
4. MASONRY
5. METALS
6. CARPENTRY
7. THERMAL AND MOISTURE PROTECTION
8. DOORS AND WINDOWS
9. FINISHES
10. SPECIALTIES
11. EQUIPMENT
12. FURNISHINGS
13. SPECIAL CONSTRUCTION
14. CONVEYING SYSTEMS
15. MECHANICAL
16. ELECTRICAL

(Outline Code 3) CSI Code Number

The CSI Code number field (Outline Code 3; Figure 10-10) is displayed as a standard column and field in most of the tables. Insert and hide this column as needed.

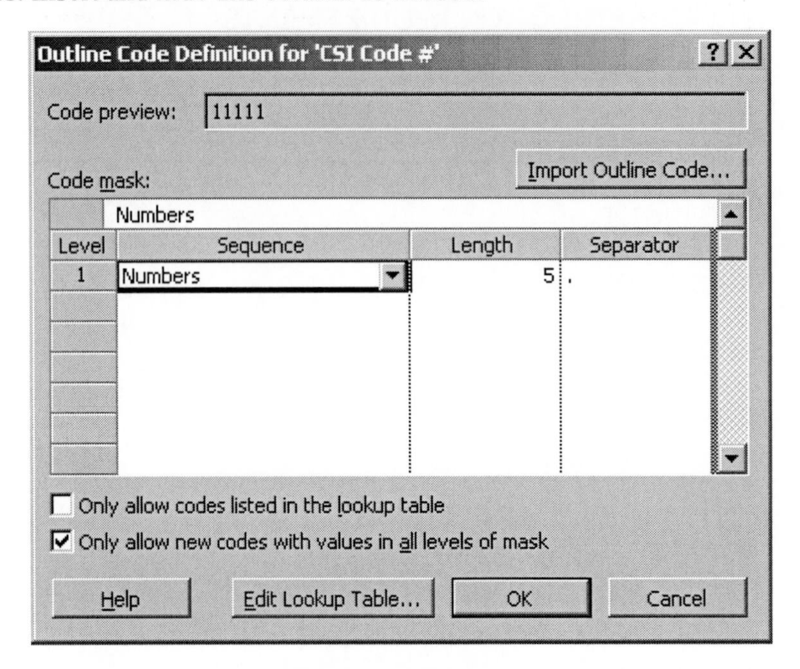

10-10: Outline Code 3

(Outline Code 4; Figure 10-11) Job Number

This is a 'user-defined' field for entering a Job # if applicable.

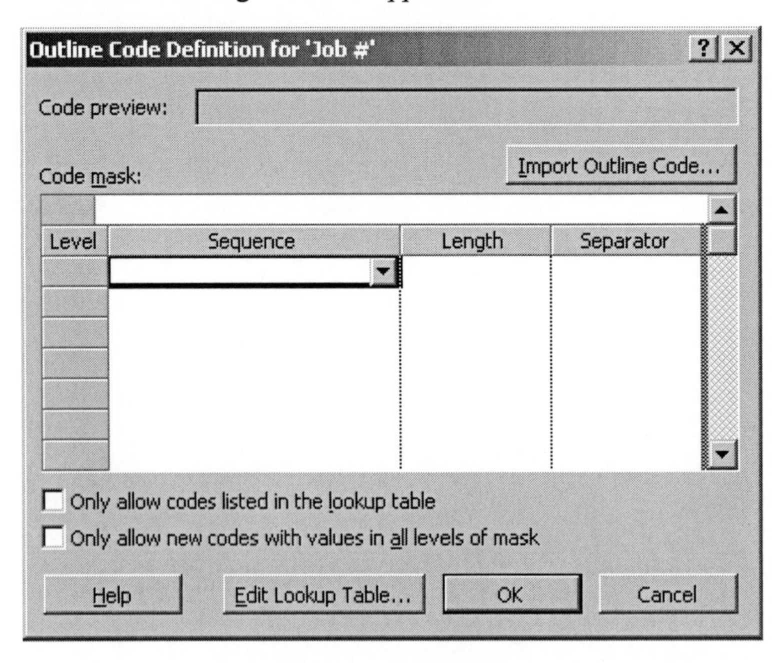

10-11: Outline Code 4

(Outline Code 5; Figure 10-12) Item

This is a 'user-defined' field for entering specific task item numbers if applicable. The default code used in all templates is the CSI Code number as defined in Outline Code 3.

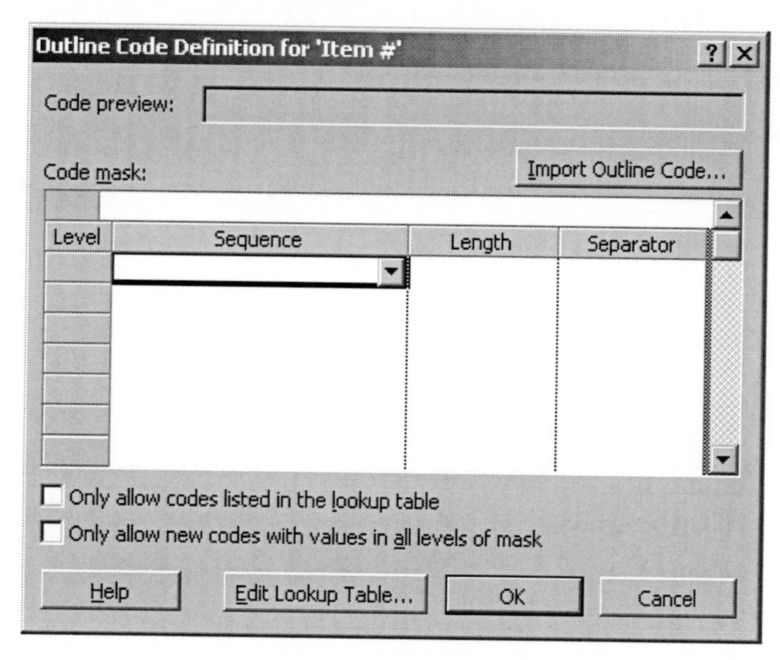

10-12: Outline Code 5

Text Fields

Text fields can be used for a variety of purposes.

Templates have pre-defined Text Fields 1-10. There are a total of 30 Text fields available in Microsoft Project.

When using Text Fields for new custom fields, use Text fields 11 – 30 to prevent any conflict in sharing schedules with others using the same scheduling structure.

(Text 1 – Text 10)

Text 1	Responsibility
Text 2	Construction Manager
Text 3	General Contractor
Text 4	Project Developer
Text 5	Project Manager
Text 6	Superintendent
Text 7	Foreman
Text 8	Subcontractor
Text 9	Draftsman
Text 10	Crew

Tables

Entry Table

The standard Entry Table (Figure 10-13) that is part of the scheduling templates is the same as the default Microsoft Project Entry Table, with the addition of the CSI Code# Column.

CSI Code #	Task Name	Duration

10-13: Entry Table

Master Entry Table

This table provides central location in which to enter and edit project specific information relative to people or entities to be tracked or filtered. The customized fields can be used, or new ones can be created. The customized fields are a predefined part of the templates as follows:

Flag 3 Subcontract Work
Outline Code 4 Job #
Outline Code 5 Item #

Text 1 Responsibility
Text 2 Construction Manager
Text 3 General Contractor
Text 4 Project Developer
Text 5 Project Manager
Text 6 Superintendent
Text 7 Foreman
Text 8 Subcontractor
Text 9 Draftsman
Text 10 Crew

Schedule of Values

The schedule of values table (Figure 10-14) allows you to enter and view the total cost for each item and the current cost to date by entering the percent complete for the tasks in the project. The current cost is automatically updated by entering the percent complete in any table or form. The first step is to enter the cost of each item in the total cost field. Note that costs cannot be entered for a summary item in the Total Cost field, as these amounts are calculated fields that represent the subtasks below it. To show the subtask cost (as in the following example for Pre-Construction Requirements), insert the fixed cost column and enter the total cost for the summary task. This will fill in the same amount in Total Cost field. Hide this column after entering a fixed cost summary item to prevent the mistake of entering other costs in the Fixed Cost field.

CSI Code #	Task Name	Total Cost	Cost To Date	Rem. Cost	% Comp.
	⊞ **PRE-CONSTRUCTION REQUIREMENTS**	**$8,300.00**	**$8,300.00**	**$0.00**	**100%**
	⊟ **PROJECT DURATION**	**$100,550.00**	**$56,445.00**	**$44,105.00**	**10%**
01000	GENERAL REQUIREMENTS	$10,000.00	$3,500.00	$6,500.00	35%
02000	SITE CONSTRUCTION	$22,500.00	$13,500.00	$9,000.00	60%
02315	EXCAVATION AND BACKFILL	$6,500.00	$6,500.00	$0.00	100%
03300	CONCRETE FOUNDATION	$8,000.00	$8,000.00	$0.00	100%
03300	CONCRETE SLABS	$4,000.00	$600.00	$3,400.00	15%
05100	⊟ **STRUCTURAL STEEL**	**$27,050.00**	**$22,370.00**	**$4,680.00**	**68%**
05100	Steel Delivery	$19,850.00	$19,850.00	$0.00	100%
05100	Structural Steel Erection	$7,200.00	$2,520.00	$4,680.00	35%
07300	ROOFING	$8,200.00	$0.00	$8,200.00	0%
08500	EXTERIOR WINDOWS	$4,000.00	$0.00	$4,000.00	0%
15500	HVAC	$4,000.00	$400.00	$3,600.00	10%

10-14: Schedule of Values

This table can be used in preparation and billing. By applying the schedule of values filter, only the tasks that have a cost entered will be viewed. By applying this table in combination with the Gantt or Tracking view, the customer or bank will be given a detailed listing of costs to date to use for billing purposes.

Tracking Table

The Tracking Table (Figure 10-15) is where the Schedule Updating is completed. The following table shows the standard view and fields that are used. If entering costs with the Schedule of Values table, complete the scheduling updating with actual starts and actual finishes first; then update the percent complete in the Schedule of Values table so that the costs can be seen as they are updated.

CSI Code #	Task Name	Act. Start	Act. Finish	% Comp.	Rem. Dur.
	⊟ PROJECT DURATION	**NA**	**NA**	**0%**	**95 d**
02000	**⊟ SITE CONSTRUCTION**	**NA**	**NA**	**0%**	**67 d**
02001	Mobilization	NA	NA	0%	2 d
02200	**⊟ Site Preparation**	**NA**	**NA**	**0%**	**15 d**
02230	Clearing and Grubbing	NA	NA	0%	5 d
02230	Strip and Stockpile Existing Soils	NA	NA	0%	2 d
02220	Building Demolition	NA	NA	0%	3 d
02300	**⊟ Earthwork**	**NA**	**NA**	**0%**	**54 d**
02310	Temporary Roads and Access	NA	NA	0%	2 d
02315	Cuts and Fills	NA	NA	0%	8 d

10-15: Tracking Table

Task Filters

All_Incomplete Subtasks

All_Incomplete Subtasks / Date Range

Construction Manager

Construction Manager / Date Range

Crew

Crew / Date Range

CSI Code #

CSI Code # / Date Range

CSI Division #

CSI Division # / Date Range

Draftsman

Draftsman / Date Range

Foreman

Foreman / Date Range

General Contractor

General Contractor / Date Range

Item #

Item # / Date Range

Job #

Job # / Date Range

Pre-Construction

Pre-Construction / Date Range

Project Developer

Project Developer / Date Range

Project Manager

Project Manager / Date Range

Resource Group_Incomplete Tasks

Resource Initials

Resource Name

Responsibility

Responsibility / Date Range

Schedule of Values

Subcontract Work

Subcontract Work / Date Range

Subcontractor

Subcontractor / Date Range

Superintendent

Superintendent / Date Range

Resource Filters

Code #

Enter the resource codes in this resource field if applicable.

Views

The views included in the templates are customized for the Gantt Chart view and the Tracking Gantt view.

In both views, the start date is to the left of the task bar, the finish date is to the right of the task bar, and the task name is on top of the task bar. In addition, gridlines are used to divide the sections and use a current date line so that past and future tasks can quickly be viewed. This format provides very specific date information for each task and lends itself to easy navigation when viewing the task information in the table and following the information in the calendar date view on the right side of the screen.

Bar Styles

Following are the bar styles customized for the Gantt Chart (Figure 10-16) and Tracking Gantt (Figures 10-17 and 10-18) Views:

Task		Normal	1	Start	Finish
Split		Normal,Split	1	Start	Finish
Critical		Normal,Critical	1	Start	Finish
Critical Split		Normal,Critical,Split	1	Start	Finish
Progress		Normal	1	Actual Start	CompleteThrough
Milestone		Milestone	1	Start	Start
Summary		Summary	1	Start	Finish
Rolled Up Task		Normal,Rolled Up,Not Summary	1	Start	Finish
Rolled Up Split		Normal,Rolled Up,Split,Not Sum	1	Start	Finish
Rolled Up Milestone		Milestone,Rolled Up,Not Summ.	1	Start	Start
Rolled Up Progress		Normal,Rolled Up,Not Summary	1	Actual Start	CompleteThrough
External Tasks		External Tasks	1	Start	Finish
Project Summary		Project Summary	1	Start	Finish
External Milestone		Milestone,External Tasks	1	Start	Start
Deadline			1	Deadline	Deadline
*Group By Summary		Group By Summary	1	Start	Finish

10-16: Gantt Bars

Layout

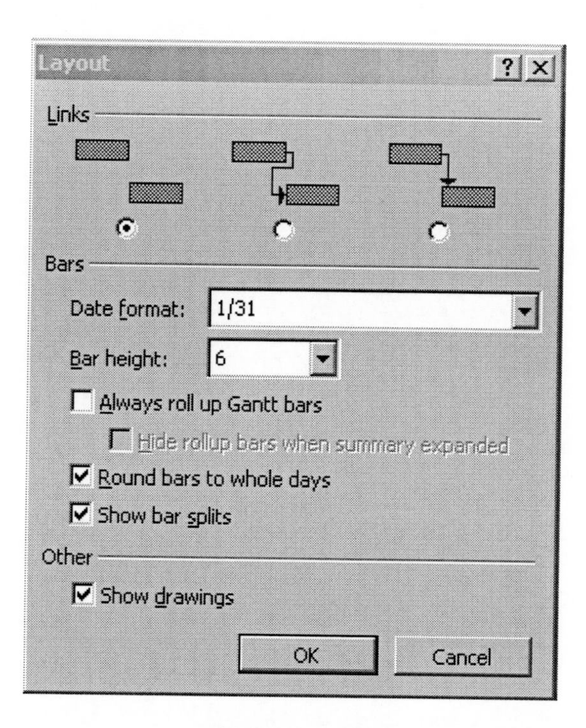

10-17: Layout

Gridlines

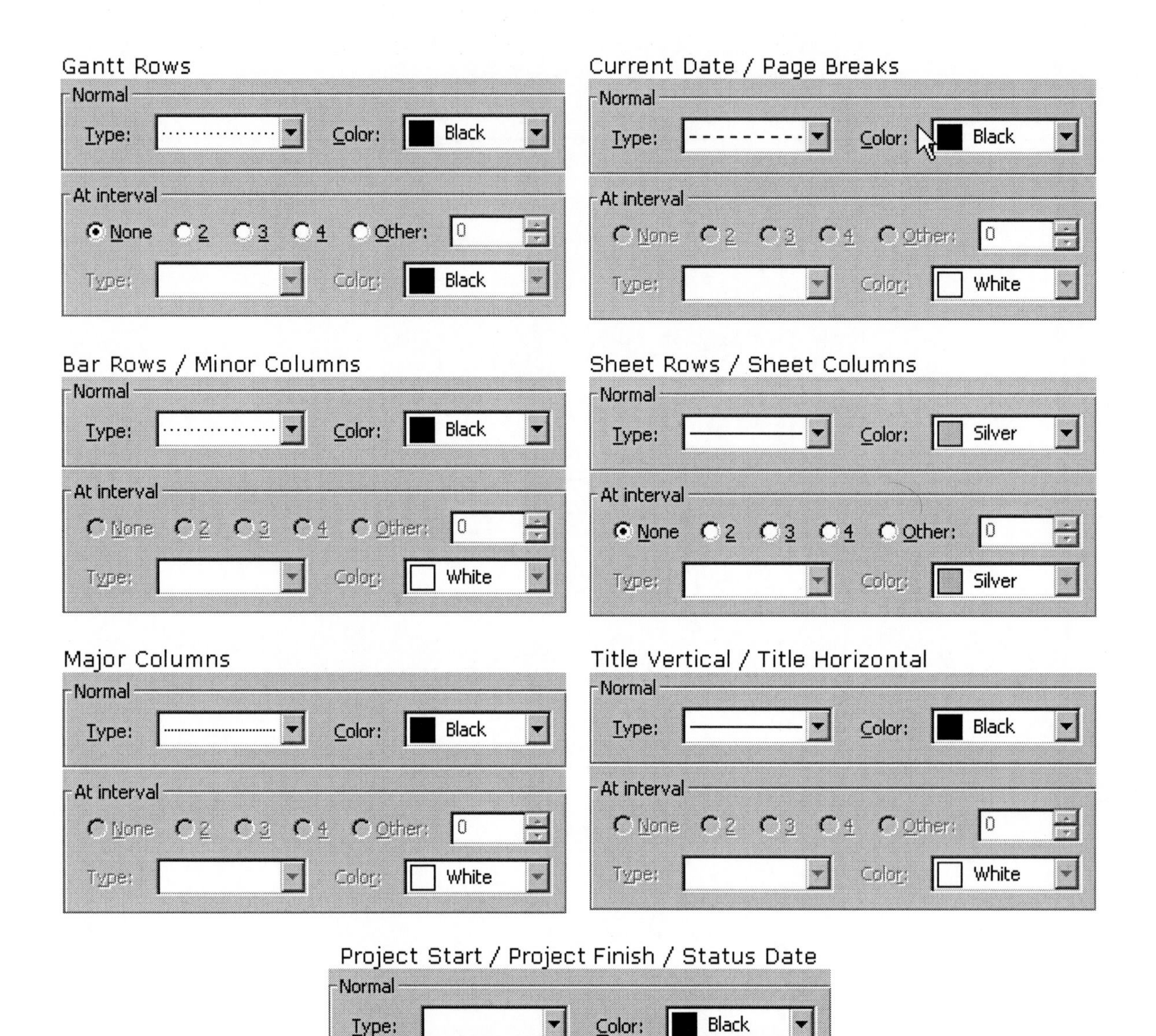

10-18: Gridlines

Summary View

This view will show the summary tasks sorted by Priority number and start date. The Summary view will keep the outline structure; showing consistent with the CSI coding structure with Pre-Construction Requirements first and then the associated summary tasks with subtasks listed under Project Duration. First make sure that the All Tasks filter is applied and then apply the sort criteria (Figure 10-19):

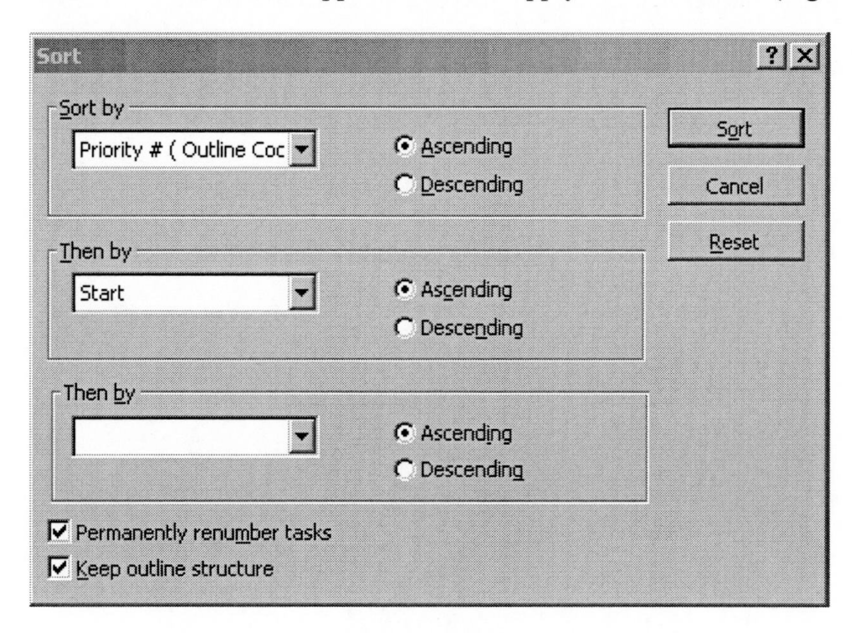

10-19: Sort By

Cascading View

This view will show the Pre-Construction Requirements and Project Duration summary tasks with all other summary tasks filtered out. This view is used when the All_Incomplete Subtasks filters are applied. The Cascading view will sort the subtasks under the Pre-Construction Requirements and Project Duration summary tasks as defined in the All_Incomplete Subtasks filter.

Make sure that the All_Incomplete Subtasks filter is active and then apply the following sort criteria (Figure 10-20):

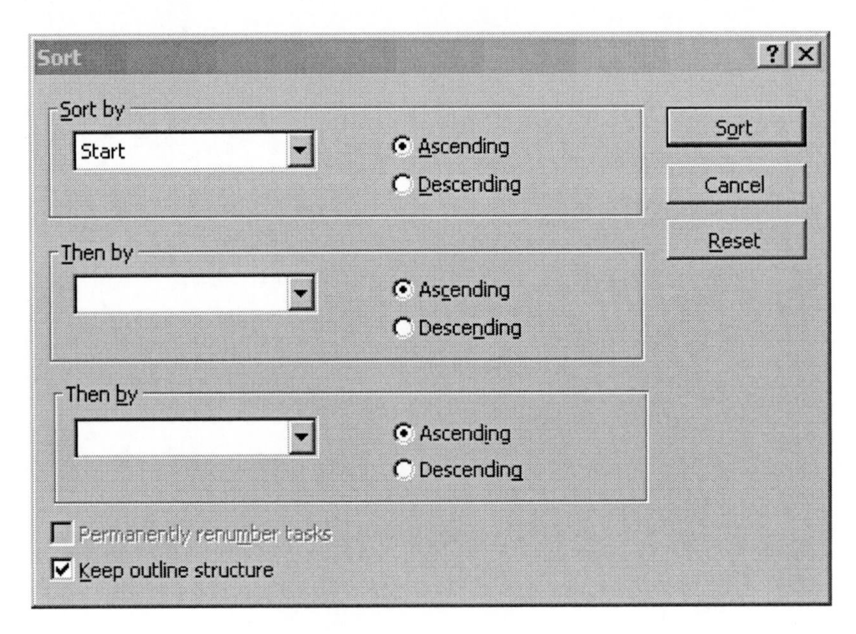

10-20: Sort By Criteria

Note: When sorting to view the Cascading View, make sure that the Keep outline structure is checked so that the related subtasks will be shown by start date under the Pre-Construction Requirements and Project Duration summary tasks.

Monthly Cash Flow (Figures 10-21 through 10-24)

	January	February	March	Total
PRE-CONSTRUCTION REQUIREMENTS	$8,300.00			$8,300.00
PROJECT DURATION				
GENERAL REQUIREMENTS	$10,000.00			$10,000.00
SITE CONSTRUCTION	$14,142.86	$8,357.14		$22,500.00
EXCAVATION AND BACKFILL	$6,500.00			$6,500.00
CONCRETE FOUNDATION	$8,000.00			$8,000.00
CONCRETE SLABS		$4,000.00		$4,000.00
STRUCTURAL STEEL				
Steel Delivery		$19,850.00		$19,850.00
Structural Steel Erection		$7,200.00		$7,200.00
ROOFING		$2,733.33	$5,466.67	$8,200.00
EXTERIOR WINDOWS			$4,000.00	$4,000.00
HVAC			$4,000.00	$4,000.00
Total	$46,942.86	$42,140.47	$13,466.67	$102,550.00

10-21: Monthly Cash Flow

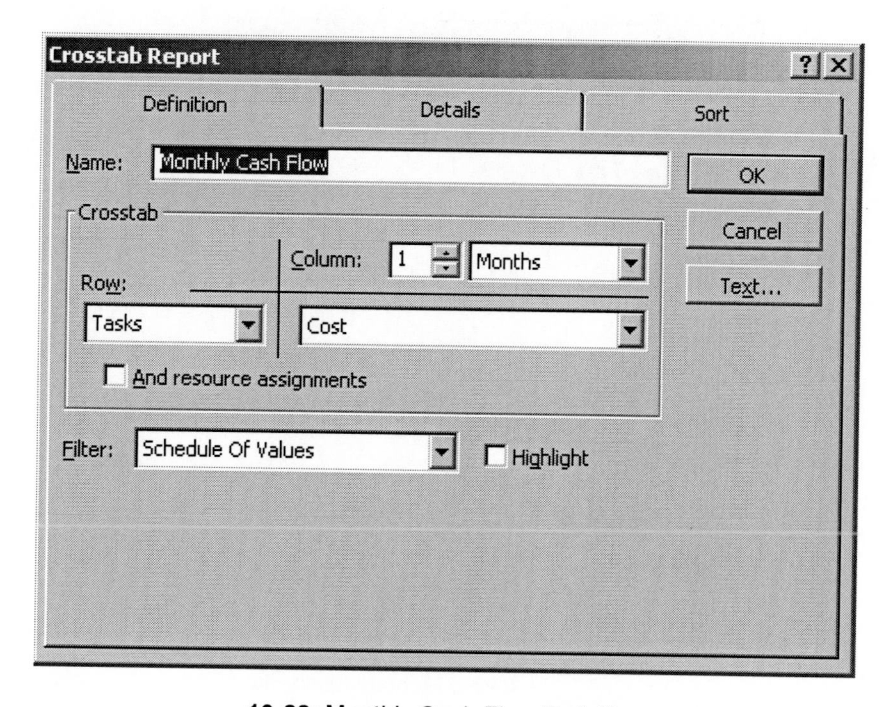

10-22: Monthly Cash Flow Definition

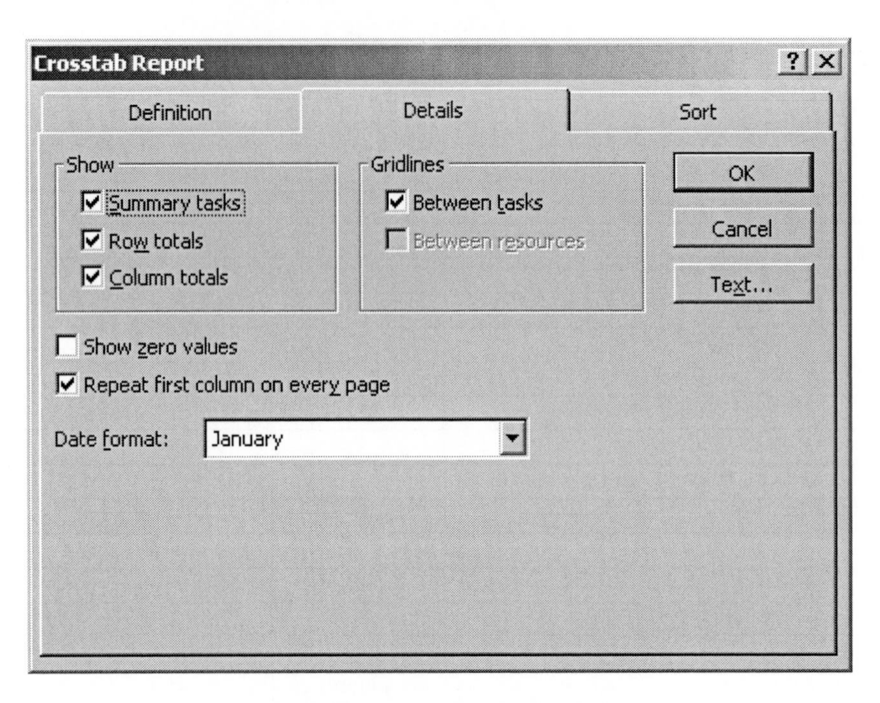

10-23: Monthly Cash Flow Details

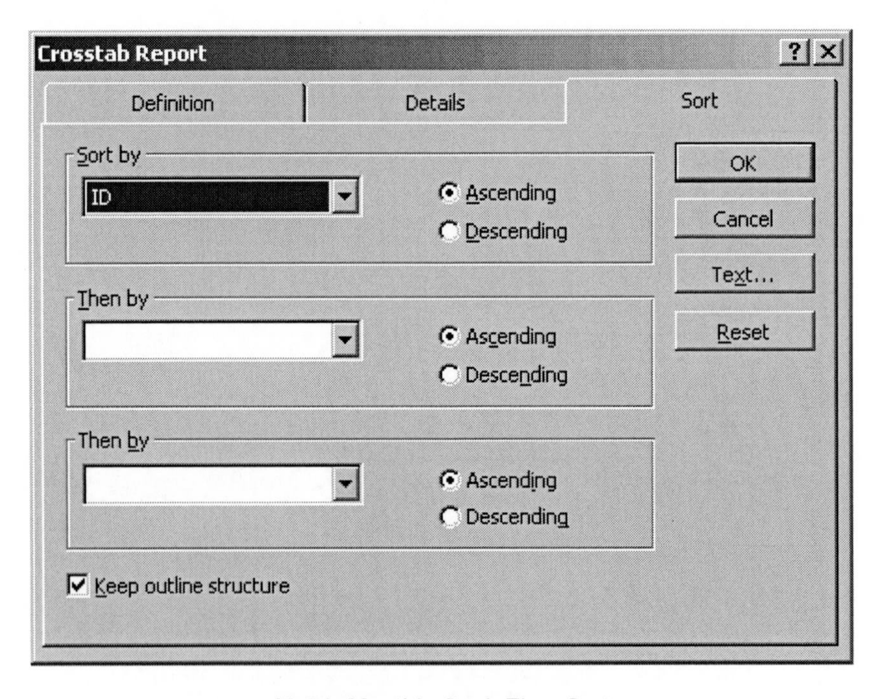

10-24: Monthly Cash Flow Sort

Weekly Cash Flow (Figures 10-25 through 10-28)

	12/31	1/7	1/14
PRE-CONSTRUCTION REQUIREMENTS	$8,300.00		
PROJECT DURATION			
GENERAL REQUIREMENTS	$1,500.00	$2,500.00	$2,500.00
SITE CONSTRUCTION	$1,928.57	$3,214.29	$3,214.29
EXCAVATION AND BACKFILL			$1,625.00
CONCRETE FOUNDATION			$1,142.86
CONCRETE SLABS			
STRUCTURAL STEEL			
Steel Delivery			
Structural Steel Erection			
ROOFING			
EXTERIOR WINDOWS			
HVAC			
Total	$11,728.57	$5,714.29	$8,482.15

10-25: Weekly Cash Flow Report

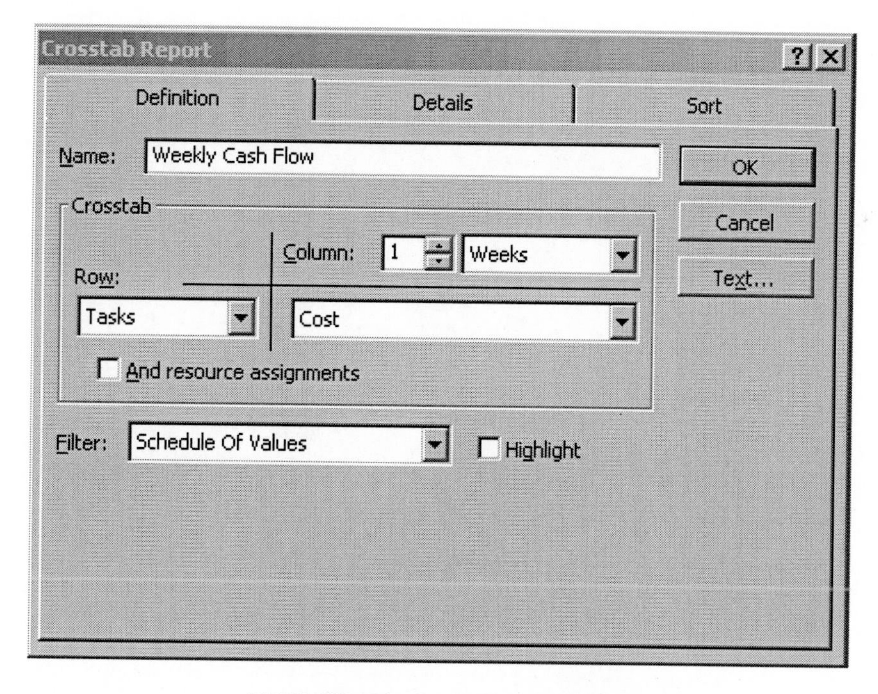

10-26: Weekly Cash Flow Definition

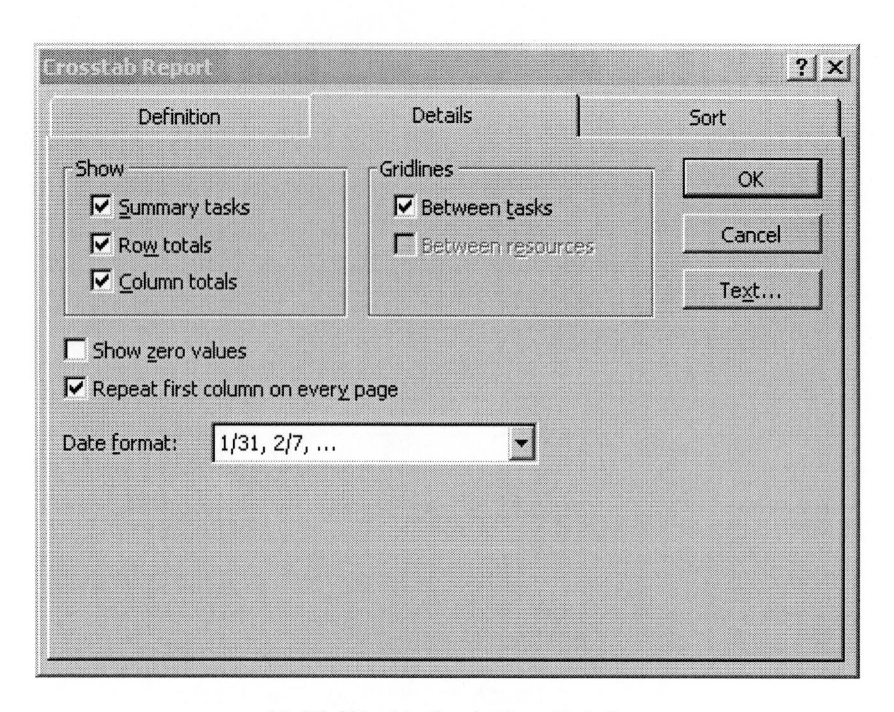

10-27: Weekly Cash Flow Details

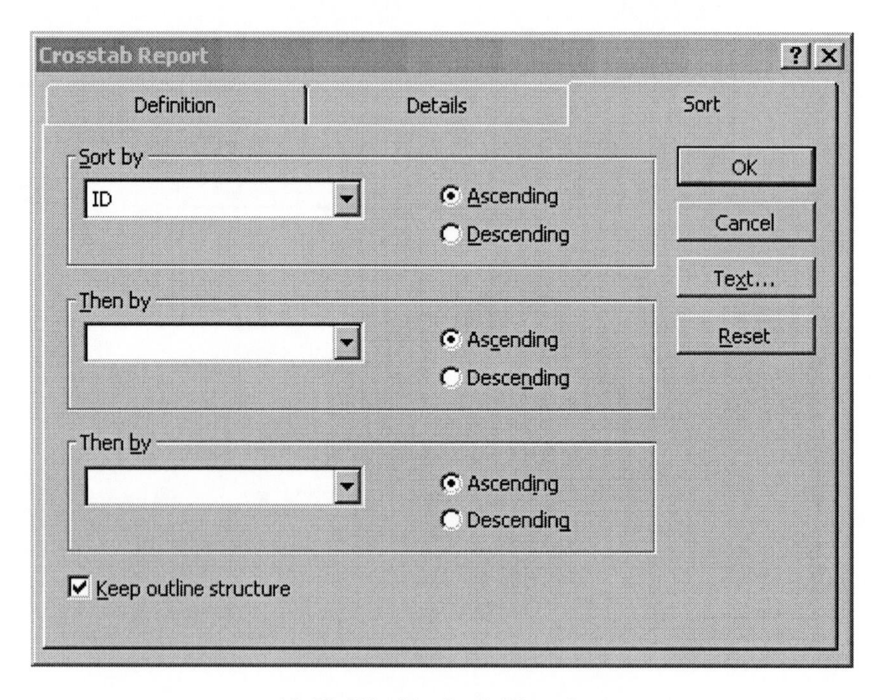

10-28: Weekly Cash Flow Sort

Project Calendar

The Project Calendar (Figure 10-29) that is set as the default calendar in all templates is a customized calendar defined as Construction.

Project calendar (Figure 10-29):

This calendar is based on a '5-day' workweek, Monday through Friday with working hours of 8:00 am to 5:00 pm, and has the following holidays programmed as non-work days:

New Year's Day

Memorial Day

Independence Day

Labor Day

Thanksgiving

Christmas Day

Option Settings

This section provides the default option settings (Figures 10-30 through 10-38) used in Microsoft Project to develop the template schedules and used as a part of the reference and views throughout this book. Settings may be changed, although the schedules may not view the same. Use this section for reference to reset the default settings if any changes are made.

view; 10-30: View

general; 10-31: General

edit; 10-32: Edit

calendar; 10-33: Calendar

schedule; 10-34: Schedule

calculation; 10-35: Calculation

spelling; 10-36: Spelling

workgroup; 10-37: Workgroup

save; 10-38: Save

Schedule Update Form

All updated schedules need to be communicated to all team members so that they can adjust their own schedules accordingly.

The following is a sample of a Schedule Update Form (Figure 10-39). Create a form similar to the form shown to promote action and responsibility.

Schedule Update

To: ____________________ **Date:** ________________
____
____________________ **Original Schedule Date:**

Attention: ____________________ ___/___/___

Project: ____________________ **Schedule Updated:**

From: ____________________ ___/___/___

Phone # ___/___/___ **Fax #** ___/___/___ **E Mail:** ____________________

Attached is a copy of the current updated schedule. All current information, changes and / or adjustments have been made. You are responsible for meeting the adjusted dates for your work, coordinating with other trades and adhering to the contract schedule requirements.

If you have any objections, concerns or have any labor, material and or equipment problems which may interfere with this schedule, please advise the project manager of the conflict in the space provided below within (2) two days of your receipt of this updated schedule.

Comments:

____
____
____
____
____
____
____
____

Signed: ____________________

Date: ___/___/___

10-39: Schedule Update Form

Appendix A

CSI Master Format – Level 1

00000 INTRODUCTORY INFORMATION

00099 BIDDING REQUIREMENTS

00499 CONTRACTING REQUIREMENTS

00980 FACILITIES AND SPACES

00990 SYSTEMS AND ASSEMBLIES

01000 GENERAL REQUIREMENTS

02000 SITE CONSTRUCTION

03000 CONCRETE

04000 MASONRY

05000 METALS

06000 WOOD AND PLASTICS

07000 THERMAL AND MOISTURE PROTECTION

08000 DOORS AND WINDOWS

09000 FINISHES

10000 SPECIALTIES

11000 EQUIPMENT

12000 FURNISHINGS

13000 SPECIAL CONSTRUCTION

14000 CONVEYING SYSTEMS

15000 MECHANICAL

16000 ELECTRICAL

Appendix B

CSI Master Format – Level 2

00000 INTRODUCTORY INFORMATION
00001 Project Title Page
00005 Certifications Page
00007 Seals Page
00010 Table of Contents
00015 List of Drawings
00020 List of Schedules
00099 BIDDING REQUIREMENTS
00100 Bid Solicitation
00200 Instruction to Bidders
00300 Information Available to Bidders
00400 Bid Forms and Supplements
00490 Bidding Agenda
00499 CONTRACTING REQUIREMENTS
00500 Agreement
00600 Bonds and Certificates
00700 General Conditions
00800 Supplementary Conditions
00900 Addenda and Modifications
00980 FACILITIES AND SPACES
00980 Facilities and Spaces
00990 SYSTEMS AND ASSEMBLIES
00990 Systems and Assemblies
01000 GENERAL REQUIREMENTS
01100 Summary
01200 Price and Payment Procedures
01300 Administrative Requirements
01400 Quality Requirements
01500 Temporary Facilities and Controls
01600 Product Requirements
01700 Execution Requirements
01800 Facility Operation
01900 Facility Decommissioning
02000 SITE CONSTRUCTION
02050 Basic Site Materials and Methods
02100 Site Remediation
02200 Site Preparation
02300 Earthwork
02400 Tunneling, Boring, and Jacking
02450 Foundation and Load-Bearing Elements
02500 Utility Services
02600 Drainage and Containment
02700 Bases, Ballasts, Pavements, and Appurtenances
02800 Site Improvements and Amenities
2900 Planting
02950 Site Restoration and Rehabilitation

03000 CONCRETE
03050 Basic Concrete Materials and Methods
03100 Concrete Forms and Accessories
03200 Concrete Reinforcement
03300 Cast-In-Place Concrete
03400 Precast Concrete
03500 Cementitious Decks and Underlayment
03600 Grouts
03700 Mass Concrete
03900 Concrete Restoration and Cleaning
04000 MASONRY
04050 Basic Masonry Materials and Methods
04200 Masonry Units
04400 Stone
04500 Refractories
04600 Corrosion-Resistant Masonry
04700 Simulated Masonry
04800 Masonry Assemblies
04900 Masonry Restoration and Cleaning
05000 METALS
05050 Basic Metal Materials and Methods
05100 Structural Metal Framing
05200 Metal Joists
05300 Metal Deck
05400 Cold-Formed Metal Framing
05500 Metal Fabrications
05600 Hydraulic Fabrications
05650 Railroad Track and Accessories
05700 Ornamental Metal
05800 Expansion Control
05900 Metal Restoration and Cleaning
06000 WOOD AND PLASTICS
06050 Basic Wood and Plastic Materials ad Methods
06100 Rough Carpentry
06200 Finish Carpentry
06400 Architectural Woodwork
06500 Structural Plastics
06600 Plastic Fabrications
06900 Wood and Plastic Restoration and Cleaning

07000 THERMAL AND MOISTURE PROTECTION
07050 Basic Thermal and Moisture Protection
07051 Materials and Methods
07100 Damp-proofing and Waterproofing
07200 Thermal Protection
07300 Shingles, Roof Tiles, and Roof Coverings
07400 Roofing and Siding Panels
07500 Membrane Roofing
07600 Flashing and Sheet Metal
07700 Roof Specialties and Accessories
07800 Fire and Smoke Protection
07900 Joint Sealers

08000 DOORS AND WINDOWS
08050 Basic Door and Window
08050 Materials and Methods
08100 Metal Doors and Frames
08200 Wood and Plastic Doors
08300 Specialty Doors
08400 Entrances and Storefronts
08500 Windows
08600 Skylights
08700 Hardware
08800 Glazing
08900 Glazed Curtain Wall

09000 FINISHES
09050 Basic Finish Materials and Methods
09100 Metal Support Assemblies
09200 Plaster and Gypsum Board
09300 Tile
09400 Terrazzo
09500 Ceilings
09600 Flooring
09700 Wall Finishes
09800 Acoustical Treatment
09900 Paints and Coatings

01000 SPECIALTIES
10100 Visual Display Boards
10150 Compartments and Cubicles
10200 Louvers and Vents
10240 Grilles and Screens
10250 Service Walls
10260 Wall and Corner Guards
10270 Access Flooring
10290 Pest Control
10300 Fireplaces and Stoves
10310 Fireplace Specialties and Accessories
10340 Manufactured Exterior Specialties
10350 Flagpoles
10400 Identification Devices
10450 Pedestrian Control Devices
10500 Lockers
10520 Fire Protection Specialties
10530 Protective Covers
10550 Postal Specialties
10600 Partitions
10670 Storage Shelving
10700 Exterior Protection
10750 Telephone Specialties
10800 Toilet, Bath, and Laundry Accessories
10880 Scales
10900 Wardrobe and Closet Specialties

11000 EQUIPMENT
11010 Maintenance Equipment
11020 Security and Vault Equipment
11030 Teller and Service Equipment
11040 Ecclesiastical Equipment
11050 Library Equipment
11060 Theater and Stage Equipment
11070 Instrumental Equipment
11080 Registration Equipment
11090 Checkroom Equipment
11100 Mercantile Equipment
11110 Commercial Laundry and Dry Cleaning Equipment
11120 Vending Equipment
11130 Audio-Visual Equipment
11140 Vehicle Service Equipment
11150 Parking Control Equipment
11160 Loading Dock Equipment
11170 Solid Waste Handling Equipment
11190 Detention Equipment
11200 Water Supply and Treatment Equipment
11300 Fluid Waste Treatment and Disposal Equipment
11400 Food Service Equipment
11450 Residential Equipment
11460 Unit Kitchens
11470 Darkroom Equipment
11480 Athletic, Recreational, and Therapeutic Equipment
11500 Industrial and Process Equipment
11600 Laboratory Equipment
11650 Planetarium Equipment
11660 Observatory Equipment
11680 Office Equipment
11700 Medical Equipment
11800 Mortuary Equipment

11850 Navigation Equipment
11870 Agricultural Equipment
11900 Exhibit Equipment

12000 FURNISHINGS
12050 Fabrics
12110 Art
12300 Manufactured Casework
12400 Furnishings and Accessories
12500 Furniture
12600 Multiple Seating
12700 Systems Furniture
12800 Interior Plants and Planters
12900 Furnishings Restoration and Repair

13000 SPECIAL CONSTRUCTION
13010 Air-Supported Structures
13020 Building Modules
13030 Special Purpose Rooms
13080 Sound, Vibration, and Seismic Control
13090 Radiation Protection
13100 Lightning Protection
13110 Cathodic Protection
13120 Pre-Engineered Structures
13150 Swimming Pools
13160 Aquariums
13165 Aquatic Park Facilities
13170 Tubs and Pools
13175 Ice Rinks
13185 Kennels and Animal Shelters
13190 Site-Constructed Incinerators
13200 Storage Tanks
13220 Filter Underdrains and Media
13230 Digester Covers and Appurtenances
13240 Oxygenation Systems
13260 Sludge Conditioning Systems
13280 Hazardous Material Remediation
13400 Measurement and Control Instrumentation
13500 Recording Instrumentation
13550 Transportation Control Instrumentation
13600 Solar and Wind Energy Equipment
13700 Security Access and Surveillance
13800 Building Automation and Control
13850 Detection and Alarm
13900 Fire Suppression

14000 CONVEYING SYSTEMS
14100 Dumbwaiters
14200 Elevators
14300 Escalators and Moving Walks
14400 Lifts
14500 Material Handling
14600 Hoists and Cranes
14700 Turntables
14800 Scaffolding
14900 Transportation

15000 MECHANICAL
15050 Basic Mechanical Materials and Methods
15100 Building Services Piping
15200 Process Piping
15300 Fire Protection Piping
15400 Plumbing Fixtures and Equipment
15500 Heat-Generation Equipment
15600 Refrigeration Equipment
15700 Heating, Ventilating, and Air Conditioning Equipment
15800 Air Distribution
15900 HVAC Instrumentation and Controls
15950 Testing, Adjusting, and Balancing

16000 ELECTRICAL
16050 Basic Electrical Materials and Methods
16100 Wiring Methods
16200 Electrical Power
16270 Transformers
16300 Transmission and Distribution
16400 Low-Voltage Distribution
16500 Lighting
16700 Communications
16800 Sound and Video

Appendix C

CSI Master Format – Level 3

00000 INTRODUCTORY INFORMATION
00001 Project Title Page
00005 Certifications Page
00007 Seals Page
00010 Table of Contents
00015 List of Drawings
20 List of Schedules
21

00099 BIDDING REQUIREMENTS
00100 Bid Solicitation
00100 Advertisement for Bids
00100 Bidder's Qualifications
00100 Invitation to Bid
00100 Request for Proposal
00200 Instruction to Bidders
00210 Supplementary Instructions to Bidders
00220 Bid Scopes
00250 Pre-Bid Meeting
00300 Information Available to Bidders
00310 Preliminary Schedules
00320 Geotechnical Data
00330 Existing Conditions
00340 Environmental Assessment Information
00350 Project Financial Information
00360 Permit Application
00400 Bid Forms and Supplements
00410 Bid Form
00430 Bid Form and Supplements
00450 Representations and Certifications
00490 Bidding Agenda

00499 CONTRACTING REQUIREMENTS
00500 Agreement
00510 Notice of Award
00520 Agreement Form
00540 Attachments to Agreement Form
00550 Notice to Proceed
00570 Definitions
00600 Bonds and Certificates
00610 Bonds
00620 Certificates
00640 Release of Liens
00650 Statutory Declaration Forms
00700 General Conditions
00800 Supplementary Conditions
00800 Anti-Pollution Measures
00800 Assigned Contracts
00800 Equal Employment Opportunity Requirements
00800 Health and Safety Criteria
00800 Insurance Requirements
00800 Labor Stabilization Agreement
00800 Letter of Assent
00800 Non-Segregated Facilities Requirements
00840 Procurement Contracts
00850 Specific Project Requirements
00860 Statutory Requirements
00870 Wage Determination Schedule
00880 Wage Rate Requirements
00890 Permits
00900 Addenda and Modifications
00910 Addenda
00920 Claims
00930 Clarifications and Proposals
00940 Modifications

00980 FACILITIES AND SPACES
00980 Facilities and Spaces

00990 SYSTEMS AND ASSEMBLIES
00990 Systems and Assemblies

Appendix

01000 GENERAL REQUIREMENTS
01100 Summary
01110 Summary of Work
01120 Multiple Contract Summary
01140 Work Restrictions
01180 Project Utility Sources
01200 Price and Payment Procedures
01210 Allowances
01230 Alternates
01240 Value Analysis
01250 Contract Modification Procedures
01270 Unit Prices
01290 Payment Procedures
01300 Administrative Requirements
01310 Project Management and Coordination
01320 Construction Progress Documentation
01330 Submittal Procedures
01350 Special Procedures
01400 Quality Requirements
01410 Regulatory Requirements
01420 References
01430 Quality Assurance
01450 Quality Control
01500 Temporary Facilities and Controls
01510 Temporary Utilities
01520 Construction Facilities
01530 Temporary Construction
01540 Construction Aids
01550 Vehicular Access and Parking
01560 Temporary Barriers and Enclosures
01570 Temporary Controls
01580 Project Identification
01600 Product Requirements
01610 Basic Product Requirements
01620 Product Options
01630 Product Substitution Procedures
01640 Owner-Furnished Products
01650 Product Delivery Requirements
01660 Product Storage and Handling Requirements
01700 Execution Requirements
01710 Examination
01720 Preparation
01730 Execution
01740 Cleaning
01750 Starting and Adjusting
01760 Protecting Installed Construction
01770 Closeout Procedures
01780 Closeout Submittals
01800 Facility Operation
01810 Commissioning
01820 Demonstration and Training
01830 Operation and Maintenance
01890 Reconstruction
01900 Facility Decommissioning
01900 Decommissioning Systems and Equipment
01900 Facility Demolition and Removal
01900 Hazardous Materials Abatement
01900 Hazardous Materials Removal and Disposal
01900 Protection of Deactivated Facilities

02000 SITE CONSTRUCTION
02050 Basic Site Materials and Methods
02055 Soils
02060 Aggregate
02065 Cement and Concrete
02070 Geosynthetics
02080 Utility Materials
02090 Joint Material
02100 Site Remediation
02105 Chemical Sampling and Analysis
02110 Excavation, Removal, and Handling of Hazardous Materials
02115 Underground Storage Tank Removal
02120 Off-Site Transportation and Disposal
02125 Drum Handling
02130 Site Decontamination
02140 Landfill Construction and Storage
02145 Groundwater Treatment Systems
02150 Hazardous Waste Recovery Processes
02160 Physical Treatment
02170 Chemical Treatment
02180 Thermal Processes
02190 Biological Processes
02195 Remediation Soil Stabilization
02200 Site Preparation
02210 Subsurface Investigation
02220 Site Demolition
02230 Site Clearing
02240 Dewatering
02250 Shoring and Underpinning
02260 Excavation Support and Protection
02280 Grade Adjustment and Abandonment of Existing Miscellaneous Structures
02285 Rebuilt Miscellaneous Structures
02290 Site Monitoring
02300 Earthwork
02310 Grading
02315 Excavation and Fill
02325 Dredging
02330 Embankment
02335 Sub-grade and Roadbed
02340 Soil Stabilization
02360 Soil Treatment
02370 Erosion and Sedimentation Control
02380 Scour Protection

02390 Shoreline Protection and Mooring Structures
02400 Tunneling, Boring, and Jacking
02410 Tunnel Excavation
02420 Initial Tunnel Support Systems
02425 Tunnel Linings
02430 Tunnel Grouting
02440 Immersed and Sunken Tube Tunnels
02441 Microtunneling
02442 Cut and Cover Tunnels
02443 Tunnel Leak Repairs
02444 Shaft Construction
02445 Boring or Jacking Conduits
02450 Foundation and Load-Bearing Elements
02455 Driven Piles
02465 Bored Piles
02475 Caissons
02480 Foundation Walls
02490 Anchors
02495 Instrumentation and Monitoring
02500 Utility Services
02510 Water Distribution
02520 Wells
02530 Sanitary Sewerage
02540 Septic Tank Systems
02550 Piped Energy Distribution
02570 Process Materials Distribution Structures
02580 Electrical and Communication Structures
02590 Site Grounding
02600 Drainage and Containment
02610 Pipe Culverts
02620 Subdrainage
02630 Storm Drainage
02640 Culverts and Manufactured Construction
02660 Ponds and Reservoirs
02670 Constructed Wetlands
02700 Bases, Ballasts, Pavements, and Appurtenances
02710 Bound Base Courses
02720 Unbound Base Courses and Ballasts
02730 Aggregate Surfacing
02740 Flexible Pavement
02750 Rigid Pavement
02755 Cement Concrete Shoulders
02760 Paving Specialties
02770 Curbs and Gutters
02775 Sidewalks
02780 Unit Pavers
02785 Flexible Pavement Coating and Micro-Surfacing
02790 Athletic and Recreational Surfaces
02795 Porous Pavement
02800 Site Improvements and Amenities
02810 Irrigation System
02815 Fountains
02820 Fences and Gates
02830 Retaining Walls

02000 SITE CONSTRUCTION

02840 Walk, Road, and Parking Appurtenances
02850 Prefabricated Bridges
02860 Screening Devices
02870 Site Furnishings
02875 Site and Street Shelters
02880 Play Field Equipment and Structures
02890 Traffic Signs and Signals
02895 Markers and Monuments
02900 Planting
02905 Plants, Planting, and Transplanting
02910 Plant Preparation
02915 Shrub and Tree Transplanting
02920 Lawns and Grasses
02930 Exterior Plants
02935 Plant Maintenance
02945 Planting Accessories
02950 Site Restoration and Rehabilitation
02995 Restoration of Underground Piping
02960 Flexible Pavement Surfacing Recovery
02965 Flexible and Bituminous Pavement Recycling
02975 Flexible and Bituminous Pavement Reinforcement and Crack and Joint Sealants
02980 Rigid Pavement Rehabilitation
02990 Structure Moving

03000 CONCRETE

03050 Basic Concrete Materials and Methods
03100 Concrete Forms and Accessories
03110 Structural Cast-in-Place Concrete Forms
03120 Architectural Cast-in-Place Concrete Forms
03130 Permanent Forms
03150 Concrete Accessories
03200 Concrete Reinforcement
03210 Reinforcing Steel
03220 Welded Wire Fabric
03230 Stressing Tendons
03240 Fibrous Reinforcing
03250 Post-Tensioning
03300 Cast-In-Place Concrete
03310 Structural Concrete
03330 Architectural Concrete

03340 Low Density Concrete
03350 Concrete Finishing
03360 Concrete Finishes
03370 Specially Placed Concrete
03380 Post-Tensioned Concrete
03390 Concrete Curing
03400 Precast Concrete
03410 Plant-Precast Structural Concrete
03420 Plant-Precast Structural Post-Tensioned Concrete
03430 Site-Precast Structural Concrete
03450 Plant-Precast Architectural Concrete
03460 Site-Precast Architectural Concrete
03470 Tilt-Up Precast Concrete
03480 Precast Concrete Specialties
03490 Glass-Fiber-Reinforced Precast Concrete
03500 Cementitious Decks and Underlayment
03510 Cementitious Roof Deck
03520 Lightweight Concrete Roof Insulation
03530 Concrete Topping
03540 Cementitious Underlayment
03600 Grouts
03700 Mass Concrete
03900 Concrete Restoration and Cleaning
03910 Concrete Cleaning
03920 Concrete Resurfacing
03930 Concrete Rehabilitation

04000 MASONRY

04050 Basic Masonry Materials and Methods
04060 Masonry Mortar
04070 Masonry Grout
04080 Masonry Anchorage and Reinforcement
04090 Masonry Accessories
04200 Masonry Units
04210 Clay Masonry Units
04220 Concrete Masonry Units
04230 Calcium Silicate Unit Masonry
04270 Glass Masonry Units
04290 Adobe Masonry Units
04400 Stone
04410 Stone Materials
04420 Collected Stone
04430 Quarried Stone
04500 Refractories
04550 Flue Liners
04560 Combustion Chambers
04570 Castable Refractories
04580 Refractory Brick
04600 Corrosion-Resistant Masonry
04610 Chemical-Resistant Brick
04620 Vitrified Clay Liner Plates
04700 Simulated Masonry
04710 Simulated Brick
04720 Cast Stone
04730 Simulated Stone

04000 MASONRY

04800 Masonry Assemblies
04810 Unit Masonry Assemblies
04820 Reinforced Unit Masonry Assemblies
04830 Non-Reinforced Unit Masonry Assemblies
04840 Prefabricated Masonry Panels
04850 Stone Assemblies
04880 Masonry Fireplaces
04900 Masonry Restoration and Cleaning
04910 Unit Masonry Restoration
04920 Stone Restoration
04930 Unit Masonry Cleaning
04940 Stone Cleaning

05000 METALS

05050 Basic Metal Materials and Methods
05060 Metal Materials
05080 Factory-Applied Metal Coatings
05090 Metal Fastenings
05100 Structural Metal Framing
05120 Structural Steel
05140 Structural Aluminum
05150 Wire Rope Assemblies
05160 Metal Framing Systems
05200 Metal Joists
05210 Steel Joists
05250 Aluminum Joists
05260 Composite Joist Assemblies
05300 Metal Deck
05310 Steel Deck
05320 Raceway Deck Systems
05330 Aluminum Deck
05340 Acoustical Metal Deck
05400 Cold-Formed Metal Framing
05410 Load-Bearing Metal Studs
05420 Cold-Formed Metal Joists
05430 Slotted Channel Framing
05450 Metal Support
05500 Metal Fabrications
05510 Metal Stairs and Ladders
05520 Handrails and Railings

05530 Gratings
05540 Floor Plates
05550 Stair Treads and Nosings
05560 Metal Castings
05580 Formed Metal Fabrications
05600 Hydraulic Fabrications
05600 Bifurcation Panels
5600 Bulkheads
05600 Manifolds
05600 Penstocks
05600 Trashracks
05650 Railroad Track and Accessories
05700 Ornamental Metal
05710 Ornamental Stairs
05715 Fabricated Spiral Stairs
05720 Ornamental Handrails and Railings
05725 Ornamental Metal Castings
05730 Ornamental Formed Metal
05740 Ornamental Forged Metal
05800 Expansion Control
05810 Expansion Joint Cover Assemblies
05820 Slide Bearings
05830 Bridge Expansion Joint Assemblies
05900 Metal Restoration and Cleaning

06000 WOOD AND PLASTICS
06050 Basic Wood and Plastic Materials ad Methods
06060 Wood Materials
06065 Plastic Materials
06070 Wood Treatment
06080 Factory-Applied Wood Coatings
06090 Wood and Plastic Fastenings
06100 Rough Carpentry
06110 Wood Framing
06120 Structural Panels
06130 Heavy Timber Construction
06140 Treated Wood Foundations
06150 Wood Decking
06160 Sheathing
06170 Prefabricated Structural Wood
06180 Glued-Laminated Construction
06200 Finish Carpentry
06220 Millwork
06250 Pre-finished Paneling
06260 Board Paneling
06270 Closet and Utility Wood Shelving
06400 Architectural Woodwork
06410 Custom Cabinets
06415 Countertops
06420 Paneling
06430 Wood Stairs and Railings
06440 Wood Ornaments
06445 Simulated Wood Ornaments
06450 Standing and Running Trim
06455 Simulated Wood Trim
06460 Wood Frames
06470 Screens, Blinds, and Shutters
06500 Structural Plastics
06510 Structural Plastic Shapes and Plates
06520 Plastic Structural Assemblies
06600 Plastic Fabrications
06600 Cultured Marble
06600 Glass-Fiber-Reinforced Plastic
06600 Plastic Handrails
06600 Plastic Paneling
06600 Solid Surfacing
06900 Wood and Plastic Restoration and Cleaning
06910 Wood Restoration and Cleaning
06920 Plastic Restoration and Cleaning

07000 THERMAL AND MOISTURE PROTECTION
07050 Basic Thermal and Moisture Protection Materials and Methods
07100 Dampproofing and Waterproofing
07110 Dampproofing
07120 Built-Up Bituminous Waterproofing
07130 Sheet Waterproofing
07140 Fluid-Applied Waterproofing
07150 Sheet Metal Waterproofing
07160 Cementitious and Reactive Waterproofing
07170 Bentonite Waterproofing
07180 Traffic Coatings
07190 Water Repellents
07200 Thermal Protection
07210 Building Insulation
07220 Roof and Deck Insulation
07240 Exterior Insulation and Finish Systems (EIFS)
07260 Vapor Retarders
07270 Air Barriers
07300 Shingles, Roof Tiles, and Roof Coverings
07310 Shingles
07320 Roof Tiles
07330 Roof Coverings
07400 Roofing and Siding Panels
07410 Metal Roof and Wall Panels
07420 Plastic Roof and Wall Panels
07430 Composite Panels
07440 Faced Panels
07450 Fiber-Reinforced Cementitious Panels
07460 Siding

07470 Wood Roof and Wall Panels
07480 Exterior Wall Assemblies
07500 Membrane Roofing
07510 Built-Up Bituminous Roofing
07520 Cold-Applied Bituminous Roofing
07530 Elastomeric Membrane Roofing
07540 Thermoplastic Membrane Roofing
07550 Modified Bituminous Membrane Roofing
07560 Fluid-Applied Roofing
07570 Coated Foamed Roofing
07580 Roll Roofing
07590 Roof Maintenance and Repairs
07600 Flashing and Sheet Metal
07610 Sheet Metal Roofing
07620 Sheet Metal Flashing and Trim
07630 Sheet Metal Roofing Specialties
07650 Flexible Flashing
07700 Roof Specialties and Accessories
07710 Manufactured Roof Specialties
07720 Roof Accessories
07760 Roof Pavers
07800 Fire and Smoke Protection
07810 Applied Fireproofing
07820 Board Fireproofing
07840 Fire-stopping
07860 Smoke Seals
07870 Smoke Containment Barriers
07900 Joint Sealers
07910 Preformed Joint Seals
07920 Joint Sealants

08000 DOORS AND WINDOWS

08050 Basic Door and Window Materials and Methods
08100 Metal Doors and Frames
08110 Steel Doors and Frames
08120 Aluminum Doors and Frames
08130 Stainless Steel Doors and Frames
08140 Bronze Doors and Frames
08150 Preassembled Metal Door and Frame Units
08160 Sliding Metal Doors and Grilles
08180 Metal Screen and Storm Doors
08190 Metal Door Restoration
08200 Wood and Plastic Doors
08210 Wood Doors
08220 Plastic Doors
08250 Preassembled Wood and Plastic Door and Frame Units
08260 Sliding Wood and Plastic Doors
08280 Wood and Plastic Storm and Screen Doors
08290 Wood and Plastic Door Restoration
08300 Specialty Doors
08310 Access Doors and Panels
08320 Detention Doors and Frames
08330 Coiling Doors and Grilles
08340 Special Function Doors
08350 Folding Doors and Grilles
08360 Overhead Doors
08370 Vertical Lift Doors
08380 Traffic Doors
08390 Pressure-Resistant Doors
08400 Entrances and Storefronts
08410 Metal-Framed Storefronts
08450 All-Glass Entrances and Storefronts
08460 Automatic Entrance Doors
08470 Revolving Entrance Doors
08480 Balanced Entrance Doors
08490 Sliding Storefronts
08500 Windows
08510 Steel Windows
08520 Aluminum Windows
08530 Stainless Steel Windows
08540 Bronze Windows
08550 Wood Windows
08560 Plastic Windows
08570 Composite Windows
08580 Special Function Windows
08590 Window Restoration and Replacement
08600 Skylights
08610 Roof Windows
08620 Unit Skylights
08630 Metal-Framed Skylights
08700 Hardware
08710 Door Hardware
08720 Weather-stripping and Seals
08740 Electro-Mechanical Hardware
08750 Window Hardware
08770 Door and Window Accessories
08780 Special Function Hardware
08790 Hardware Restoration
08800 Glazing
08810 Glass
08830 Mirrors
08840 Plastic Glazing
08850 Glazing Accessories
08890 Glazing Restoration
08900 Glazed Curtain Wall
08910 Metal Framed Curtain Wall
08950 Translucent Wall and Roof Assemblies
08960 Sloped Glazing Assemblies
08970 Structural Glass Curtain Walls
08990 Glazed Curtain Wall Restoration

09000 FINISHES
09050 Basic Finish Materials and Methods
09100 Metal Support Assemblies
09110 Non-Load-Bearing Wall Framing
09120 Ceiling Suspension
09130 Acoustical Suspension
09190 Metal Frame Restoration
09200 Plaster and Gypsum Board
09205 Furring and Lathing
09210 Gypsum Plaster
09220 Portland Cement Plaster
09230 Plaster Fabrications
09250 Gypsum Board
09260 Gypsum Board Assemblies
09270 Gypsum Board Accessories
09280 Plaster Restoration
09300 Tile
09305 Tile Setting Materials and Accessories
09310 Ceramic Tile
09330 Quarry Tile
09340 Paver Tile
09350 Glass Mosaics
09360 Plastic Tile
09370 Metal Tile
09380 Cut Natural Stone Tile
09390 Tile Restoration
09400 Terrazzo
09410 Portland Cement Terrazzo
09420 Precast Terrazzo
09430 Conductive Terrazzo
09440 Plastic Matrix Terrazzo
09490 Terrazzo Restoration
09500 Ceilings
09510 Acoustical Ceilings
09545 Specialty Ceilings
09550 Mirror Panel Ceilings
09560 Textured Ceilings
09570 Linear Wood Ceilings
09580 Suspended Decorative Grids
09590 Ceiling Assembly Restoration
09600 Flooring
09610 Floor Treatment
09620 Specialty Flooring
09630 Masonry Flooring
09640 Wood Flooring
09650 Resilient Flooring
09660 Static Control Flooring
09670 Fluid-Applied Flooring
09680 Carpet
09690 Flooring Restoration
09700 Wall Finishes
09710 Acoustical Wall Treatment
09720 Wall Covering
09730 Wall Carpet
09740 Flexible Wood Sheets
09750 Stone Facing
09760 Plastic Blocks
09770 Special Wall Surfaces
09790 Wall Finish Restoration
09800 Acoustical Treatment
09810 Acoustical Space Units
09820 Acoustical Insulation and Sealants
09830 Acoustical Barriers
09840 Acoustical Wall Treatment
09900 Paints and Coatings
09910 Paints
09930 Stains and Transparent Finishes
09940 Decorative Finishes
09960 High-Performance Coatings
09970 Coatings for Steel
09980 Coatings for Concrete and Masonry
9990 Paint Restoration

10000 SPECIALTIES
10100 Visual Display Boards
10110 Chalkboards
10115 Markerboards
10120 Tackboard and Visual Aid Boards
10130 Operable Board Units
10140 Display Track Assemblies
10145 Visual Aid Board Units
10150 Compartments and Cubicles
10160 Metal Toilet Compartments
10165 Plastic Laminate Toilet Compartments
10170 Plastic Toilet Compartments
10175 Particleboard Toilet Compartments
10180 Stone Toilet Compartments
10185 Shower and Dressing Compartments
10190 Cubicles
10200 Louvers and Vents
10210 Wall Louvers
10220 Louvered Equipment Enclosures
10225 Door Louvers
10230 Vents
10240 Grilles and Screens
10250 Service Walls
10260 Wall and Corner Guards
10260 Bumper Guards
10260 Corner Guards
10260 Impact-Resistant Wall Protection
10270 Access Flooring
10270 Rigid Grid Assemblies
10270 Snap-on Stringer Assemblies
10270 Stringerless Assemblies
10290 Pest Control
10290 Bird Control
10290 Insect Control
10290 Rodent Control

10300 Fireplaces and Stoves
10305 Manufactured Fireplaces
10310 Fireplace Specialties and Accessories
10130 Fireplace Dampers
10130 Fireplace Inserts
10130 Fireplace Screens and Doors
10130 Fireplace Water Heaters
10320 Stoves
10330 Fireplace and Stove Restoration
10340 Manufactured Exterior Specialties
10340 Clocks
10340 Cupolas
10340 Spires
10340 Steeples
10340 Weathervanes
10345 Exterior Specialties Restoration
10350 Flagpoles
10350 Automatic Flagpoles
10350 Ground-Set Flagpoles
10350 Nautical Flagpoles
10350 Wall-Mounted Flagpoles
10350 Flags (usually)
10400 Identification Devices
10410 Directories
10420 Plaques
10430 Exterior Signage
10440 Interior Signage
10450 Pedestrian Control Devices
10450 Detection Specialties
10450 Portable Posts and Railings
10450 Rotary Gates
10450 Turnstiles
10500 Lockers
10500 Coin Operated Lockers
10500 Glass Lockers
10500 Metal Lockers
10500 Plastic-Laminate-Faced Lockers
10500 Plastic Lockers
10500 Recycled Plastic Lockers
10500 Wood Lockers
10520 Fire Protection Specialties
10520 Fire Blankets and Cabinets
10520 Fire Extinguisher Accessories
10520 Fire Extinguisher Cabinets
10520 Fire Extinguishers
10520 Wheeled Fire Extinguisher Units
10530 Protective Covers
10530 Awnings
10530 Canopies
10530 Car Shelters
10530 Walkway Coverings
10550 Postal Specialties
10550 Central Mail Delivery Boxes
10550 Collection Boxes
10550 Mail Boxes
10550 Mail Chutes
10600 Partitions
10605 Wire Mesh Partitions
10610 Folding Gates
10615 Demountable Partitions
10630 Portable Partitions, Screens, and Panels
10650 Operable Partitions
10670 Storage Shelving
10670 Metal Storage Shelving
10670 Mobile Storage Units
10670 Prefabricated Wood Storage Shelving
10670 Recycled Plastic Storage Shelving
10670 Wire Storage Shelving
10700 Exterior Protection
10705 Exterior Sun Control Devices
10710 Exterior Shutters
10715 Storm Panels
10720 Exterior Louvers
10750 Telephone Specialties
10750 Telephone Directory Units
10750 Telephone Enclosures
10750 Telephone Shelving
10800 Toilet, Bath, and Laundry Accessories
10810 Toilet Accessories
10820 Bath Accessories
10830 Laundry Accessories
10880 Scales
10880 Fixed Scales
10900 Wardrobe and Closet Specialties
11010 Maintenance Equipment
11010 Floor and Wall Cleaning Equipment
11010 Housekeeping Carts
11010 Vacuum Cleaning Systems
11010 Window Washing Systems
11020 Security and Vault Equipment
11020 Safe Deposit Boxes
11020 Safes
11020 Vault Doors and Day Gates
11030 Teller and Service Equipment
11030 Automatic Banking Systems
11030 Money Cart Pass-Through
11030 Package Transfer Units
11030 Service and Teller Window Units
11030 Teller Equipment Systems
11040 Ecclesiastical Equipment
11040 Baptisteries
11040 Chancel Fittings

11050 Library Equipment
11050 Automated Book Storage and Retrieval Systems
11050 Book Depositories
11050 Book Theft Protection Equipment
11050 Library Stack Systems
11060 Theater and Stage Equipment
11060 Acoustical Shells
11060 Folding and Portable Stages
11060 Rigging Systems and Controls
11060 Stage Curtains
11070 Instrumental Equipment
11070 Bells
11070 Carillons
11070 Organs
11080 Registration Equipment
11090 Checkroom Equipment
11100 Mercantile Equipment
11100 Barber and Beauty Shop Equipment
11100 Cash Registers and Checking Equipment
11100 Display Cases
11100 Food Processing Equipment
11100 Food Weighing and Wrapping Equipment
11110 Commercial Laundry and Dry Cleaning Equipment
11110 Dry Cleaning Equipment
11110 Drying and Conditioning Equipment
11110 Finishing Equipment
11110 Ironing Equipment
11110 Washers and Extractors
11120 Vending Equipment
11120 Money-Changing Machines
11120 Vending Machines
11130 Audio-Visual Equipment
11130 Learning Laboratories
11130 Projection Screens
11130 Projectors
11140 Vehicle Service Equipment
11140 Compressed Air Equipment
11140 Fuel Dispensing Equipment
11140 Lubrication Equipment
11140 Tire Changing Equipment
11140 Vehicle Washing Equipment
11150 Parking Control Equipment
11150 Coin Machine Units
11150 Key and Card Control Units
11150 Parking Gates
11150 Ticket Dispensers
11160 Loading Dock Equipment
11160 Dock Bumpers
11160 Dock Levelers
11160 Dock Lifts
11160 Portable Ramps, Bridges, and Platforms
11160 Seals and Shelters
11160 Truck Restraints
11170 Solid Waste Handling Equipment
11170 Bins
11170 Chutes and Collectors
11170 Packaged Incinerators
11170 Pneumatic Waste Equipment
11170 Pulping Machines
11170 Recycling Equipment
11170 Waste Compactors and Destructors
11190 Detention Equipment
11200 Water Supply and Treatment Equipment
11210 Supply and Treatment Pumps
11220 Mixers and Flocculators
11225 Clarifiers
11230 Water Aeration Equipment
11240 Chemical Feed Equipment
11250 Water Softening Equipment
11260 Disinfectant Feed Equipment
11270 Fluoridation Equipment
11285 Hydraulic Gates
11295 Hydraulic Valves
11300 Fluid Waste Treatment and Disposal Equipment
11310 Sewage and Sludge Pumps
11320 Grit Collecting Equipment
11330 Screening and Grinding Equipment
11335 Sedimentation Tank Equipment
11340 Scum Removal Equipment
11345 Chemical Equipment
11350 Sludge Handling and Treatment Equipment
11360 Filter Press Equipment
11365 Trickling Filter Equipment
11370 Compressors
11375 Aeration Equipment
11380 Sludge Digestion Equipment
11385 Digester Mixing Equipment
11390 Package Sewage Treatment Plants
11400 Food Service Equipment
11405 Food Storage Equipment
11410 Food Preparation Equipment
11415 Food Delivery Carts and Conveyors
11420 Food Cooking Equipment
11425 Hood and Ventilation Equipment
11430 Food Dispensing Equipment
11435 Ice Machines
11440 Cleaning and Disposal Equipment
11450 Residential Equipment
11450 Residential Appliances
11450 Residential Kitchen Equipment
11450 Retractable Stairs
11460 Unit Kitchens
11470 Darkroom Equipment
11470 Darkroom Processing Equipment
11470 Revolving Darkroom Doors
11470 Transfer Cabinets
11480 Athletic, Recreational, and Therapeutic Equipment
11480 Backstops
11480 Bowling Alleys

11480 Exercise Equipment
11480 Gym Dividers
11480 Gymnasium Equipment
11480 Scoreboards
11480 Shooting Ranges
11480 Therapy Equipment
11500 Industrial and Process Equipment
11600 Laboratory Equipment
11650 Planetarium Equipment
11660 Observatory Equipment
11680 Office Equipment
11700 Medical Equipment
11710 Medical Sterilizing Equipment
11720 Examination and Treatment Equipment
11730 Patient Care Equipment
11740 Dental Equipment
11750 Optical Equipment
11760 Operating Room Equipment
11770 Radiology Equipment
11780 Mortuary Equipment
11850 Navigation Equipment
11870 Agricultural Equipment
11900 Exhibit Equipment

12000 FURNISHINGS
12050 Fabrics
12110 Art
12110 Murals
12120 Wall Decorations
12140 Sculptures
12170 Art Glass
12190 Ecclesiastical Art
12300 Manufactured Casework
12310 Manufactured Metal Casework
12320 Manufactured Wood Casework
12350 Specialty Casework
12400 Furnishings and Accessories
12410 Office Accessories
12420 Table Accessories
12430 Portable Lamps
12440 Bath Furnishings
12450 Bedroom Furnishings
12460 Furnishing Accessories
12480 Rugs and Mats
12490 Window Treatments
12500 Furniture
12510 Office Furniture
12520 Seating
12540 Hospitality Furniture
12560 Institutional Furniture
12580 Residential Furniture
12600 Multiple Seating
12610 Fixed Audience Seating
12620 Portable Audience Seating
12630 Stadium and Arena Seating
12640 Booths and Tables
12650 Multiple-Used Fixed Seating
12660 Telescoping Stands
12670 Pews and Benches
12680 Seat and Table Assemblies
12700 Systems Furniture
12710 Panel-Hung Component System Furniture
12720 Free-Standing Component System Furniture
12730 Beam System Furniture
12740 Desk System Furniture
12800 Interior Plants and Planters
12810 Interior Live Plants
12820 Interior Artificial Plants
12830 Interior Planters
12840 Interior Landscape Accessories
12850 Interior Plant Maintenance
12900 Furnishings Restoration and Repair

13000 SPECIAL CONSTRUCTION
13010 Air-Supported Structures
13020 Building Modules
13020 Prison Cells
13020 Hotel and Dormitory Units
13030 Special Purpose Rooms
13030 Athletic Rooms
13030 Clean Rooms
13030 Cold Storage Rooms
13030 Hyperbaric Rooms
13030 Insulated Rooms
13030 Office Shelters and Booths
13030 Planetariums
13030 Prefabricated Rooms
13030 Saunas
13030 Sound-Conditioned Rooms
13030 Steam Baths
13030 Vaults
13080 Sound, Vibration, and Seismic Control
13090 Radiation Protection
13100 Lightning Protection
13110 Cathodic Protection
13120 Pre-Engineered Structures
13120 Cable-Supported Structures
13120 Fabric Structures
13120 Glazed Structures
13120 Grandstands and Bleachers
13120 Metal Building Systems
13120 Modular Mezzanines
13120 Observatories
13120 Portable and Mobile Buildings
13120 Pre-Engineered Buildings
13120 Prefabricated Control Booths
13120 Prefabricated Dome Structures
13150 Swimming Pools
13150 Below Grade Swimming Pools
13150 Elevated Swimming Pools
13150 On-Grade Swimming Pools
13150 Re-circulating Gutter Systems

13150 Swimming Pool Accessories
13150 Swimming Pool Cleaning Systems
13160 Aquariums
13165 Aquatic Park Facilities
13165 Water Slides
13165 Wave Pools
13170 Tubs and Pools
13170 Hot Tubs
13170 Therapeutic Pools
13170 Whirlpool Tubs
13175 Ice Rinks
13185 Kennels and Animal Shelters
13190 Site-Constructed Incinerators
13190 Sludge Incinerators
13190 Solid Waste Incinerators
13190 Waste Disposal Incinerators
13200 Storage Tanks
13200 Elevated Storage Tanks
13200 Ground Storage Tanks
13200 Tank Cleaning Procedures
13200 Tank Lining
13200 Underground Storage Tanks
13220 Filter Underdrains and Media
13220 Filter Bottoms
13220 Filter Media
13220 Package Filters
13230 Digester Covers and Appurtenances
13230 Fixed Covers
13230 Floating Covers
13230 Gasholder Covers
13240 Oxygenation Systems
13240 Oxygen Dissolution System
13240 Oxygen Generators
13240 Oxygen Storage Facility
13260 Sludge Conditioning Systems
13280 Hazardous Material Remediation
13400 Measurement and Control Instrumentation
13410 Basic Measurement and Control Instrumentation Materials and Methods
13420 Instruments
13430 Boxes, Panels, and Control Centers
13440 Indicators, Recorders, and Controllers
13450 Central Control
13480 Instrument Lists and Reports
13490 Measurement and Control Commissioning
13500 Recording Instrumentation
13510 Stress Instrumentation
13520 Seismic Instrumentation
13530 Meteorological Instrumentation
13550 Transportation Control Instrumentation
13560 Airport Control Instrumentation
13570 Railroad Control Instrumentation
13580 Subway Control Instrumentation
13590 Transit Vehicle Control Instrumentation
13600 Solar and Wind Energy Equipment
13610 Solar Flat Plate Collectors
13620 Solar Concentrating Collectors
13625 Solar Vacuum Tube Collectors
13630 Solar Collector Components
13640 Packaged Solar Equipment
13650 Photovoltaic Collectors
13660 Wind Energy Equipment
13700 Security Access and Surveillance
13700 Door Answering
13700 Intrusion Detection
13700 Security Access
13700 Video Surveillance
13800 Building Automation and Control
13800 Clock Control
13800 Door Control
13800 Elevator Monitoring and Control
13800 Energy Monitoring and Control
13800 Environmental Control
13800 Escalator and Moving Walks
13800 Lighting Control
13850 Detection and Alarm
13850 Fire Alarm
13850 Gas Detection
13850 Leak Detection
13850 Smoke Alarm
13900 Fire Suppression
13910 Fire Protection Basic Materials and Methods
13920 Fire Pumps
13930 Wet-Pipe Fire Suppression Sprinklers
13935 Dry-Pipe Fire Suppression Sprinklers
13940 Pre-Action Fire Suppression Sprinklers
13945 Combination Dry-Pipe and Pre-Action Fire Suppression Sprinklers
13950 Deluge Fire Suppression Sprinklers
13955 Foam Fire Extinguishing
13960 Carbon Dioxide Fire Extinguishing
13965 Alternative Fire Extinguishing Systems
13970 Dry Chemical Fire Extinguishing
13975 Standpipes and Hoses

14000 CONVEYING SYSTEMS
14100 Dumbwaiters
14110 Manual Dumbwaiters
14120 Electric Dumbwaiters
14140 Hydraulic Dumbwaiters
14200 Elevators
14210 Electric Traction Elevators
14240 Hydraulic Elevators
14270 Custom Elevator Cabs
14280 Elevator Equipment and Controls
14290 Elevator Renovation
14300 Escalators and Moving Walks
14400 Lifts
14410 People Lifts
14420 Wheelchair Lifts
14430 Platform Lifts
14440 Sidewalk Lifts
14450 Vehicle Lifts
14500 Material Handling
14510 Material Transport
14530 Postal Conveying
14540 Baggage Conveying and Dispensing
14550 Conveyors
14560 Chutes
14570 Feeder Equipment
14580 Pneumatic Tube Systems
14600 Hoists and Cranes
14605 Crane Rails
14610 Fixed Hoists
14620 Trolley Hoists
14630 Bridge Cranes
14640 Gantry Cranes
14650 Jib Cranes
14670 Tower Cranes
14680 Mobile Cranes
14690 Derricks
14700 Turntables
14800 Scaffolding
14810 Suspended Scaffolding
14820 Rope Climbers
14830 Telescoping Platforms
14840 Powered Scaffolding
14900 Transportation
14910 People Movers
14920 Monorails
14930 Funiculars
14940 Aerial Tramways
14950 Aircraft Passenger Loading

15000 MECHANICAL
15050 Basic Mechanical Materials and Methods
15060 Hangers and Supports
15070 Mechanical Sound, Vibration, and Seismic Control
15075 Mechanical Identification
15080 Mechanical Insulation
15090 Mechanical Restoration and Retrofit
15100 Building Services Piping
15105 Pipes and Tubes
15110 Valves
15120 Piping Specialties
15130 Pumps
15140 Domestic Water Piping
15150 Sanitary Waste and Vent Piping
15160 Storm Drainage Piping
15170 Swimming Pool and Fountain Piping
15180 Heating and Cooling Piping
15190 Fuel Piping
15200 Process Piping
15210 Process Air and Gas Piping
15220 Process Water and Waste Piping
15230 Industrial Process Piping
15300 Fire Protection Piping
15400 Plumbing Fixtures and Equipment
15410 Plumbing Fixtures
15440 Plumbing Pumps
15450 Potable Water Storage Tanks
15460 Domestic Water Conditioning Equipment
15470 Domestic Water Filtrating Equipment
15480 Domestic Water Heaters
15490 Pool and Fountain Equipment
15500 Heat-Generation Equipment
15510 Heating Boilers and Accessories
15520 Feedwater Equipment
15530 Furnaces
15540 Fuel-Fired Heaters
15550 Breechings, Chimneys, and Stacks
15600 Refrigeration Equipment
15610 Refrigeration Compressors
15620 Packaged Water Chillers
15630 Refrigerant Monitoring Systems
15640 Packaged Cooling Towers
15650 Field-Erected Cooling Towers
15660 Liquid Coolers and Evaporative Condensers
15670 Refrigerant Condensing Units
15700 Heating, Ventilating, and Air Conditioning Equipment
15710 Heat Exchangers
15720 Air Handling Units
15730 Unitary Air Conditioning Equipment
15740 Heat Pumps
15750 Humidity Control Equipment

15760 Terminal Heating and Cooling Units
15770 Floor-Heating and Snow-Melting Equipment
15780 Energy Recovery Equipment
15800 Air Distribution
15810 Ducts
15820 Duct Accessories
15830 Fans
15840 Air Terminal Units
15850 Air Outlets and Inlets
15860 Air Cleaning Devices
15900 HVAC Instrumentation and Controls
15905 HVAC Instrumentation
15910 Direct Digital Controls
15915 Electric and Electronic Control
15920 Pneumatic Controls
15925 Pneumatic and Electric Controls
15930 Self-Powered Controls
15935 Building Systems Controls
15940 Sequence of Operation
15950 Testing, Adjusting, and Balancing
15950 Demonstration of Mechanical Equipment
15950 Duct Testing, Adjusting, and Balancing
15950 Equipment Testing, Adjusting, and Balancing
15950 Mechanical Equipment Starting/Commissioning
15950 Pipe Testing, Adjusting, and Balancing

16000 ELECTRICAL
16050 Basic Electrical Materials and Methods
16060 Grounding and Bonding
16070 Hangers and Supports
16075 Electrical Identification
16080 Electrical Testing
16090 Restoration and Repair
16100 Wiring Methods
16120 Conductors and Cables
16130 Raceway and Boxes
16140 Wiring Devices
16150 Wiring Connections
16200 Electrical Power
16210 Electrical Utility Services
16220 Motors and Generators
16230 Generator Assemblies
16240 Battery Equipment
16260 Static Power Converters
16270 Transformers
16270 Distribution Transformers
16270 Network Transformers
16270 Pad-Mounted Transformers
16270 Power Transformers
16270 Substation Transformers
16280 Power Filters and Conditioners
16300 Transmission and Distribution
16310 Transmission and Distribution Accessories
16320 High-Voltage Switching and Protection
16330 Medium-Voltage Switching and Protection
16340 Medium-Voltage Switching and Protection Assemblies
16360 Unit Substations
16400 Low-Voltage Distribution
16410 Enclosed Switches and Circuit Breakers
16420 Enclosed Controllers
16430 Low-Voltage Switchgear
16440 Switchboards, Panelboards, and Control Centers
16450 Enclosed Bus Assemblies
16460 Low-Voltage Transformers
16470 Power Distribution Units
16490 Components and Accessories
16500 Lighting
16510 Interior Luminaires
16520 Exterior Luminaires
16530 Emergency Lighting
16540 Classified Location Lighting
16550 Special-Purpose Lighting
16560 Signal Lighting
16570 Dimming Control
16580 Lighting Accessories
16590 Lighting Restoration and Repair
16700 Communications
16710 Communications Circuits
16720 Telephone and Intercommunication Equipment
16740 Communication and Data Processing Equipment
16770 Cable Transmission and Reception Equipment
16780 Broadcast Transmission and Reception Equipment
16790 Microwave Transmission and Reception Equipment
16800 Sound and Video
16810 Sound and Video Circuits
16820 Sound Reinforcement
16830 Broadcast Studio Audio Equipment
16840 Broadcast Studio Video Equipment
16850 Television Equipment
16880 Multimedia Equipment

Appendix D

Master Format – Level 4

00000 INTRODUCTORY INFORMATION
00001 Project Title Page
00005 Certifications Page
00007 Seals Page
00010 Table of Contents
00015 List of Drawings
00020 List of Schedules
00099 BIDDING REQUIREMENTS
00100 Bid Solicitation
00100 Advertisement for Bids
00100 Bidder's Qualifications
00100 Invitation to Bid
00100 Request for Proposal
00200 Instruction to Bidders
00210 Supplementary Instructions to Bidders
00220 Bid Scopes
00250 Pre-Bid Meeting
00300 Information Available to Bidders
00310 Preliminary Schedules
00320 Geotechnical Data
00330 Existing Conditions
00340 Environmental Assessment Information
00350 Project Financial Information
00360 Permit Application
00400 Bid Forms and Supplements
00410 Bid Form
00410 Bid Form, Construction Management
00410 Bid Form, Cost-Plus Fee
00410 Bid Form, Procurement
00410 Bid Form, Stipulated Sum
00410 Bid Form, Unit Price
00430 Bid Form and Supplements
00430 Allowances
00430 Alternates
00430 Bid Security
00430 Bid Submittal Checklist
00430 Estimated Quantities
00430 Proposed Products
00430 Proposed Subcontractors
00430 Unit Prices
00430 Wage Rates
00430 Work Plan and Equipment Schedule
00450 Representations and Certifications
00490 Bidding Agenda

00499 CONTRACTING REQUIREMENTS
00500 Agreement
00510 Notice of Award
00520 Agreement Form
00520 Agreement, Construction Management
00520 Agreement, Cost-Plus Fee
00520 Agreement, Procurement
00520 Agreement, Stipulated Sum
00520 Agreement, Unit Price
00540 Attachments to Agreement Form
00540 Allowance Amounts
00540 Supplementary Scope Statement
00540 Unit Price Schedule
00550 Notice to Proceed
00570 Definitions
00600 Bonds and Certificates
00610 Bonds
00610 Consent of Surety
00610 Lien Bonds
00610 Maintenance Bonds
00610 Performance Bonds
00610 Payment Bonds
00610 Special Bonds
00610 Warranty Bonds
00620 Certificates
00620 Acceptance Certificate
00620 Application for Payment Certificate
00620 Certificates of Insurance
00620 Certificates of Compliance
00620 Certificates of Substantial Performance and Completion
00620 Certificates of Final Performance and Completion
00620 Final Settlement Certificate
00640 Release of Liens
00650 Statutory Declaration Forms
00700 General Conditions
00800 Supplementary Conditions
00800 Anti-Pollution Measures
00800 Assigned Contracts
00800 Equal Employment Opportunity Requirements
00800 Health and Safety Criteria
00800 Insurance Requirements
00800 Labor Stabilization Agreement
00800 Letter of Assent

00800 Non-Segregated Facilities Requirements
00800 Procurement Contracts
00800 Specific Project Requirements
00800 Statutory Requirements
00800 Wage Determination Schedule
00800 Wage Rate Requirements
00890 Permits
00890 Notices
00900 Addenda and Modifications
00910 Addenda
00920 Claims
00930 Clarifications and Proposals
00930 Change Order Proposals
00930 Change Order Requests
00930 Clarification Notices
00930 Minor Changes in the Work
00930 Proposal Requests
00930 Proposal Worksheet Summaries
00930 Requests for Interpretation
00940 Modifications
00940 Amendments
00940 Architect's Supplemental Instructions
00940 Change Orders
00940 Construction Change Directives
00940 Field Order
00940 Supplemental Instructions
00940 Work Change Directives
00940 Written Amendments

00980 FACILITIES AND SPACES
00980 Facilities and Spaces

00990 SYSTEMS AND ASSEMBLIES
00990 Systems and Assemblies

01000 GENERAL REQUIREMENTS
01100 Summary
01110 Summary of Work
01110 Work By Owner
01110 Work Covered By Contract Documents
01120 Multiple Contract Summary
01120 Construction Sequence
01120 Construction By Owner
01120 Contract Interface
01120 Summary of Contracts
01140 Work Restrictions
01140 Access to Site
01140 Coordination With Occupants
01140 Use of Site
01140 Use of Premises
01180 Project Utility Sources
01200 Price and Payment Procedures
01210 Allowances
01210 Cash Allowances
01210 Contingency Allowances
01210 Inspecting and Testing Allowances
01210 Installation Allowances
01210 Product Allowances
01210 Quantity Allowances
01230 Alternates
01240 Value Analysis
01250 Contract Modification Procedures
01270 Unit Prices
01290 Payment Procedures
01290 Applications and Certificates for Payment
01290 Schedule of Values
01300 Administrative Requirements
01310 Project Management and Coordination
01310 Project Coordination
01310 Project Meetings
01310 Project Site Administration
01320 Construction Progress Documentation
01320 Construction Photographs
01320 Contract Progress Reporting
01320 Network Analysis Schedules
01320 Periodic Site Observation
01320 Progress Schedules and Reports
01320 Purchase Order Tracking
01320 Scheduling of Construction
01320 Survey and Layout Data
01330 Submittal Procedures
01330 Certificates
01330 Design Data
01330 Field Test Reports
01330 Shop Drawings, Product Data, and Samples
01330 Source Quality Control Reports
01350 Special Procedures
01400 Quality Requirements
01410 Regulatory Requirements
01420 References
01420 Abbreviations and Acronyms
01420 Reference Standards
01420 Symbols
01430 Quality Assurance
01430 Fabricator Qualifications
01430 Installer Qualifications
01430 Manufacturer Qualifications
01430 Manufacturers Field Services
01430 Supplier Qualifications
01430 Testing Agency Qualifications
01450 Quality Control

01450 Contractor's Quality Control
01450 Field Quality Control
01450 Field Samples
01450 Mock-up Requirements
01450 Plant Inspection
01450 Source Quality Control
01450 Testing and Inspection Services
01450 Testing Laboratory Services
01500 Temporary Facilities and Controls
01510 Temporary Utilities
01510 Temporary Electricity
01510 Temporary Fire Protection
01510 Temporary Fuel Oil
01510 Temporary Heating, Cooling, and Ventilating
01510 Temporary Lighting
01510 Temporary Natural Gas
01510 Temporary Telephone
01510 Temporary Water
01520 Construction Facilities
01520 Field Offices and Sheds
01520 First Aid
01520 Sanitary Facilities
01530 Temporary Construction
01530 Temporary Bridges
01530 Temporary Decking
01530 Temporary Overpasses
01530 Temporary Ramps
01530 Temporary Runarounds
01540 Construction Aids
01540 Construction Elevators, Hoists, and Cranes
01540 Scaffolding and Platform
01540 Swing Staging
01550 Vehicular Access and Parking
01550 Access Roads
01550 Haul Routes
01550 Parking Areas
01550 Temporary Roads
01550 Traffic Control
01560 Temporary Barriers and Enclosures
01560 Air Barriers
01560 Barricades
01560 Dust Barriers
01560 Fences
01560 Noise Barriers
01560 Pollution Control
01560 Security Measures
01560 Tree and Plant Protection
01570 Temporary Controls
01570 Erosion and Sediment Control
01570 Pest Control
01580 Project Identification
01580 Project Signs
01600 Product Requirements
01610 Basic Product Requirements
01620 Product Options
01630 Product Substitution Procedures
01630 Substitution Procedures During Bidding
01630 Substitution Procedures During Construction
01640 Owner-Furnished Products
01650 Product Delivery Requirements
01660 Product Storage and Handling Requirements
01700 Execution Requirements
01710 Examination
01710 Acceptance of Conditions
01710 Existing Conditions
01720 Preparation
01720 Construction Layout
01720 Field Engineering
01720 Protection of Adjacent Construction
01720 Surveying
01730 Execution
01730 Application
01730 Cutting and Patching
01730 Erection
01730 Existing Products
01730 Installation
01730 Selective Demolition
01740 Cleaning
01740 Final Cleaning
01740 Progress Cleaning
01740 Site Maintenance
01750 Starting and Adjusting
01750 Checkout Procedures
01760 Protecting Installed Construction
01770 Closeout Procedures
01780 Closeout Submittals
01780 Final Site Survey
01780 Maintenance Contracts
01780 Maintenance Data
01780 Maintenance Materials
01780 Operation Data
01780 Preventive Maintenance Instructions
01780 Product Warranties
01780 Product Bonds
01780 Project Record Documents
01780 Spare Parts
01800 Facility Operation
01810 Commissioning

01810 Commissioning Summary
01810 System Performance Evaluation
01810 Testing, Adjusting, and Balancing Procedures
01820 Demonstration and Training
01830 Operation and Maintenance
01890 Reconstruction
01900 Facility Decommissioning
01900 Decommissioning Systems and Equipment
01900 Facility Demolition and Removal
01900 Hazardous Materials Abatement
01900 Hazardous Materials Removal and Disposal
1900 Protection of Deactivated Facilities

02000 SITE CONSTRUCTION
02050 Basic Site Materials and Methods
02055 Soils
02060 Aggregate
02065 Cement and Concrete
02065 Asphalt Cement
02065 Hydraulic Cement
02065 Plant-Mixed Bituminous Concrete
02065 Recycled Plant-Mixed Bituminous Concrete
02070 Geosynthetics
02070 Geocomposite Edge Drains
02070 Geocomposite In-Place Wall Drains
02070 Geogrids
02070 Geotextiles
02080 Utility Materials
02080 Hydrants
02080 Manholes
02080 Meters
02080 Utility Boxes
02080 Valves
02090 Joint Material
02100 Site Remediation
02105 Chemical Sampling and Analysis
02110 Excavation, Removal, and Handling of Hazardous Materials
02115 Underground Storage Tank Removal
02120 Off-Site Transportation and Disposal
02125 Drum Handling
02130 Site Decontamination
02140 Landfill Construction and Storage
02145 Groundwater Treatment Systems
02150 Hazardous Waste Recovery Processes
02150 Air and Steam Stripping
02150 Soil Vapor Extraction
02150 Soil Washing and Flushing
02160 Physical Treatment
02160 Coagulation and Flocculation
02160 Reverse Osmosis
02160 Solidification and Stabilization
02170 Chemical Treatment
02170 Chemical Precipitation
02170 Ion Exchange
02170 Neutralization
02180 Thermal Processes
02180 Incineration
02180 Thermal Desorption
02180 Vitrification
02190 Biological Processes
02190 Aerobic Processes
02190 Anaerobic Processes
02190 Bio-Remediation
02195 Remediation Soil Stabilization
02200 Site Preparation
02210 Subsurface Investigation
02210 Boring and Exploratory Drilling
02210 Core Drilling
02210 Geophysical Investigations
02210 Groundwater Monitoring
02210 Seismic Investigation
02210 Standard Penetration Testing
02210 Test Pits
02220 Site Demolition
02220 Building Demolition
02220 Minor Site Demolition for Remodeling
02220 Selective Site Demolition
02230 Site Clearing
02230 Clearing and Grubbing
02230 Selective Clearing
02230 Selective Tree Removal and Trimming
02230 Sod Stripping
02230 Stripping and Stockpiling of Soil
02230 Tree Pruning
02240 Dewatering
02250 Shoring and Underpinning
02250 Foundation Grouting
02250 Grillages
02250 Needle Beams
02250 Sheetpiling
02250 Shoring
02250 Slabjacking
02250 Soil Stabilization
02250 Underpinning
02250 Vibroflotation and Densification
02260 Excavation Support and Protection
02260 Anchor Tiebacks
02260 Cofferdams
02260 Cribbing and Walers
02260 Ground Freezing
02260 Reinforced Earth
02260 Slurry Wall Construction

02260 Soil and Rock Anchors
02280 Grade Adjustment and Abandonment of Existing Miscellaneous Structures
02290 Site Monitoring
02300 Earthwork
02310 Grading
02310 Finish Grading
02310 Rough Grading
02315 Excavation and Fill
02315 Backfill
02315 Borrow Excavation
02315 Compaction
02315 Excavation
02315 Fill
02315 Trenching
02325 Dredging
02330 Embankment
02330 Armoring
02330 Earth Dams
02330 Soil Embankment
02335 Subgrade and Roadbed
02335 Pre-watering of Excavation Areas
02335 Reconditioning
02335 Subgrade Modification
02340 Soil Stabilization
02340 Asphalt Soil Stabilization
02340 Cement Soil Stabilization
02340 Geotextile Soil Stabilization and Layer Separation
02340 Lime Slurry Soil Stabilization
02340 Lime Soil Stabilization
02340 Pressure Grouting Soil Stabilization
02360 Soil Treatment
02360 Rodent Control
02360 Termite Control
02360 Vegetation Control
02370 Erosion and Sedimentation Control
02370 Cement Concrete Paving for Stream Beds
02370 Erosion Control Blankets and Mats
02370 Gabions
02370 Geogrids
02370 Geotextile Sedimentation and Erosion Control
02370 Mulch Control Netting
02370 Paved Energy Dissipators
02370 Riprap and Rock Lining
02370 Rock Barriers
02370 Rock Basins
02370 Rock Energy Dissipators
02370 Slope Paving
02370 Synthetic Erosion Control and Revegetation Mats
02370 Turf Reinforcement Mats
02370 Water Course and Slope Erosion Protection
02380 Scour Protection
02390 Shoreline Protection and Mooring Structures
02390 Breakwaters
02390 Groins
02390 Jetties
02390 Moles
02390 Revetments
02390 Seawalls
02400 Tunneling, Boring, and Jacking
02410 Tunnel Excavation
02410 Compressed Air Tunneling
02410 Muck Disposal
02410 Rock Excavation - Drill and Blast
02410 Rock Excavation - Tunnel Boring Machine (TBM)
02410 Soft Ground Shield-Driven Tunneling
02420 Initial Tunnel Support Systems
02420 Prefabricated Steel Tunnel Linings
02420 Rock Bolting
02420 Steel Ribs and Lagging
02425 Tunnel Linings
02425 Cast-in-Place Concrete Tunnel Linings
02425 Precast Concrete Tunnel Lining
02430 Tunnel Grouting
02430 Earth Stabilization Chemical Grouting
02430 Rock Seam Pressure Grouting
02430 Tunnel Liner Grouting
02440 Immersed and Sunken Tube Tunnels
02441 Microtunneling
02442 Cut and Cover Tunnels
02443 Tunnel Leak Repairs
02444 Shaft Construction
02445 Boring or Jacking Conduits
02450 Foundation and Load-Bearing Elements
02455 Driven Piles
02455 Cast-in-Place Concrete Piles
02455 Composite Piles
02455 Concrete Displacement Piles
02455 Concrete-Filled Steel Piles
02455 Driven Pile Load Tests
02455 Driven Pile Repairs
02455 Precast Concrete Piles
02455 Prestressed Concrete Piles
02455 Sheet Piles
02455 Steel H Piles
02455 Timber Piles
02455 Unfilled Tubular Steel Piles
02465 Bored Piles
02465 Auger Cast Grout Piles
02465 Bored and Augered Pile Load Tests

02465 Bored and Augered Pile Repairs
02465 Bored and Augered Test Piles
02465 Bored and Belled Concrete Piles
02465 Bored and Socketed Piles
02465 Bored Friction Concrete Piles
02465 Drilled Caissons
02465 Drilled Concrete Piers and Shafts
02465 Uncased Cast-in-Place Concrete Piles
02475 Caissons
02475 Box Caissons
02475 Excavated Caissons
02475 Floating Caissons
02475 Open Caissons
02475 Pneumatic Caissons
02475 Sheeted Caissons
02480 Foundation Walls
02480 Anchored Walls
02480 Concrete Cribbing
02480 Manufactured Modular Walls
02480 Mechanically Stabilized Earth Walls
02480 Metal Cribbing
02480 Permanently Anchored Soldier-Beam Walls
02480 Slurry Diaphragm Foundation Walls
02480 Soldier-Beam Walls
02490 Anchors
02490 Rock Anchors
02495 Instrumentation and Monitoring
02500 Utility Services
02510 Water Distribution
02510 Cisterns
02510 Disinfection of Water Distribution
02510 Fire Protection
02510 Hydrants
02510 Site Water
02510 Valves
02510 Water Supply
02520 Wells
02520 Extraction Wells
02520 Monitoring Wells
02520 Recharge Wells
02520 Test Wells
02520 Water Supply Wells
02520 Well Abandonment
02530 Sanitary Sewerage
02530 Gauging Stations
02530 Packaged Pumping Stations
02530 Packaged Lift Stations
02530 Sanitary Cleanouts
02530 Sanitary Sewage Systems
02530 Sanitary Sewer Manholes, Frames, and Covers
02530 Sewage Collection Lines
02530 Sewage Force Mains
02530 Site Sanitary Sewage Lines
02540 Septic Tank Systems
02540 Drainage Field
02540 Grease Interceptor
02540 Sand Filter
02540 Septic Tank
02540 Siphon Tank
02550 Piped Energy Distribution
02550 Chilled Water Distribution
02550 Hot Water Distribution
02550 Liquid Petroleum Gas Distribution
02550 Natural Gas Distribution
02550 Oil Distribution
02550 Steam Distribution
02570 Process Materials Distribution Structures
02580 Electrical and Communication Structures
02580 Antenna Towers
02580 Lighting Poles and Standards
02580 Transmission Towers
02580 Underground Ducts and Manholes
02580 Utility Poles
02590 Site Grounding
02600 Drainage and Containment
02610 Pipe Culverts
02620 Subdrainage
02620 Foundation Drainage Piping
02620 Geocomposite Drains
02620 Geotextile Subsurface Drainage Filtration
02620 Pipe Underdrain and Pavement Base Drain
02620 Retaining Wall Drainage Piping
02620 Subgrade Drains
02620 Tunnel Drainage Piping
02620 Underslab Drainage Piping
02630 Storm Drainage
02630 Catch Basins, Grates, and Frames
02630 Combination Storm Drain and Underdrain
02630 Inlets
02630 Storm Drainage Manholes, Frames, and Covers
02630 Storm Drainage Pipe and Fittings
02630 Water Detention Chambers
02640 Culverts and Manufactured Construction
02640 Concrete Arch Buried Bridge
02640 Metal Pipe - Arch Culverts
02640 Metal Plate Culverts
02640 Precast Reinforced Concrete Arch Culverts
02640 Precast Reinforced Concrete Box Culverts
02640 Precast Reinforced Concrete Rigid Frame Culverts
02660 Ponds and Reservoirs

02660 Cooling Water Ponds
02660 Distributions Reservoirs
02660 Fire Protection Reservoirs
02660 Leaching Pits
02660 Pond and Reservoir Covers
02660 Pond and Reservoir Liners
02660 Retention Basins
02660 Sewage Lagoons
02660 Stabilization Ponds
02670 Constructed Wetlands
02700 Bases, Ballasts, Pavements, and Appurtenances
02710 Bound Base Courses
02710 Aggregate-Bituminous Base Course
02710 Aggregate-Cement Base Course
02710 Asphalt-Treated Permeable Base Course
02710 Bituminous Concrete Base Course
02710 Cement Stabilized Open Graded Base Course
02710 Cement Treated Courses
02710 Cold-Recycled Bituminous Base Course
02710 Hydraulic Cement Concrete Base Course
02710 Lean Concrete Base Course
02710 Lime Treated Courses
02710 Lime-Fly-Ash Treated Courses
02710 Plain Cement Concrete Base Course
02720 Unbound Base Courses and Ballasts
02720 Aggregate Sub-base
02720 Aggregate Base Course
02720 Cracked and Seated Portland Cement Concrete Base Course
02720 Rubberized Portland Cement Concrete Base Course
02720 Sub-Ballast
02720 Top Ballast
02730 Aggregate Surfacing
02730 Cinder Surfacing
02730 Crushed Stone Surfacing
02740 Flexible Pavement
02740 Asphalt-Rubber and Rubber Modified Bituminous Pavement
02740 Athletic Bituminous Pavement
02740 Bituminous Concrete Pavement
02740 Cold Mix Bituminous Pavement
02740 Fiber-Modified Bituminous Pavement
02740 Polymer-Modified Bituminous Pavement and Rubberized Asphalt
02740 Road Mix Bituminous Pavement
02740 Stone and Split-Mastic Bituminous Pavement
02750 Rigid Pavement
02750 Continuous Reinforced Cement Concrete Pavement
02750 Exposed Aggregate Pavement
02750 Plain Cement Concrete Pavement
02750 Power Compacted Concrete Pavement
02750 Pre-stressed Reinforced Cement Concrete Pavement
02750 Reinforced Cement Concrete Pavement
02750 Roller-Compacted Concrete Pavement
02755 Cement Concrete Shoulders
02760 Paving Specialties
02760 Cold Plastic Pavement Markings and Legends
02760 Curb Cut Ramps
02760 Painted Traffic Lines and Markings
02760 Pavement Joint Sealant
02760 Pavement Markings
02760 Raised Pavement Markers
02760 Snow Melting Cables and Mats
02760 Stamped Pattern Concrete Pavement
02760 Tactile Warning Surfaces
02770 Curbs and Gutters
02770 Bituminous Concrete Curbs
02770 Cement Concrete Curbs
02770 Cement Concrete Gutters and Curbs
02770 Gutters
02770 Stone Curbs
02770 Simulated Stone Curbs
02775 Sidewalks
02780 Unit Pavers
02780 Asphalt Block Pavers
02780 Brick Pavers
02780 Interlocking Precast Concrete Pavers
02780 Precast Concrete Pavers
02780 Pressed Pavers
02780 Stone Pavers
02785 Flexible Pavement Coating and Micro-Surfacing
02785 Cape Seal
02785 Chip Seal
02785 Fog Seal
02785 Latex-Modified Emulsion
02785 Micro-Surfacing
02785 Sand Seal
02785 Sandwich Seal
02785 Slurry Seal
02785 Surface Treatment
02790 Athletic and Recreational Surfaces
02790 Baseball Field Surfacing
02790 Multi-Purpose Court Surfacing
02790 Resilient Matting
02790 Synthetic Grass Surfacing
02790 Synthetic Running Track Surfacing
02790 Tennis Court Surfacing
02795 Porous Pavement
02800 Site Improvements and Amenities

02810 Irrigation System
02810 Agricultural Irrigation System
02810 Drip Irrigation System
02810 Lawn Sprinkler System
02815 Fountains
02820 Fences and Gates
02820 Chain Link Fences and Gates
02820 Ornamental Metal Fences and Gates
02820 Plastic Fences and Gates
02820 Wire Fences and Gates
02820 Wood Fences and Gates
02830 Retaining Walls
02830 Cast-in-Place Concrete Retaining Walls
02830 Interlocking Block Retaining Walls
02830 Masonry Retaining Walls
02830 Precast Concrete Retaining Walls
02830 Timber Retaining Walls
02840 Walk, Road, and Parking Appurtenances
02840 Concrete Median Barrier
02840 Crash Barriers
02840 Delineators
02840 Fenders
02840 Guide Rail
02840 Impact Attenuating Devices
02840 Metal Median Barriers
02840 Parking Bumpers
02840 Traffic Barriers
02850 Prefabricated Bridges
02860 Screening Devices
02860 Jet Blast Barriers
02860 Screens and Louvers
02860 Sound Barriers
02870 Site Furnishings
02870 Bicycle Racks
02870 Prefabricated Planters
02870 Seating
02870 Tables
02870 Trash and Litter Receptors
02875 Site and Street Shelters
02875 Bus Stop Shelters
02880 Play Field Equipment and Structures
02880 Athletic or Recreational Screening
02880 Playground Equipment
02880 Play Structures
02880 Tennis Court Windbreakers
02890 Traffic Signs and Signals
02890 Post-Mounted Signs
02890 Structure-Mounted Signs
02890 Traffic Signal Supports and Equipment
02895 Markers and Monuments
02895 Boundary and Survey Markers
02900 Planting
02905 Plants, Planting, and Transplanting
02910 Plant Preparation
02910 Blankets
02910 Forms and Stabilizers
02910 Hydro-Punching
02910 Mats
02910 Mulching
02910 Netting
02910 Soil Preparation
02910 Stakes
02910 Topsoil
02915 Shrub and Tree Transplanting
02920 Lawns and Grasses
02920 Hydro-Mulching
02920 Plugging
02920 Seeding and Soil Supplements
02920 Sodding
02920 Sprigging
02920 Stolonizing
02930 Exterior Plants
02930 Ground Covers
02930 Plants and Bulbs
02930 Shrubs
02930 Trees
02935 Plant Maintenance
02935 Fertilizing
02935 Liming
02935 Mowing
02935 Pruning
02935 Watering
02945 Planting Accessories
02945 Landscape Edging
02945 Landscape Timbers
02945 Planters
02945 Tree Grates
02945 Tree Grids
02950 Site Restoration and Rehabilitation
02995 Restoration of Underground Piping
02995 Grouting Underground Piping and Units
02995 Relining Underground Piping and Units
02995 Sealing Underground Piping and Units
02960 Flexible Pavement Surfacing Recovery
02960 Pavement Milling and Pavement Cold Planing
02965 Flexible and Bituminous Pavement Recycling
02965 Cement-Based Bituminous Pavement Base Courses
02965 Cold In-Place Recycled Bituminous Pavement Courses

02965 Full-Depth Reclaimed Bituminous Pavement Base Courses
02965 Hot In-Place Recycled Bituminous Pavement Courses
02965 Hot In-Place Recycled Surface Course Overlays
02965 Roadbed Modification
02975 Flexible and Bituminous Pavement Reinf. and Crack and Joint Sealants
02975 Fabric Reinforcement for Bituminous Pavement
02975 Polymer Modified Asphalt Joint and Crack Sealing
02975 Stress-Absorbing Membrane
02975 Stress-Absorbing Membrane Interlayer
02980 Rigid Pavement Rehabilitation
02980 Cement Concrete Bonded Overlays
02980 Cement Concrete Direct Partially Bonded Overlays
02980 Cement Concrete Pavement Recycling
02980 Cement Concrete Unbonded Overlays
02980 Concrete Pavement Jacking and Slabjacking
02980 Concrete Pavement Patching
02980 Full Depth Patching
02980 Grinding of Concrete Pavement
02980 Grooving of Concrete Pavement
02980 Joint Cleaning and Resealing
02980 Joint Rehabilitation
02980 Partial Depth Patching
02980 Pavement-Shoulder Joint Resealing
02980 Slab Stabilization
02980 Sub-sealing and Stabilization
02990 Structure Moving

03000 CONCRETE
03050 Basic Concrete Materials and Methods
03100 Concrete Forms and Accessories
03110 Structural Cast-in-Place Concrete Forms
03120 Architectural Cast-in-Place Concrete Forms
03130 Permanent Forms
03130 Permanent Steel Forms
03130 Prefabricated Stair Forms
03150 Concrete Accessories
03200 Concrete Reinforcement
03210 Reinforcing Steel
03220 Welded Wire Fabric
03230 Stressing Tendons
03240 Fibrous Reinforcing
03250 Post-Tensioning
03300 Cast-In-Place Concrete
03310 Structural Concrete
03330 Architectural Concrete
03340 Low Density Concrete
03350 Concrete Finishing
03360 Concrete Finishes
03360 High-Tolerance Floor
03360 Special Concrete Finishes
03370 Specially Placed Concrete
03380 Post-Tensioned Concrete
03390 Concrete Curing
03400 Precast Concrete
03410 Plant-Precast Structural Concrete
03420 Plant-Precast Structural Post-Tensioned Concrete
03430 Site-Precast Structural Concrete
03430 Lift-Slab Concrete
03430 Precast Post-Tensioned Concrete
03430 Structural Precast Pretensioned Concrete
03450 Plant-Precast Architectural Concrete
03450 Faced Architectural Precast Concrete
03450 Architectural Concrete Cast Offsite
03460 Site-Precast Architectural Concrete
03470 Tilt-Up Precast Concrete
03480 Precast Concrete Specialties
03490 Glass-Fiber-Reinforced Precast Concrete
03500 Cementitious Decks and Underlayment
03510 Cementitious Roof Deck
03510 Cementitious Wood Fiber Plank
03510 Gypsum Concrete Roof Deck
03520 Lightweight Concrete Roof Insulation
03520 Composite Concrete and Insulation
03520 Lightweight Cellular Concrete
03520 Lightweight Insulating Concrete
03530 Concrete Topping
03540 Cementitious Underlayment
03540 Gypsum Underlayment
03540 Portland Cement Underlayment
03600 Grouts
03700 Mass Concrete
03900 Concrete Restoration and Cleaning
03910 Concrete Cleaning
03920 Concrete Resurfacing
03930 Concrete Rehabilitation

04000 MASONRY
04050 Basic Masonry Materials and Methods
04060 Masonry Mortar
04060 Cement and Lime Mortar
04060 Chemical-Resistant Mortar
04060 Epoxy Mortar
04060 Mortar for Engineered Masonry
04060 Mortar for Masonry Restoration
04060 Mortar Pigments
04060 Premixed Mortar
04060 Refractory Mortar
04060 Surface Bonding Mortar
04070 Masonry Grout
04070 Chemical-Resistant Grout
04070 Grout for Engineered Masonry
04080 Masonry Anchorage and Reinforcement
04080 Continuous Joint Reinforcing
04080 Flexible Masonry Ties
04080 Masonry Veneer Ties
04080 Reinforcing Bars
04080 Rigid Masonry Ties
04080 Stone Anchors
04090 Masonry Accessories
04090 Control Joint Materials
04090 Embedded Flashing
04090 Expansion Joint Materials
04090 Weepholes
04200 Masonry Units
04210 Clay Masonry Units
04210 Brick
04210 Building Brick
04210 Ceramic Glazed Clay Masonry Units
04210 Clay Tile
04210 Facing Brick
04210 Glazed Structural Clay Tile
04210 Structural Clay Tile
04210 Terra Cotta
04220 Concrete Masonry Units
04220 Concrete Brick
04220 Exposed Aggregate Concrete Masonry Units
04220 Fluted Concrete Masonry Units
04220 Interlocking Concrete Masonry Units
04220 Molded-Face Concrete Masonry Units
04220 Prefaced Concrete Masonry Units
04220 Preinsulated Concrete Masonry Units
04220 Sound-Absorbing Concrete Masonry Units
04220 Split-Face Concrete Masonry Units
04230 Calcium Silicate Unit Masonry
04270 Glass Masonry Units
04290 Adobe Masonry Units
04400 Stone
04410 Stone Materials
04410 Bluestone
04410 Granite
04410 Limestone
04410 Marble
04410 Quartzite
04410 Sandstone
04410 Slate
04420 Collected Stone
04420 Excavated Stone
04420 Field Stone
04420 Riverbed Stone
04420 Natural Stone
04430 Quarried Stone
04430 Cleft-Face Stone
04430 Crushed Stone
04430 Dimension Stone
04430 Cut stone
04500 Refractories
04550 Flue Liners
04560 Combustion Chambers
04570 Castable Refractories
04580 Refractory Brick
04600 Corrosion-Resistant Masonry
04610 Chemical-Resistant Brick
04620 Vitrified Clay Liner Plates
04700 Simulated Masonry
04710 Simulated Brick
04720 Cast Stone
04730 Simulated Stone
04800 Masonry Assemblies
04810 Unit Masonry Assemblies
04810 Cavity Walls
04810 Composite Unit Masonry
04810 Masonry Veneer
04810 Multiple-Wythe Unit Masonry
04810 Single-Wythe Unit Masonry
04810 Surface-Bonded Masonry
04810 Thin Brick Veneer
04820 Reinforced Unit Masonry Assemblies
04830 Non-Reinforced Unit Masonry Assemblies
04840 Prefabricated Masonry Panels
04850 Stone Assemblies
04850 Dry-Placed Stone Assemblies
04850 Metal-Supported Stone Assemblies
04850 Mortar-Placed Stone Assemblies

04850 Stone-Clad Precast Concrete
04850 Stone Curtain Wall Assemblies
04880 Masonry Fireplaces
04900 Masonry Restoration and Cleaning
04910 Unit Masonry Restoration
04910 Unit Masonry Repair
04910 Unit Masonry Replacement
04910 Unit Masonry Repointing
04920 Stone Restoration
04920 Stone Repair
04920 Stone Replacement
04920 Stone Repointing
04930 Unit Masonry Cleaning
04930 Brick Cleaning
04930 Terra Cotta Cleaning
04940 Stone Cleaning

05000 METALS
05050 Basic Metal Materials and Methods
05060 Metal Materials
05080 Factory-Applied Metal Coatings
05090 Metal Fastenings
05100 Structural Metal Framing
05120 Structural Steel
05120 Architecturally Exposed Structural Steel
05120 Fabricated Fireproofed Steel Columns
05120 Tubular Steel
05140 Structural Aluminum
05140 Architecturally-Exposed Structural Aluminum
05150 Wire Rope Assemblies
05150 Aluminum Wire Rope Assemblies
05150 Steel Wire Rope Assemblies
05160 Metal Framing Systems
05160 Geodesic Structures
05160 Space Frames
05200 Metal Joists
05210 Steel Joists
05210 Deep Longspan Steel Joists
05210 Longspan Steel Joists
05210 Open Web Steel Joists
05210 Steel Joist Girders
05250 Aluminum Joists
05260 Composite Joist Assemblies
05300 Metal Deck
05310 Steel Deck
05310 Composite Metal Deck
05310 Steel Floor Deck
05310 Steel Roof Deck
05320 Raceway Deck Systems
05330 Aluminum Deck
05330 Aluminum Floor Deck
05330 Aluminum Roof Deck
05340 Acoustical Metal Deck
05400 Cold-Formed Metal Framing
05410 Load-Bearing Metal Studs
05420 Cold-Formed Metal Joists
05430 Slotted Channel Framing
05450 Metal Support
05450 Electrical Supports
05450 Mechanical Supports
05450 Medical Supports
05500 Metal Fabrications
05510 Metal Stairs and Ladders
05520 Handrails and Railings
05520 Pipe and Tube Railings
05520 Tempered Glass Railing Assemblies
05530 Gratings
05540 Floor Plates
05550 Stair Treads and Nosings
05560 Metal Castings
05580 Formed Metal Fabrications
05580 Column Covers
05580 Formed Metal Enclosures
05580 Heating/Cooling Unit Enclosures
05600 Hydraulic Fabrications
05600 Bifurcation Panels
05600 Bulkheads
05600 Manifolds
05600 Penstocks
05600 Trashracks
05650 Railroad Track and Accessories
05700 Ornamental Metal
05710 Ornamental Stairs
05715 Fabricated Spiral Stairs
05720 Ornamental Handrails and Railings
05725 Ornamental Metal Castings
05730 Ornamental Formed Metal
05740 Ornamental Forged Metal
05800 Expansion Control
05810 Expansion Joint Cover Assemblies
05820 Slide Bearings
05830 Bridge Expansion Joint Assemblies
05900 Metal Restoration and Cleaning

06000 WOOD AND PLASTICS
06050 Basic Wood and Plastic Materials ad Methods
06060 Wood Materials
06065 Plastic Materials
06070 Wood Treatment
06070 Fire-Retardant Treatment
06070 Preservative Treatment
06080 Factory-Applied Wood Coatings
06090 Wood and Plastic Fastenings
06100 Rough Carpentry
06110 Wood Framing
06110 Engineered Lumber Products
06110 Stress-Rated Lumber
06120 Structural Panels
06120 Cementitious Reinforced Panels
06120 Stressed Skin Panels
06130 Heavy Timber Construction
06130 Log Structures
06130 Mill-Framed Structures
06130 Pole Construction
06130 Railroad Ties
06130 Timber Bridges and Trestles
06130 Timber Trusses
06140 Treated Wood Foundations
06150 Wood Decking
06160 Sheathing
06160 Insulating Sheathing
06160 Subflooring
06160 Underlayment
06160 Wood Board Sheathing
06160 Wood Panel Product Sheathing
06170 Prefabricated Structural Wood
06170 Laminated Veneer Lumber
06170 Metal-Web Wood Joists
06170 Parallel Strand Lumber
06170 Wood I Joists
06170 Wood Trusses
06180 Glued-Laminated Construction
06200 Finish Carpentry
06220 Millwork
06250 Prefinished Paneling
06260 Board Paneling
06270 Closet and Utility Wood Shelving
06400 Architectural Woodwork
06410 Custom Cabinets
06410 Cabinet Hardware
06410 Laminate-Clad Wood Cabinets
06410 Opaque-Finish Wood Cabinets
06410 Transparent-Finish Wood Cabinets
06415 Countertops
06420 Paneling
06430 Wood Stairs and Railings
06430 Ornamental Wood Stairs
06440 Wood Ornaments
06440 Custom Wood Turnings
06440 Fluted Wood Pilasters
06440 Ornamental Wood Grilles
06440 Wood Balusters
06440 Wood Corbels
06440 Wood Cupolas
06440 Wood Finials
06440 Wood Mantels
06440 Wood Pediment Heads
06440 Wood Posts and Columns
06445 Simulated Wood Ornaments
06450 Standing and Running Trim
06450 Door and Window Wood Casings
06450 Wood Aprons
06450 Wood Base and Shoe Molding
06450 Wood Chair Rail
06450 Wood Cornices
06450 Wood Facias and Soffits
06450 Wood Handrails and Guard Rails
06450 Wood Stops, Stools, and Sills
06455 Simulated Wood Trim
06460 Wood Frames
06460 Exterior Door Wood Frames and Jambs
06460 Fire-Rated Door Wood Frames
06460 Interior Door Wood Frames and Jambs
06460 Ornamental Wood Frames
06460 Stick-Built Wood Windows
06460 Wood Veneer Frames
06470 Screens, Blinds, and Shutters
06500 Structural Plastics
06510 Structural Plastic Shapes and Plates
06510 Plastic Lumber
06510 Structural Tubular Plastic
06520 Plastic Structural Assemblies
06600 Plastic Fabrications
06600 Cultured Marble
06600 Glass-Fiber-Reinforced Plastic
06600 Plastic Handrails
06600 Plastic Paneling
06600 Solid Surfacing
06900 Wood and Plastic Restoration and Cleaning
06910 Wood Restoration and Cleaning
06910 Architectural Woodwork Cleaning

06910 Architectural Woodwork Refinishing
06910 Architectural Woodwork Restoration
06910 Wood Framing Restoration
6920 Plastic Restoration and Cleaning

07000 THERMAL AND MOISTURE PROTECTION
07050 Basic Thermal and Moisture Protection Materials and Methods
07100 Dampproofing and Waterproofing
07110 Dampproofing
07110 Bituminous Dampproofing
07110 Cementitious Dampproofing
07120 Built-Up Bituminous Waterproofing
07120 Built-Up Asphalt Waterproofing
07120 Built-Up Coal Tar Waterproofing
07130 Sheet Waterproofing
07130 Bituminous Sheet Waterproofing
07130 Elastomeric Sheet Waterproofing
07130 Modified Bituminous Sheet Waterproofing
07130 Thermoplastic Sheet Waterproofing
07140 Fluid-Applied Waterproofing
07150 Sheet Metal Waterproofing
07160 Cementitious and Reactive Waterproofing
07160 Acrylic Modified Cement Waterproofing
07160 Crystalline Waterproofing
07160 Metal Oxide Waterproofing
07170 Bentonite Waterproofing
07170 Bentonite Panel Waterproofing
07170 Bentonite Sheet Waterproofing
07180 Traffic Coatings
07180 Pedestrian Traffic Coating
07180 Vehicular Traffic Coating
07190 Water Repellents
07190 Acrylic Water Repellents
07190 Silane Water Repellents
07190 Silicone Water Repellents
07190 Siloxane Water Repellents
07190 Stearate Water Repellents
07200 Thermal Protection
07210 Building Insulation
07220 Roof and Deck Insulation
07220 Asphaltic Perlite Concrete Deck
07220 Roof Board Insulation
07240 Exterior Insulation and Finish Systems (EIFS)
07240 Exterior Finish Assemblies
07260 Vapor Retarders
07270 Air Barriers
07300 Shingles, Roof Tiles, and Roof Coverings
07310 Shingles
07310 Asphalt Shingles
07310 Fiberglass Reinforced Shingles
07310 Metal Shingles
07310 Mineral Fiber Cement Shingles
07310 Plastic Shakes
07310 Porcelain Enamel Shingles
07310 Slate Shingles
07310 Wood Shingles
07310 Wood Shakes
07320 Roof Tiles
07320 Clay Roof Tiles
07320 Concrete Roof Tiles
07320 Metal Roof Tiles
07320 Mineral Fiber Cement Roof Tiles
07320 Plastic Roof Tiles
07330 Roof Coverings
07330 Sod Roofing
07330 Thatched Roofing
07400 Roofing and Siding Panels
07410 Metal Roof and Wall Panels
07410 Metal Roof Panels
07410 Metal Wall Panels
07420 Plastic Roof and Wall Panels
07420 Plastic Roof Panels
07420 Plastic Wall Panels
07430 Composite Panels
07440 Faced Panels
07440 Aggregate Coated Panels
07440 Porcelain Enameled Faced Panels
07440 Tile Faced Panels
07450 Fiber-Reinforced Cementitious Panels
07450 Glass-Fiber-Reinforced Cementitious Panels
07450 Mineral-Fiber-Reinforced Cementitious Panels
07460 Siding
07460 Aluminum Siding
07460 Composition Siding
07460 Hardboard Siding
07460 Mineral Fiber Cement Siding
07460 Plastic Siding
07460 Plywood Siding
07460 Steel Siding
07460 Wood Siding
07470 Wood Roof and Wall Panels
07470 Wood Roof Panels
07470 Wood Wall Panels
07480 Exterior Wall Assemblies
07500 Membrane Roofing
07510 Built-Up Bituminous Roofing
07510 Built-Up Asphalt Roofing
07510 Built-Up Coal-Tar Roofing
07520 Cold-Applied Bituminous Roofing
07520 Cold-Applied Mastic Roof Membrane

07520 Glass-Fiber-Reinforced Asphalt Emulsion Roofing
07530 Elastomeric Membrane Roofing
07530 Chlorinated Polyethylene (CPE) Roofing
07530 Chlorosulfonated Polyethylene (CSPE) Roofing
07530 Copolymer Alloy (CPA) Roofing
07530 Ethylene Propylene Diene Monomer (EPDM) Roofing
07530 Nitrile Butadiene Polymer (NBP) Roofing
07530 Polyisobutylene (PIB) Roofing
07540 Thermoplastic Membrane Roofing
07540 Ethylene Interpolymer (EIP) Roofing
07540 Polyvinyl Chloride (PVC) Roofing
07540 Thermoplastic Alloy (TPA) Roofing
07550 Modified Bituminous Membrane Roofing
07560 Fluid-Applied Roofing
07570 Coated Foamed Roofing
07580 Roll Roofing
07590 Roof Maintenance and Repairs
07590 Preparation for Reroofing
07590 Roof Maintenance Program
07590 Roof Moisture Survey
07590 Roof Removal
07590 Roofing Restaurants
07590 Roofing Restoration
07600 Flashing and Sheet Metal
07610 Sheet Metal Roofing
07620 Sheet Metal Flashing and Trim
07630 Sheet Metal Roofing Specialties
07650 Flexible Flashing
07650 Laminated Sheet Flashing
07650 Modified Bituminous Sheet Flashing
07650 Plastic Sheet Flashing
07650 Rubber Sheet Flashing
07650 Self-Adhering Sheet Flashing
07700 Roof Specialties and Accessories
07710 Manufactured Roof Specialties
07710 Copings
07710 Counterflashing Systems
07710 Gravel Stops and Facias
07710 Gutters and Downspouts
07710 Reglets
07710 Roof Expansion Assemblies
07710 Scuppers
07720 Roof Accessories
07720 Manufactured Curbs
07720 Relief Vents
07720 Ridge Vents
07720 Roof Hatches
07720 Roof Walk Boards
07720 Roof Walkways
07720 Smoke Vents
07720 Snow Guards
07720 Waste Containment Assemblies
07760 Roof Pavers
07760 Roof Ballast Pavers
07760 Roof Decking Pavers
07800 Fire and Smoke Protection
07810 Applied Fireproofing
07810 Cement Aggregate Fireproofing
07810 Cementitious Fireproofing
07810 Foamed Magnesium Oxychloride Fireproofing
07810 Intumescent Mastic Fireproofing
07810 Magnesium Cement Fireproofing
07810 Mineral Fiber Cementitious Fireproofing
07810 Mineral Fiber Fireproofing
07820 Board Fireproofing
07820 Calcium Silicate Board Fireproofing
07820 Slag Fiber Board Fireproofing
07840 Firestopping
07840 Annular Space Protection
07840 Fibrous Fire Safing
07840 Fire-Resistant Joint Sealants
07840 Intumescent Firestopping Foams
07840 Silicone Firestopping Foams
07840 Thermal Barriers for Plastics
07840 Through Penetration Firestopping Mortars
07840 Through Penetration Firestopping Pillows
07860 Smoke Seals
07870 Smoke Containment Barriers
07900 Joint Sealers
07910 Preformed Joint Seals
07910 Compression Seals
07910 Joint Gaskets
07920 Joint Sealants
07920 Backer Rods
07920 Calking
07920 Joint Fillers
7920 Sealants

08000 DOORS AND WINDOWS
08050 Basic Door and Window Materials and Methods
08100 Metal Doors and Frames
08110 Steel Doors and Frames
08110 Custom Steel Doors and Frames
08110 Standard Steel Doors and Frames
08120 Aluminum Doors and Frames
08130 Stainless Steel Doors and Frames
08140 Bronze Doors and Frames
08150 Preassembled Metal Door and Frame Units
08160 Sliding Metal Doors and Grilles
08160 Sliding Metal Doors
08160 Sliding Metal Grilles
08180 Metal Screen and Storm Doors
08180 Aluminum Screen and Storm Doors
08180 Metal Screen Doors
08180 Metal Storm Doors
08180 Steel Screen and Storm Doors
08190 Metal Door Restoration
08200 Wood and Plastic Doors
08210 Wood Doors
08210 Carved Wood Doors
08210 Flush Wood Doors
08210 Metal-Faced Wood Doors
08210 Plastic-Faced Wood Doors
08210 Prefinished Wood Doors
08210 Stile and Rail Wood Doors
08220 Plastic Doors
08220 Laminated Plastic Doors
08220 Solid Plastic Doors
08250 Preassembled Wood and Plastic Door and Frame Units
08250 Preassembled Plastic Door and Frame Units
08250 Preassembled Wood Door and Frame Units
08260 Sliding Wood and Plastic Doors
08260 Sliding Plastic Doors
08260 Sliding Wood Doors
08280 Wood and Plastic Storm and Screen Doors
08280 Plastic Screen Doors
08280 Plastic Storm and Screen Doors
08280 Plastic Storm Doors
08280 Wood Screen Doors
08280 Wood Storm and Screen Doors
08280 Wood Storm Doors
08290 Wood and Plastic Door Restoration
08300 Specialty Doors
08310 Access Doors and Panels
08310 Access Doors and Frames
08310 Access Panels and Frames
08320 Detention Doors and Frames
08320 Detention Door Frame Protection
08320 Steel Detention Doors and Frames
08320 Steel Plate Detention Doors and Frames
08330 Coiling Doors and Grilles
08330 Coiling Counter Doors
08330 Coiling Counter Grilles
08330 Overhead Coiling Doors
08330 Overhead Coiling Grilles
08330 Side Coiling Doors
08330 Side Coiling Grilles
08340 Special Function Doors
08340 Cold Storage Doors
08340 Hangar Doors
08340 Lightproof Doors
08340 Radiation Protection Doors
08340 Radio-Frequency-Interference (RFI) Shielding Doors
08340 Sound Control Doors
08340 Industrial Doors
08350 Folding Doors and Grilles
08350 Accordion Folding Doors
08350 Accordion Folding Grilles
08350 Folding Fire Doors
08350 Panel Folding Doors
08360 Overhead Doors
08360 Sectional Overhead Doors
08360 Single-Panel Overhead Doors
08370 Vertical Lift Doors
08370 Multi-Leaf Vertical Lift Doors
08370 Telescoping Vertical Lift Doors
08380 Traffic Doors
08380 Flexible Strip Doors
08380 Flexible Traffic Doors
08380 Rigid Traffic Doors
08390 Pressure-Resistant Doors
08390 Airtight Doors
08390 Blast-Resistant Doors
08390 Watertight Doors
8400 Entrances and Storefronts
08410 Metal-Framed Storefronts
08410 Aluminum-Framed Storefronts
08410 Bronze-Framed Storefronts
08410 Stainless Steel-Framed Storefronts
08410 Steel-Framed Storefronts
08450 All-Glass Entrances and Storefronts
08460 Automatic Entrance Doors
08470 Revolving Entrance Doors
08480 Balanced Entrance Doors

08490 Sliding Storefronts
08500 Windows
08510 Steel Windows
08520 Aluminum Windows
08530 Stainless Steel Windows
08540 Bronze Windows
0550 Wood Windows
08550 Metal-Clad Wood Windows
08550 Plastic-Clad Wood Windows
08560 Plastic Windows
08570 Composite Windows
08580 Special Function Windows
08580 Detention Windows
08580 Detention Window Screens
08580 Pass Windows
08580 Radiation Protection Windows
08580 Radio-Frequency-Interference (RFI) Shielding Windows
08580 Security Window Screens
08580 Sound Control Windows
08580 Storm Windows
08590 Window Restoration and Replacement
08590 Window Restoration
08590 Window Replacement
08600 Skylights
08610 Roof Windows
08610 Metal Roof Windows
08610 Wood Roof Windows
08620 Unit Skylights
08620 Domed Unit Skylights
08620 Pyramidal Unit Skylights
08620 Vaulted Unit Skylights
08630 Metal-Framed Skylights
08630 Domed Metal-Framed Skylights
08630 Motorized Metal-Framed Skylights
08630 Pyramidal Metal-Framed Skylights
08630 Ridge Metal-Framed Skylights
08630 Vaulted Metal-Framed Skylights
08700 Hardware
08710 Door Hardware
08710 Controlling Hardware
08710 Door Trim
08710 Hanging Hardware
08710 Latching Hardware
08710 Locking Hardware
08720 Weatherstripping and Seals
08720 Door Weatherstripping and Seals
08720 Thresholds
08720 Window Weatherstripping and Seals
08740 Electro-Mechanical Hardware
08740 Card Key Locking Hardware
08740 Electrical Locking Control
08740 Electromagnetic Door Holders
08750 Window Hardware
08750 Automatic Window Equipment
08750 Window Locks
08750 Window Lifts
08750 Window Operators
08770 Door and Window Accessories
08770 Key Storage and Control Equipment
08780 Special Function Hardware
08780 Detention Hardware
08790 Hardware Restoration
08800 Glazing
08810 Glass
08810 Bent Glass
08810 Bullet-Resistant Glass
08810 Chemically-Strengthened Glass
08810 Coated Glass
08810 Composite Glass
08810 Decorative Glass
08810 Fire Rated Glass
08810 Float Glass
08810 Heat-Strengthened Glass
08810 Impact-Resistant Glass
08810 Insulating Glass
08810 Laminated Glass
08810 Mirrored Glass
08810 Rolled Glass
08810 Tempered Glass
08810 Wired Glass
08830 Mirrors
08840 Plastic Glazing
08840 Bullet-Resistant Plastic Glazing
08840 Decorative Plastic Glazing
08840 Insulating Plastic Glazing
08840 Translucent Plastic Glazing
08840 Transparent Plastic Glazing
08850 Glazing Accessories
08890 Glazing Restoration
08900 Glazed Curtain Wall
08910 Metal Framed Curtain Wall
08910 Glazed Aluminum Curtain Wall
08910 Glazed Bronze Curtain Wall
08910 Glazed Stainless Steel Curtain Wall
08950 Translucent Wall and Roof Assemblies
08960 Sloped Glazing Assemblies
08970 Structural Glass Curtain Walls

09000 FINISHES

09050 Basic Finish Materials and Methods
09100 Metal Support Assemblies
09110 Non-Load-Bearing Wall Framing
09120 Ceiling Suspension
09130 Acoustical Suspension
09190 Metal Frame Restoration
09200 Plaster and Gypsum Board
09205 Furring and Lathing
09205 Gypsum Lath
09205 Lead-Lined Lath
09205 Metal Lath
09205 Plaster Accessories
09205 Veneer Plaster Base
09210 Gypsum Plaster
09210 Acoustical Gypsum Plaster
09210 Fireproofing Gypsum Plaster
09210 Gypsum Veneer Plaster
09220 Portland Cement Plaster
09220 Adobe Finish
09220 Portland Cement Veneer Plaster
09230 Plaster Fabrications
09250 Gypsum Board
09250 Cementitious Backing Board
09250 Factory-Finished Gypsum Board
09250 Gypsum Sheathing
09250 Shaft Wall Liner
09260 Gypsum Board Assemblies
09260 Gypsum Board Area Separation Walls
09260 Shaft Wall Assemblies
09270 Gypsum Board Accessories
09280 Plaster Restoration
09300 Tile
09305 Tile Setting Materials and Accessories
09310 Ceramic Tile
09310 Ceramic Mosaics
09310 Conductive Tile
09330 Quarry Tile
09330 Chemical-Resistant Quarry Tile
09340 Paver Tile
09340 Terra Cotta Tile
09350 Glass Mosaics
09360 Plastic Tile
09370 Metal Tile
09380 Cut Natural Stone Tile
09380 Agglomerate Tile
09380 Stone Floor and Wall Covering
09390 Tile Restoration
09400 Terrazzo
09410 Portland Cement Terrazzo
09420 Precast Terrazzo
09420 Terrazzo Tile
09430 Conductive Terrazzo
09440 Plastic Matrix Terrazzo
09490 Terrazzo Restoration
09500 Ceilings
09510 Acoustical Ceilings
09510 Acoustical Metal Pan Ceilings
09510 Acoustical Panel Ceilings
09510 Acoustical Tile Ceilings
09545 Specialty Ceilings
09545 Integrated Ceilings
09545 Linear Metal Ceilings
09545 Luminous Ceilings
09550 Mirror Panel Ceilings
09560 Textured Ceilings
09560 Gypsum Panel
09560 Metal Panel
09570 Linear Wood Ceilings
09580 Suspended Decorative Grids
09590 Ceiling Assembly Restoration
9600 Flooring
09610 Floor Treatment
09610 Slip-Resistant Floor Treatment
09610 Static-Resistant Floor Treatment
09620 Specialty Flooring
09620 Asphalt Plank Flooring
09620 Athletic Flooring
09620 Plastic-Laminate Flooring
09630 Masonry Flooring
09630 Brick Flooring
09630 Chemical-Resistant Brick Flooring

09000 FINISHES

09630 Flagstone Flooring
09630 Granite Flooring
09630 Marble Flooring
09630 Slate Flooring
09630 Stone Flooring
09640 Wood Flooring
09640 Cushioned Wood Flooring Assemblies
09640 Mastic Set Wood Flooring Assemblies
09640 Resilient Wood Flooring Assemblies
09640 Wood Athletic Flooring
09640 Wood Block Flooring
09640 Wood Composition Flooring
09640 Wood Parquet Flooring
09640 Wood Strip Flooring
09650 Resilient Flooring
09650 Resilient Base and Accessories
09650 Resilient Sheet Flooring
09650 Resilient Tile Flooring
09660 Static Control Flooring
09660 Conductive Elastomeric Liquid Flooring
09660 Conductive Resilient Flooring
09660 Static Dissipative Resilient Flooring
09660 Static-Resistant Resilient Flooring
09670 Fluid-Applied Flooring
09670 Elastomeric Liquid Flooring
09670 Epoxy-Marble Chip Flooring
09670 Magnesium Oxychloride Flooring
09670 Mastic Fills
09670 Resinous Flooring
09670 Seamless Quartz Flooring
09680 Carpet
09680 Carpet Cushion
09680 Carpet Tile
09680 Indoor and Outdoor Carpet
09680 Sheet Carpet
09690 Flooring Restoration
09700 Wall Finishes
09710 Acoustical Wall Treatment
09720 Wall Covering
09720 Cork Wall Covering
09720 Vinyl-Coated Fabric Wall Covering
09720 Vinyl Wall Covering
09720 Wall Fabrics
09720 Wallpaper
09730 Wall Carpet
09740 Flexible Wood Sheets
09740 Flexible Wood Veneers
09750 Stone Facing
09760 Plastic Blocks
09770 Special Wall Surfaces
09790 Wall Finish Restoration
09800 Acoustical Treatment
09810 Acoustical Space Units
09820 Acoustical Insulation and Sealants
09820 Acoustical Insulation
09820 Acoustical Sealants
09830 Acoustical Barriers
09840 Acoustical Wall Treatment
09900 Paints and Coatings
09910 Paints
09910 Exterior Paints
09910 Interior Paints
09930 Stains and Transparent Finishes
09930 Exterior Opaque Stains
09930 Exterior Transparent Finishes
09930 Exterior Transparent Stains
09930 Interior Opaque Stains
09930 Interior Transparent Finishes
09930 Interior Transparent Stains
09930 Transparent Coatings
09940 Decorative Finishes
09960 High-Performance Coatings
09960 Abrasion-Resistant Coatings
09960 Chemical-Resistant Coatings
09960 Elastomeric Coatings
09960 Fire-Resistant Paints
09960 Graffiti-Resistant Coatings
09960 High-Build Glazed Coatings
09960 Intumescent Paints
09960 Marine Coatings
09960 Textured Plastic Coatings
09970 Coatings for Steel
09980 Coatings for Concrete and Masonry
09990 Paint Restoration
09990 Maintenance Coatings
09990 Paint Cleaning

10000 SPECIALTIES
10100 Visual Display Boards
10110 Chalkboards
10110 Fixed Chalkboards
10110 Operable Chalkboards
10110 Portable Chalkboards
10115 Markerboards
10115 Electronic Markerboards
10115 Fixed Markerboards
10115 Operable Markerboards
10115 Portable Markerboards
10120 Tackboard and Visual Aid Boards
10120 Fixed Tackboard
10120 Magnetic Display Boards
10120 Operable Display Boards
10120 Portable Tackboard
10130 Operable Board Units
10140 Display Track Assemblies
10145 Visual Aid Board Units
10150 Compartments and Cubicles
10160 Metal Toilet Compartments
10165 Plastic Laminate Toilet Compartments
10170 Plastic Toilet Compartments
10175 Particleboard Toilet Compartments
10180 Stone Toilet Compartments
10185 Shower and Dressing Compartments
10190 Cubicles
10190 Cubicle Curtains
10190 Cubicle Track and Hardware
10200 Louvers and Vents
10210 Wall Louvers
10210 Motorized Louvers
10210 Operable Louvers
10210 Stationary Louvers
10220 Louvered Equipment Enclosures
10225 Door Louvers
10230 Vents
10230 Soffit Vents
10230 Wall Vents
10240 Grilles and Screens
10250 Service Walls
10260 Wall and Corner Guards
10260 Bumper Guards
10260 Corner Guards
10260 Impact-Resistant Wall Protection
10270 Access Flooring
10270 Rigid Grid Assemblies
10270 Snap-on Stringer Assemblies
10270 Stringerless Assemblies
10290 Pest Control
10290 Bird Control
10290 Insect Control
10290 Rodent Control
10300 Fireplaces and Stoves
10305 Manufactured Fireplaces
10305 Manufactured Fireplace Chimneys
10305 Manufactured Fireplace Forms
10310 Fireplace Specialties and Accessories
10310 Fireplace Dampers
10310 Fireplace Inserts
10310 Fireplace Screens and Doors
10310 Fireplace Water Heaters
10320 Stoves
10330 Fireplace and Stove Restoration
10340 Manufactured Exterior Specialties
10340 Clocks
10340 Cupolas
10340 Spires
10340 Steeples
10340 Weathervanes
10345 Exterior Specialties Restoration
10350 Flagpoles
10350 Automatic Flagpoles
10350 Ground-Set Flagpoles
10350 Nautical Flagpoles
10350 Wall-Mounted Flagpoles
10350 Flags (usually)
10400 Identification Devices
10410 Directories
10410 Electronic Directories
10420 Plaques
10430 Exterior Signage
10430 Dimensional Letter Signage
10430 Electronic Message Signage
10430 Illuminated Exterior Signage
10430 Non-Illuminated Exterior Signage
10430 Post and Panel/Pylon Exterior Signage
10440 Interior Signage
10440 Dimensional Letters
10440 Door Signs
10440 Electronic Message Signs
10440 Engraved Signs
10440 Illuminated Interior Signs
10440 Non-Illuminated Interior Signs
10450 Pedestrian Control Devices
10450 Detection Specialties
10450 Portable Posts and Railings
10450 Rotary Gates

10450 Turnstiles
10500 Lockers
10500 Coin Operated Lockers
10500 Glass Lockers
10500 Metal Lockers
10500 Plastic-Laminate-Faced Lockers
10500 Plastic Lockers
10500 Recycled Plastic Lockers
10500 Wood Lockers
10520 Fire Protection Specialties
10520 Fire Blankets and Cabinets
10520 Fire Extinguisher Accessories
10520 Fire Extinguisher Cabinets
10520 Fire Extinguishers
10520 Wheeled Fire Extinguisher Units
10530 Protective Covers
10530 Awnings
10530 Canopies
10530 Car Shelters
10530 Walkway Coverings
10550 Postal Specialties
10550 Central Mail Delivery Boxes
10550 Collection Boxes
10550 Mail Boxes
10550 Mail Chutes
10600 Partitions
10605 Wire Mesh Partitions
10610 Folding Gates
10615 Demountable Partitions
10615 Central Mail Delivery Boxes
10615 Collection Boxes
10615 Mail Boxes
10615 Mail Chutes
10630 Portable Partitions, Screens, and Panels
10650 Operable Partitions
10650 Accordion Folding Partitions
10650 Coiling Partitions
10650 Folding Panel Partitions
10650 Sliding Partitions
10670 Storage Shelving
10670 Metal Storage Shelving
10670 Mobile Storage Units
10670 Prefabricated Wood Storage Shelving
10670 Recycled Plastic Storage Shelving
10670 Wire Storage Shelving
10700 Exterior Protection
10705 Exterior Sun Control Devices
10710 Exterior Shutters
10715 Storm Panels
10720 Exterior Louvers
10750 Telephone Specialties
10750 Telephone Directory Units
10750 Telephone Enclosures
10750 Telephone Shelving
10800 Toilet, Bath, and Laundry Accessories
10810 Toilet Accessories
10810 Commercial Toilet Accessories
10810 Detention Toilet Accessories
10810 Hospital Toilet Accessories
10820 Bath Accessories
10820 Residential Bath Accessories
10820 Shower and Tub Doors
10830 Laundry Accessories
10830 Built-In Ironing Boards
10830 Clothes Drying Racks
10880 Scales
10880 Fixed Scales
10900 Wardrobe and Closet Specialties

10000 EQUIPMENT

11010 Maintenance Equipment
11010 Floor and Wall Cleaning Equipment
11010 Housekeeping Carts
11010 Vacuum Cleaning Systems
11010 Window Washing Systems
11020 Security and Vault Equipment
11020 Safe Deposit Boxes
11020 Safes
11020 Vault Doors and Day Gates
11030 Teller and Service Equipment
11030 Automatic Banking Systems
11030 Money Cart Pass-Through
11030 Package Transfer Units
11030 Service and Teller Window Units
11030 Teller Equipment Systems
11040 Ecclesiastical Equipment
11040 Baptisteries
11040 Chancel Fittings
11050 Library Equipment
11050 Automated Book Storage and Retrieval Systems
11050 Book Depositories
11050 Book Theft Protection Equipment
11050 Library Stack Systems
11060 Theater and Stage Equipment
11060 Acoustical Shells
11060 Folding and Portable Stages
11060 Rigging Systems and Controls

11060 Stage Curtains
11070 Instrumental Equipment
11070 Bells
11070 Carillons
11070 Organs
11080 Registration Equipment
11090 Checkroom Equipment
11100 Mercantile Equipment
11100 Barber and Beauty Shop Equipment
11100 Cash Registers and Checking Equipment
11100 Display Cases
11100 Food Processing Equipment
11100 Food Weighing and Wrapping Equipment
11110 Commercial Laundry and Dry Cleaning Equipment
11110 Dry Cleaning Equipment
11110 Drying and Conditioning Equipment
11110 Finishing Equipment
11110 Ironing Equipment
11110 Washers and Extractors
11120 Vending Equipment
11120 Money-Changing Machines
11120 Vending Machines
11130 Audio-Visual Equipment
11130 Learning Laboratories
11130 Projection Screens
11130 Projectors
11140 Vehicle Service Equipment
11140 Compressed Air Equipment
11140 Fuel Dispensing Equipment
11140 Lubrication Equipment
11140 Tire Changing Equipment
11140 Vehicle Washing Equipment
11150 Parking Control Equipment
11150 Coin Machine Units
11150 Key and Card Control Units
11150 Parking Gates
11150 Ticket Dispensers
11160 Loading Dock Equipment
11160 Dock Bumpers
11160 Dock Levelers
11160 Dock Lifts
11160 Portable Ramps, Bridges, and Platforms
11160 Seals and Shelters
11160 Truck Restraints
11170 Solid Waste Handling Equipment
11170 Bins
11170 Chutes and Collectors
11170 Packaged Incinerators
11170 Pneumatic Waste Equipment
11170 Pulping Machines
11170 Recycling Equipment
11170 Waste Compactors and Destructors
11190 Detention Equipment
11200 Water Supply and Treatment Equipment
11210 Supply and Treatment Pumps
11220 Mixers and Flocculators
11225 Clarifiers
11230 Water Aeration Equipment
11240 Chemical Feed Equipment
11250 Water Softening Equipment
11250 Base-Exchange or Zeolite Equipment
11250 Lime-Soda Process Equipment
11260 Disinfectant Feed Equipment
11260 Chlorination Equipment
11260 Hydrogen Power (pH) Equipment
11270 Fluoridation Equipment
11285 Hydraulic Gates
11285 Bulkhead Gates
11285 High-Pressure Gates
11285 Hinged-Leaf Gates
11285 Radial Gates
11285 Slide Gates
11285 Sluice Gates
11285 Spillway Crest Gates
11285 Vertical-Lift Gates
11295 Hydraulic Valves
11295 Butterfly Valves
11295 Regulating Valves
11300 Fluid Waste Treatment and Disposal Equipment
11310 Sewage and Sludge Pumps
11310 Oil-Water Separators
11310 Packaged Pump Stations
11310 Sewage Ejectors
11320 Grit Collecting Equipment
11330 Screening and Grinding Equipment
11335 Sedimentation Tank Equipment
11340 Scum Removal Equipment
11345 Chemical Equipment
11350 Sludge Handling and Treatment Equipment
11360 Filter Press Equipment
11365 Trickling Filter Equipment
11370 Compressors
11375 Aeration Equipment
11380 Sludge Digestion Equipment
11385 Digester Mixing Equipment
11390 Package Sewage Treatment Plants
11400 Food Service Equipment

11405 Food Storage Equipment
11410 Food Preparation Equipment
11415 Food Delivery Carts and Conveyors
11420 Food Cooking Equipment
11425 Hood and Ventilation Equipment
11425 Surface Fire Suppression Equipment
11430 Food Dispensing Equipment
11430 Bar Equipment
11430 Service Line Equipment
11430 Soda Fountain Equipment
11435 Ice Machines
11440 Cleaning and Disposal Equipment
11450 Residential Equipment
11450 Residential Appliances
11450 Residential Kitchen Equipment
11450 Retractable Stairs
11460 Unit Kitchens
11470 Darkroom Equipment
11470 Darkroom Processing Equipment
11470 Revolving Darkroom Doors
11470 Transfer Cabinets
11480 Athletic, Recreational, and Therapeutic Equipment
11480 Backstops
11480 Bowling Alleys
11480 Exercise Equipment
11480 Gym Dividers
11480 Gymnasium Equipment
11480 Scoreboards
11480 Shooting Ranges
11480 Therapy Equipment
11500 Industrial and Process Equipment
11600 Laboratory Equipment
11650 Planetarium Equipment
11660 Observatory Equipment
11680 Office Equipment
11700 Medical Equipment
11710 Medical Sterilizing Equipment
11720 Examination and Treatment Equipment
11730 Patient Care Equipment
11740 Dental Equipment
11750 Optical Equipment
11760 Operating Room Equipment
11770 Radiology Equipment
11780 Mortuary Equipment
11850 Navigation Equipment
11870 Agricultural Equipment
11900 Exhibit Equipment

12000 FURNISHINGS
12050 Fabrics
12110 Art
12110 Murals
12110 Commissioned Paintings
12110 Faux Finishes
12110 Photo Murals
12120 Wall Decorations
12140 Sculptures
12170 Art Glass
12170 Etched Glass
12170 Stained Glass
12190 Ecclesiastical Art
12300 Manufactured Casework
12310 Manufactured Metal Casework
12320 Manufactured Wood Casework
12320 Wood-Veneer-Faced Casework
12320 Plastic-Laminate-Faced Casework
12350 Specialty Casework
12350 Bank Casework
12350 Dental Casework
12350 Display Casework
12350 Dormitory Casework
12350 Educational Facility Casework
12350 Hospital Casework
12350 Kitchen Casework
12350 Laboratory Casework
12350 Library Casework
12350 Medical Casework
12350 Nurse Station Casework
12350 Religious Casework
12350 Residential Casework
12350 Study Carrels
12400 Furnishings and Accessories
12410 Office Accessories
12410 Desk Accessories
12420 Table Accessories
12420 China
12420 Flatware
12420 Glassware
12420 Hollowware
12420 Napery
12420 Silverware
12420 Table Linens
12430 Portable Lamps
12430 Desk Lamps
12430 Floor Lamps
12430 Table Lamps
12440 Bath Furnishings

12440 Bath Linens
12440 Bath Mats
12440 Bath Towels
12440 Shower Curtains
12450 Bedroom Furnishings
12450 Bed Linens
12450 Blankets
12450 Comforters
12450 Pillows
12460 Furnishing Accessories
12460 Ash Receptacles
12460 Bowls
12460 Clocks
12460 Decorative Crafts
12460 Vases
12460 Waste Receptacles
12480 Rugs and Mats
12480 Floor Mats
12480 Rugs
12480 Runners
12490 Window Treatments
12490 Blinds
12490 Curtains and Drapes
12490 Interior Shutters
12490 Shades
12490 Solar-Control Film
12490 Window Treatment Hardware
12500 Furniture
12510 Office Furniture
12510 Custom Office Furniture
12510 Office Case Goods
12520 Seating
12520 Chairs
12520 Custom Upholstered Seating
12520 Upholstered Seating
12540 Hospitality Furniture
12540 Hotel Furniture
12540 Motel Furniture
12540 Restaurant Furniture
12560 Institutional Furniture
12560 Book Shelves
12560 Classroom Furniture
12560 Dormitory Furniture
12560 Health Care Furniture
12560 Index Card File Cabinets
12560 Laboratory Furniture
12560 Library Furniture
12560 Study Carrels
12580 Residential Furniture
12600 Multiple Seating
12610 Fixed Audience Seating
12620 Portable Audience Seating
12620 Folding Chairs
12620 Interlocking Chairs
12620 Stacking Chairs
12630 Stadium and Arena Seating
12640 Booths and Tables
12650 Multiple-Used Fixed Seating
12660 Telescoping Stands
12660 Telescoping Bleachers
12660 Telescoping Chair Platforms
12670 Pews and Benches
12680 Seat and Table Assemblies
12700 Systems Furniture
12710 Panel-Hung Component System Furniture
12720 Free-Standing Component System Furniture
12730 Beam System Furniture
12740 Desk System Furniture
12800 Interior Plants and Planters
12810 Interior Live Plants
12820 Interior Artificial Plants
12830 Interior Planters
12840 Interior Landscape Accessories
12850 Interior Plant Maintenance
12900 Furnishings Restoration and Repair

13000 SPECIAL CONSTRUCTION
13010 Air-Supported Structures
13020 Building Modules
13020 Prison Cells
13020 Hotel and Dormitory Units
13030 Special Purpose Rooms
13030 Athletic Rooms
13030 Clean Rooms
13030 Cold Storage Rooms
13030 Hyperbaric Rooms
13030 Insulated Rooms
13030 Office Shelters and Booths
13030 Planetariums
13030 Prefabricated Rooms
13030 Saunas
13030 Sound-Conditioned Rooms
13030 Steam Baths
13030 Vaults
13080 Sound, Vibration, and Seismic Control
13090 Radiation Protection
13100 Lightning Protection
13110 Cathodic Protection

13120 Pre-Engineered Structures
13120 Cable-Supported Structures
13120 Fabric Structures
13120 Glazed Structures
13120 Grandstands and Bleachers
13120 Metal Building Systems
13120 Modular Mezzanines
13120 Observatories
13120 Portable and Mobile Buildings
13120 Pre-Engineered Buildings
13120 Prefabricated Control Booths
13120 Prefabricated Dome Structures
13150 Swimming Pools
13150 Below Grade Swimming Pools
13150 Elevated Swimming Pools
13150 On-Grade Swimming Pools
13150 Recirculating Gutter Systems
13150 Swimming Pool Accessories
13150 Swimming Pool Cleaning Systems
13160 Aquariums
13165 Aquatic Park Facilities
13165 Water Slides
13165 Wave Pools
13170 Tubs and Pools
13170 Hot Tubs
13170 Therapeutic Pools
13170 Whirlpool Tubs
13175 Ice Rinks
13185 Kennels and Animal Shelters
13190 Site-Constructed Incinerators
13190 Sludge Incinerators
13190 Solid Waste Incinerators
13190 Waste disposal incinerators
13200 Storage Tanks
13200 Elevated Storage Tanks
13200 Ground Storage Tanks
13200 Tank Cleaning Procedures
13200 Tank Lining
13200 Underground Storage Tanks
13220 Filter Underdrains and Media
13220 Filter Bottoms
13220 Filter Media
13220 Package Filters
13230 Digester Covers and Appurtenances
13230 Fixed Covers
13230 Floating Covers
13230 Gasholder Covers
13240 Oxygenation Systems
13240 Oxygen Dissolution System
13240 Oxygen Generators
13240 Oxygen Storage Facility
13260 Sludge Conditioning Systems
13280 Hazardous Material Remediation
13400 Measurement and Control Instrumentation
13410 Basic Measurement and Control Instrumentation Materials and Methods
13410 Electronic Wire and Cable
13410 Instrument Air Accessories
13410 Instrument Air Tubing
13410 Optical Fiber Cable
13420 Instruments
13420 Control Valves
13420 Differential Pressure Instruments
13420 Flowmeters
13420 Gage Glasses and Cocks
13420 Level Instruments
13420 Orifice Plates and Flanges
13420 Potentiometric Instruments
13420 Pressure Instruments
13420 Receiver Instruments
13420 Temperature Instruments
13420 Traps and Drainers
13430 Boxes, Panels, and Control Centers
13440 Indicators, Recorders, and Controllers
13440 Annunciators
13440 Controllers
13440 Indicators
13440 Recorders
13450 Central Control
13480 Instrument Lists and Reports
13480 Instrument Diagrams
13480 Instrument Layouts
13480 Instrument Lists
13490 Measurement and Control Commissioning
13500 Recording Instrumentation
13510 Stress Instrumentation
13520 Seismic Instrumentation
13530 Meteorological Instrumentation
13550 Transportation Control Instrumentation
13560 Airport Control Instrumentation
13570 Railroad Control Instrumentation
13580 Subway Control Instrumentation
13590 Transit Vehicle Control Instrumentation
13600 Solar and Wind Energy Equipment
13610 Solar Flat Plate Collectors
13610 Air Collectors
13610 Liquid Collectors
13620 Solar Concentrating Collectors
13625 Solar Vacuum Tube Collectors

13630 Solar Collector Components
13630 Solar Absorber Plates or Tubing
13630 Solar Coatings and Surface Treatment
13630 Solar Collector Insulation
13630 Solar Glazing
13630 Solar Housing and Framing
13630 Solar Reflectors
13640 Packaged Solar Equipment
13650 Photovoltaic Collectors
13660 Wind Energy Equipment
13700 Security Access and Surveillance
13700 Door Answering
13700 Intrusion Detection
13700 Security Access
13700 Video Surveillance
13800 Building Automation and Control
13800 Clock Control
13800 Door Control
13800 Elevator Monitoring and Control
13800 Energy Monitoring and Control
13800 Environmental Control
13800 Escalator and Moving Walks
13800 Lighting Control
13850 Detection and Alarm
13850 Fire Alarm
13850 Gas Detection
13850 Leak Detection
13850 Smoke Alarm
13900 Fire Suppression
13910 Fire Protection Basic Materials and Methods
13910 Hangers and Supports
13910 Pipe and Fittings
13910 Piping Specialties
13910 Valves
13920 Fire Pumps
13930 Wet-Pipe Fire Suppression Sprinklers
13935 Dry-Pipe Fire Suppression Sprinklers
13940 Pre-Action Fire Suppression Sprinklers
13945 Combination Dry-Pipe and Pre-Action Fire Suppression Sprinklers
13950 Deluge Fire Suppression Sprinklers
13955 Foam Fire Extinguishing
13960 Carbon Dioxide Fire Extinguishing
13965 Alternative Fire Extinguishing Systems
13965 Halon Alternative
13965 Water Mist
13970 Dry Chemical Fire Extinguishing
13975 Standpipes and Hoses

14000 CONVEYING SYSTEMS
14100 Dumbwaiters
14110 Manual Dumbwaiters
14120 Electric Dumbwaiters
14140 Hydraulic Dumbwaiters
14200 Elevators
14210 Electric Traction Elevators
14210 Electric Traction Freight Elevators
14210 Electric Traction Passenger Elevators
14210 Electric Traction Residential Elevators
14210 Electric Traction Service Elevators
14240 Hydraulic Elevators
14240 Hydraulic Freight Elevators
14240 Hydraulic Passenger Elevators
14240 Hydraulic Residential Elevators
14240 Hydraulic Service Elevators
14270 Custom Elevator Cabs
14270 Cab Finishes
14280 Elevator Equipment and Controls
14280 Elevator Doors
14280 Elevator Controls
14290 Elevator Renovation
14300 Escalators and Moving Walks
14400 Lifts
14410 People Lifts
14410 Counterbalanced People Lifts
14410 Endless Belt People Lifts
14420 Wheelchair Lifts
14420 Inclined Wheelchair Lifts
14420 Vertical Wheelchair Lifts
14430 Platform Lifts
14430 Orchestra Lifts
14430 Stage Lifts
14440 Sidewalk Lifts
14450 Vehicle Lifts
14500 Material Handling
14510 Material Transport
14510 Guided Vehicle Material Handling
14510 Track Vehicle Material Handling
14530 Postal Conveying
14540 Baggage Conveying and Dispensing
14550 Conveyors
14550 Belt Conveyors
14550 Bucket Conveyors
14550 Container Conveyors
14550 Hopper and Track Conveyors
14550 Monorail Conveyors
14550 Oscillating Conveyors
14550 Pneumatic Conveyors

14550 Roller Conveyors
14550 Scoop Conveyors
14550 Screw Conveyors
14550 Selective Vertical Conveyors
14560 Chutes
14560 Coal Chutes
14560 Dry Bulk Material Chutes
14560 Escape Chutes
14560 Laundry and Linen Chutes
14560 Package Chutes
14560 Refuse Chutes
14570 Feeder Equipment
14570 Apron Feeders
14570 Reciprocating Plate Feeders
14570 Rotary Airlock Feeders
14570 Rotary Flow Feeders
14570 Vibratory Feeders
14580 Pneumatic Tube Systems
14600 Hoists and Cranes
14605 Crane Rails
14610 Fixed Hoists
14610 Air-Powered Fixed Hoists
14610 Electric Fixed Hoists
14610 Manual Fixed Hoists
14620 Trolley Hoists
14620 Air-Powered Trolley Hoists
14620 Electric Trolley Hoists
14620 Manual Trolley Hoists
14630 Bridge Cranes
14630 Top-Running Overhead Cranes
14630 Underslung Overhead Cranes
14640 Gantry Cranes
14650 Jib Cranes
14670 Tower Cranes
14680 Mobile Cranes
14690 Derricks
14700 Turntables
14800 Scaffolding
14810 Suspended Scaffolding
14810 Beam Scaffolding
14810 Carriage Scaffolding
14810 Hook Scaffolding
14820 Rope Climbers
14820 Manual Rope Climbers
14820 Powered Rope Climbers
14830 Telescoping Platforms
14830 Electric and Battery Telescoping Platforms
14830 Pneumatic Telescoping Platforms
14840 Powered Scaffolding
14900 Transportation
14910 People Movers
14920 Monorails
14930 Funiculars
14940 Aerial Tramways
14950 Aircraft Passenger Loading

15000 MECHANICAL
15050 Basic Mechanical Materials and Methods
15060 Hangers and Supports
15070 Mechanical Sound, Vibration, and Seismic Control
15075 Mechanical Identification
15080 Mechanical Insulation
15080 Duct Insulation
15080 Equipment Insulation
15080 Piping Insulation
15090 Mechanical Restoration and Retrofit
15100 Building Services Piping
15105 Pipes and Tubes
15110 Valves
15120 Piping Specialties
15130 Pumps
15140 Domestic Water Piping
15140 Disinfecting Potable Water Piping
15140 Non-Potable Water Piping
15140 Potable Water Piping
15150 Sanitary Waste and Vent Piping
15150 Interceptors
15150 Sanitary Piping
15150 Sanitary Piping Separators
15160 Storm Drainage Piping
15160 Retrofit Roof Drains
15160 Roof Drains
15160 Roof Drain Specialties
15160 Storm Drainage Piping
15170 Swimming Pool and Fountain Piping
15170 Reflecting Pool and Fountain Piping
15170 Reflecting Pool and Fountain Specialties
15170 Swimming Pool Piping
15170 Swimming Pool Specialties
15180 Heating and Cooling Piping
15180 Condensate Drain Piping
15180 Heating and Cooling Pumps
15180 Hydronic Piping
15180 Refrigerant Piping
15180 Steam and Condensate Piping
15180 Steam Condensate Pumps
15180 Water Treatment Equipment

15190 Fuel Piping
15190 Fuel Oil Piping
15190 Fuel Oil Pumps
15190 Fuel Oil Specialties
15190 Liquid Petroleum Gas Piping
15190 Natural Gas and Liquid Petroleum Gas Specialties
15190 Natural Gas Piping
15200 Process Piping
15210 Process Air and Gas Piping
15210 Air Compressors
15210 Compressed-Air Piping
15210 Gas Equipment
15210 Gas Piping
15210 Helium Piping
15210 Nitrogen Piping
15210 Nitrous Oxide Gas Piping
15210 Oxygen Gas Piping
15210 Vacuum Pumps
15210 Vacuum Piping
15220 Process Water and Waste Piping
15220 Deionized Water Piping
15220 Distilled Water Piping
15220 Interceptors
15220 Laboratory Acid Waste and Vent Piping
15220 Process Piping Interceptors
15220 Reverse Osmosis Water Piping
15230 Industrial Process Piping
15230 Dry Product Piping
15230 Fluid Product Piping
15300 Fire Protection Piping
15400 Plumbing Fixtures and Equipment
15410 Plumbing Fixtures
15440 Plumbing Pumps
15440 Base-Mounted Pumps
15440 Compact Circulators
15440 Inline Pumps
15440 Packaged Booster Pumping Station
15440 Sewage Ejectors
15440 Sump Pumps
15450 Potable Water Storage Tanks
15460 Domestic Water Conditioning Equipment
15460 Water Conditioners
15460 Water Softeners
15470 Domestic Water Filtrating Equipment
15470 Disposable Filters
15470 Rechargeable Filters
15480 Domestic Water Heaters
15480 Domestic Water Heat Exchangers
15480 Packaged Domestic Water Heaters
15490 Pool and Fountain Equipment
15490 Fountain Equipment
15490 Reflecting Pool Equipment
15490 Swimming Pool Equipment
15500 Heat-Generation Equipment
15510 Heating Boilers and Accessories
15510 Cast-Iron Boilers
15510 Condensing Boilers
15510 Finned Water-Tube Boilers
15510 Firebox Heating Boilers
15510 Flexible Water-Tube Boilers
15510 Pulse Combustion Boilers
15510 Scotch Marine Boilers
15510 Steel Water-Tube Boilers
15520 Feedwater Equipment
15520 Boiler Feedwater Pumps
15520 Deaerators
15520 Packaged Deaerator and Feedwater Equipment
15530 Furnaces
15530 Electric-Resistance Furnaces
15530 Fuel-Fired Furnaces
15540 Fuel-Fired Heaters
15540 Fuel-Fired Duct Heaters
15540 Fuel-Fired Radiant Heaters
15540 Fuel-Fired Unit Heaters
15550 Breechings, Chimneys, and Stacks
15550 Draft Control Devices
15550 Fabricated Breechings and Accessories
15550 Fabricated Stacks
15550 Gas Vents
15550 Insulated Sectional Chimneys
15600 Refrigeration Equipment
15610 Refrigeration Compressors
15610 Centrifugal Refrigerant Compressors
15610 Reciprocating Refrigerant Compressors
15610 Rotary-Screw Refrigerant Compressors
15620 Packaged Water Chillers
15620 Absorption Water Chillers
15620 Centrifugal Water Chillers
15620 Reciprocating Water Chillers
15620 Rotary-Screw Water Chillers
15630 Refrigerant Monitoring Systems
15640 Packaged Cooling Towers
15640 Mechanical-Draft Cooling Towers
15640 Natural-Draft Cooling Towers
15650 Field-Erected Cooling Towers
15660 Liquid Coolers and Evaporative Condensers
15670 Refrigerant Condensing Units

15670 Packaged Refrigerant Condensing Coils and Fan Units
15670 Refrigerant Condensing Coils
15700 Heating, Ventilating, and Air Conditioning Equipment
15710 Heat Exchangers
15710 Steam-to-Water Heat Exchangers
15710 Water-to-Water Heat Exchangers
15720 Air Handling Units
15720 Built-Up Indoor Air Handling Units
15720 Customized Rooftop Air Handling Units
15720 Modular Indoor Air Handling Units
15720 Modular Rooftop Air Handling Units
15730 Unitary Air Conditioning Equipment
15730 Packaged Air Conditioners
15730 Packaged Rooftop Air Conditioning Units
15730 Packaged Rooftop Make-Up Air Conditioning Units
15730 Packaged Terminal Air Conditioning Units
15730 Room Air Conditioners
15730 Split System Air Conditioning Units
15740 Heat Pumps
15740 Air-Source Heat Pumps
15740 Rooftop Heat Pumps
15740 Water-Source Heat Pumps
15750 Humidity Control Equipment
15750 Dehumidifiers
15750 Humidifiers
15750 Swimming Pool Dehumidification Units
15760 Terminal Heating and Cooling Units
15760 Air Coils
15760 Convectors
15760 Fan Coil Units
15760 Finned-Tube Radiation
15760 Induction Units
15760 Infrared Heaters
15760 Unit Heaters
15760 Unit Ventilators
15770 Floor-Heating and Snow-Melting Equipment
15780 Energy Recovery Equipment
15780 Energy Storage Tanks
15780 Heat Pipe
15780 Heat Wheels
15800 Air Distribution
15810 Ducts
15810 Duct Hangers and Supports
15810 Fibrous Glass Ducts
15810 Flexible Ducts
15810 Glass-Fiber-Reinforced Plastic Ducts
15810 Metal Ducts
15820 Duct Accessories
15830 Fans
15830 Axial Fans
15830 Ceiling Fans
15830 Centrifugal Fans
15830 Industrial Ventilating Equipment
15830 Power Ventilators
15840 Air Terminal Units
15840 Constant Volume
15840 Variable Volume Units
15850 Air Outlets and Inlets
15850 Diffusers, Registers, and Grilles
15850 Gravity Ventilators
15850 Intake and Relief Ventilators
15850 Louvers and Vents
15850 Penthouse Ventilators
15860 Air Cleaning Devices
15860 Air Filters
15860 Dust Collectors
15860 Electronic Air Cleaners
15860 High-Efficiency Air Filters
15860 ULPA Filters
15900 HVAC Instrumentation and Controls
15905 HVAC Instrumentation
15910 Direct Digital Controls
15915 Electric and Electronic Control
15915 Central Control Equipment
15915 Control Panels
15915 Terminal Boxes
15920 Pneumatic Controls
15925 Pneumatic and Electric Controls
15930 Self-Powered Controls
15935 Building Systems Controls
15940 Sequence of Operation
15950 Testing, Adjusting, and Balancing
15950 Demonstration of Mechanical Equipment
15950 Duct Testing, Adjusting, and Balancing
15950 Equipment Testing, Adjusting, and Balancing
15950 Mechanical Equipment Starting/Commissioning
15950 Pipe Testing, Adjusting, and Balancing

16000 ELECTRICAL
16050 Basic Electrical Materials and Methods
16060 Grounding and Bonding
16070 Hangers and Supports
16075 Electrical Identification
16080 Electrical Testing
16090 Restoration and Repair

16100 Wiring Methods
16120 Conductors and Cables
16130 Raceway and Boxes
16130 Cabinets
16130 Conduit and Tubing
16130 Cutout Boxes
16130 Enclosures
16130 Indoor Service Poles
16130 Junction Boxes
16130 Multi-Outlet Assemblies
16130 Outlet Boxes
16130 Pull Boxes
16130 Surface Raceway
16130 Wireway and Auxiliary Gutters
16140 Wiring Devices
16140 Floor Boxes
16140 Receptacles
16140 Remote-Control Switching Devices
16140 Wall Plates
16140 Wall Switches and Dimmers
16150 Wiring Connections
16200 Electrical Power
16210 Electrical Utility Services
16220 Motors and Generators
16230 Generator Assemblies
16230 Engine Generators
16230 Frequency Changers
16230 Motor Generators
16230 Rotary Converters
16230 Rotary Uninterruptible Power Units
16240 Battery Equipment
16240 Batteries
16240 Battery Racks
16240 Battery Units
16240 Central Battery Equipment
16260 Static Power Converters
16260 Battery Chargers
16260 Direct Current (DC) Drive Controllers
16260 Slip Controllers
16260 Static Frequency Converters
16260 Static Uninterruptible Power Supplies
16260 Variable Frequency Controllers
16270 Transformers
16270 Distribution Transformers
16270 Network Transformers
16270 Pad-Mounted Transformers
16270 Power Transformers
16270 Substation Transformers
16280 Power Filters and Conditioners
16280 Capacitors
16280 Chokes and Inductors
16280 Electromagnetic-Interference (EMI) Filters
16280 Harmonic Filters
16280 Power Factor Controllers
16280 Radio-Frequency-Interference (RFI) Filters
16280 Surge Suppressors
16280 Voltage Regulators
16300 Transmission and Distribution
16310 Transmission and Distribution Accessories
16310 Arresters
16310 Cutouts
16310 Insulators
16310 Line Materials
16310 Supports
16320 High-Voltage Switching and Protection
16320 High-Voltage Circuit Breakers
16320 High-Voltage Cutouts
16320 High-Voltage Fuses
16320 High-Voltage Lightning Arresters
16320 High-Voltage Surge Arresters
16320 High-Voltage Reclosers
16330 Medium-Voltage Switching and Protection
16330 Medium-Voltage Circuit Protection Devices
16330 Medium-Voltage Cutouts
16330 Medium-Voltage Fuses
16330 Medium-Voltage Lightning Arresters
16330 Medium-Voltage Reclosers
16330 Medium-Voltage Surge Arresters
16340 Medium-Voltage Switching and Protection Assemblies
16340 Medium-Voltage Circuit Breaker Switchgear
16340 Medium-Voltage Enclosed Bus
16340 Medium-Voltage Enclosed Fuse Cutouts
16340 Medium-Voltage Enclosed Fuses
16340 Medium-Voltage Fusible Interrupter Switchgear
16340 Medium-Voltage Motor Controllers
16340 Medium-Voltage Vacuum Interrupter Switchgear
16360 Unit Substations
16400 Low-Voltage Distribution
16410 Enclosed Switches and Circuit Breakers
16420 Enclosed Controllers
16430 Low-Voltage Switchgear
16440 Switchboards, Panelboards, and Control Centers
16450 Enclosed Bus Assemblies
16460 Low-Voltage Transformers
16470 Power Distribution Units
16490 Components and Accessories

16500 Lighting
16510 Interior Luminaires
16520 Exterior Luminaires
16520 Area Lighting
16520 Aviation Lighting
16520 Flood Lighting
16520 Navigation Lighting
16520 Parking Lighting
16520 Roadway Lighting
16520 Site Lighting
16520 Sports Lighting
16520 Walkway Lighting
16530 Emergency Lighting
16540 Classified Location Lighting
16550 Special-Purpose Lighting
16550 Detention Lighting
16550 Display Lighting
16550 Medical Lighting
16550 Outline Lighting
16550 Theatrical Lighting
16550 Underwater Lighting
16550 Medical Lighting
16550 Security Lighting
16560 Signal Lighting
16560 Hazard Warning Lighting
16560 Obstruction Lighting
16570 Dimming Control
16580 Lighting Accessories
16590 Lighting Restoration and Repair
16700 Communications
16710 Communications Circuits
16720 Telephone and Intercommunication Equipment
16740 Communication and Data Processing Equipment
16770 Cable Transmission and Reception Equipment
16780 Broadcast Transmission and Reception Equipment
16790 Microwave Transmission and Reception Equipment
16800 Sound and Video
16810 Sound and Video Circuits
16820 Sound Reinforcement
16830 Broadcast Studio Audio Equipment
16840 Broadcast Studio Video Equipment
16850 Television Equipment
16880 Multimedia Equipment

Appendix E

Residential Master

Available at www.ConstructionScheduling.com

01000	**GENERAL REQUIREMENTS**
01110	Work by Owner
01290	Monthly Requisition for Payment
01300	Administration Requirements
01310	Project Supervision
01310	Monthly Project Meetings
01310	Weekly Project Meetings
01310	Meetings with Customer
01310	Safety Meetings
01310	Check Job Site
01320	Construction Photographs
01320	Contract Progress Reporting
01320	Schedule Updates
01320	Periodic Site Observation
01320	Progress Reports
01320	Survey and Layout
01330	Submittal Procedures
01330	Shop Drawings, Product Data, and Samples
01330	Quality Control Reports
01400	Quality Requirements
01450	Contractors Quality Control
01450	Field Quality Control
01450	Owners Quality Control
01450	Testing and Inspection Services
01450	Testing Laboratory Services
01450	Soils Testing
01450	Soils Compaction Testing
01450	Concrete Testing
01450	Asbestos Testing & Certification
01450	Engineer Inspection
01450	Architectural Inspection
01450	Building Department Inspection
01450	Utility Department Inspection
01450	Planning Department Inspection
01450	Construction Photos
01500	Temporary Facilities and Controls
01510	Temporary Electricity
01510	Temporary Fire Protection
01510	Temporary Fuel Oil
01510	Temporary Heating, Cooling, and Ventilating
01510	Temporary Lighting
01510	Temporary Natural Gas
01510	Temporary Telephone
01510	Temporary Water
01520	Field Office Trailer
01520	Storage Trailer
01520	Waste Disposal
01520	Portable Toilets
01530	Temporary Construction
01550	Access Roads
01550	Parking Areas
01550	Temporary Roads
01560	Security Control
01560	Temporary Barriers and Enclosures
01560	Temporary Fencing
01580	Project Signs
01700	Execution Requirements
01720	Construction Layout / Preparation
01720	Field Engineering
01720	Surveying
01720	As Built Foundation Location Survey
01730	Winter Protection
01740	Progress Cleaning
01740	Final Cleaning
01740	Site Maintenance
01740	Snow & Ice Removal

Appendix

02000	**SITE CONSTRUCTION**
02001	Mobilization
02200	Site Preparation
02210	Subsurface Investigation
02210	Soil Borings
02210	Core Drilling
02210	Groundwater Monitoring
02210	Test Pits
02220	Site Demolition
02220	Building Demolition
02220	Selective Site Demolition
02220	Demolition Waste Removal
02230	Site Clearing
02230	Clearing and Grubbing
02230	Selective Clearing
02230	Selective Tree Removal and Trimming
02230	Sod Stripping
02230	Strip and Stockpile Existing Soils
02230	Tree Pruning
02240	Dewatering
02250	Shoring and Underpinning
02250	Shoring
02250	Slabjacking
02250	Soil Stabilization
02250	Underpinning
02300	Earthwork
02310	Temporary Roads and Access
02310	Rough Grading
02310	Finish Grading at Pavement Areas
02310	Pavement Sub Base
02310	Slab Sub Base
02310	Sidewalk Sub Base
02310	Respread Topsoil
02310	Finish Grading Topsoil
02315	Cuts and Fills
02315	Site Bulk Excavation
02315	Site Bulk Fill
02315	Site Excavation
02315	Site Backfill
02315	Foundation Excavation
02315	Foundation Backfill
02315	Foundation Excavation / Backfill
02315	Rough Grade Backfilled Material
02315	Underslab Trenching and Backfill
02315	Borrow Excavation
02315	Off Site Hauling
02315	Site Compaction
02315	Site Trenching
02315	Site Lighting Excavation and Backfill
02370	Erosion Control
02370	Erosion Control Blankets and Mats
02370	Riprap
02370	Erosion Control Fencing
02410	Rock Excavation
02500	Utility Services
02510	Fire Protection Supply
02510	Hydrants
02510	Water Supply Line
02520	Monitoring Wells
02520	Drilled Well
02520	Test Wells
02520	Water Supply Wells
02520	Well Points
02530	Sanitary Sewage Systems
02530	Sanitary Sewer Manholes, Frames, and Covers
02530	Sewage Force Mains
02540	Septic Tank Systems
02540	Septic Drainage Field
02540	Septic Tank
02580	Underground Ducts and Manholes
02580	Utility Pole Excavation and Backfill
.02585	Electrical / Telephone Entrance Excavation / Backfill
02585	Electrical Service Entrance Excavation / Backfill
02585	Telephone Service Entrance Excavation / Backfill
02585	Cable Service Entrance Excavation / Backfill
02585	Gas Service Entrance Excavation / Backfill
02590	Site Grounding
02600	Drainage and Containment
02610	Pipe Culverts
02620	Foundation Drainage Piping
02620	Pipe Underdrain and Pavement Base Drain
02620	Retaining Wall Drainage Piping
02620	Subgrade Drains
02620	Underslab Drainage Piping
02630	Storm Drainage Systems
02630	Storm Drainage Catch Basins, Grates, and Frames
02630	Storm Drainage Manholes, Frames, and Covers
02640	Culverts

02670	Excavated Wetlands
02700	Paving and Surfacing
02740	Asphalt Pavement
02750	Concrete Pavement
02760	Pavement Marking
02760	Stamped Concrete
02760	Tack Coating
02770	Gutters
02770	Curbing
02775	Concrete Sidewalks
02775	Concrete Pads
02775	Concrete Stairs
02780	Pavers
02785	Sealcoating
02800	Site Improvements
02810	Lawn Sprinkler System
02815	Fountains
02820	Fences and Gates
02830	Retaining Walls
02840	Guide / Guard Rails
02840	Parking Bumpers
02870	Site Furnishings
02900	Landscaping
02910	Mulching
02910	Soil Preparation
02910	Topsoil
02910	Fine Grade Topsoil
02915	Shrub and Tree Transplanting
02920	Lawns and Grasses
02920	Seeding
02920	Hydroseeding
02920	Sodding
02930	Exterior Plants
02930	Ground Covers
02930	Shrubs and Trees
02935	Plant Maintenance
02935	Landscape Maintenance
02935	Fertilizing
02935	Mowing
03000	**CONCRETE**
03300	Concrete Foundations
03310	Concrete Footings
03310	Concrete Piers / Columns
03310	Concrete Foundation Walls
03310	Concrete Form Removal
03310	Concrete Rubbing
03310	Transformer Pad
03300	Concrete Slabs
03310	Concrete Slab Preparation
03310	Concrete Slab Placement
03310	Concrete Slab Sawcutting
03310	Concrete Slab Sealer
03310	Concrete Cure Time
03300	Exterior Concrete (If Not Covered Under Site)
03310	Concrete Paving
03310	Concrete Sidewalks
03310	Concrete Pads
03310	Concrete Stairs
03310	Stamped Concrete
03310	Structural Concrete
03330	Architectural Concrete
03360	Special Concrete Finishes
03400	Precast Concrete
03410	Precast Structural Concrete
03480	Pre-Cast Lintels
03480	Pre-Cast Sills
04000	**MASONRY**
04005	Lintels (Delivery)
04050	Scaffolding
04070	Masonry Grouting
04080	Masonry Reinforcement
04210	Brick (Delivery)
04210	Brick Masonry
04220	Concrete Block Foundation Wall
04220	Concrete Block (Delivery)
04220	Concrete Block Masonry
04220	Concrete Sills
04220	Glazed Concrete Block
04220	Split-Face Block
04270	Glass Block
04290	Adobe Masonry Units
04400	Stone Masonry
04400	Marble
04400	Limestone
04400	Granite
04400	Sandstone
04400	Slate
04400	Stone Sills
04700	Simulated Masonry
04710	Simulated Brick
04730	Simulated Stone
04810	Thin Brick Veneer

04840 Prefabricated Masonry Panels
04880 Masonry Fireplaces
04880 Chimneys
04900 Masonry Restoration and Cleaning
0491 Masonry Restoration
04910 Masonry Cleaning
04910 Masonry Repair
04910 Masonry Replacement
04910 Masonry Repointing
04910 Masonry Parging
04910 Toothing Masonry
04910 Masonry Cutting / Patching
04920 Stone Restoration
04920 Stone Repair
04920 Stone Replacement
04920 Stone Repointing
04930 Masonry Cleaning

05000 METALS
05100 Structural Steel
05100 Steel Delivery
05100 Steel Lintels / Angles (Delivery)
05100 Structural Steel Erection
05400 Light Gauge Metal Framing
05500 Miscellaneous Metals
05700 Ornamental Metal

06000 CARPENTRY
06100 Rough Carpentry
06110 Wood Framing
06110 Wood Truss Delivery
06110 Furring and Blocking
06110 Ceiling Hatches
06130 Heavy Timber Construction
06150 Wood Decking
06160 Sheathing
06160 Wall Sheathing
06160 Roof Sheathing
06160 Insulating Sheathing
06160 Subflooring
06160 Underlayment
06170 Wood Joists
06170 Wood Truss Erection
06180 Glued-Laminated Construction
06200 Finish Carpentry
06220 Millwork
06250 Prefinished Paneling
06260 Board Paneling
06270 Closet and Utility Shelving
06400 Architectural Woodwork
06410 Custom Cabinets
06410 Cabinetry
06415 Countertops
06520 Paneling
06430 Wood Stairs and Railings
06430 Ornamental Wood Stairs
06450 Wood Trim
06450 Wood Casings
06450 Wood Aprons
06450 Window Sills
06450 Wood Base
06450 Wood Chair Rail
06450 Wood Cornices
06450 Wood Facias and Soffits
06450 Wood Handrails and Guard Rails
06450 Wood Stops, Stools, and Sills
06445 Simulated Wood Trim
06450 Wood Frames
06450 Exterior Door Wood Frames and Jambs
06450 Fire-Rated Door Wood Frames
06450 Interior Door Wood Frames and Jambs
06450 Ornamental Wood Frames
06470 Screens, Blinds, and Shutters
06600 Plastic Laminate
06910 Wood Restoration and Cleaning
06910 Architectural Woodwork Cleaning
06910 Architectural Woodwork Refinishing
06910 Architectural Woodwork Restoration
06910 Wood Framing Restoration

07000 THERMAL AND MOISTURE PROTECTION
07050 Drainage Systems
07100 Dampproofing and Waterproofing
07110 Foundation Coating
07130 Sheet Waterproofing
07130 Membrane Waterproofing
07190 Water Repellents
07200 Thermal Protection
07210 Foundation Insulation
07210 Blown-In Insulation
07210 Fiberglass Insulation

07210 Ceiling Insulation
07210 Sound Insulation
07210 Rigid Insulation
07210 Sprayed Insulation
07220 Roof Deck Insulation
07240 Exterior Insulation Finish Systems (EIFS)
07260 Vapor Retarders
07300 Roofing
07310 Roofing Shingles
07310 Asphalt Shingles
07310 Fiberglass Shingles
07310 Metal Shingles
07310 Mineral Fiber Cement Shingles
07310 Plastic Shakes
07310 Porcelain Enamel Shingles
07310 Slate Shingles
07310 Wood Shingles
07310 Wood Shakes
07320 Roof Tiles
07330 Roof Coverings
07400 Roofing Panels
07410 Metal Roof
07410 Metal Roof Panels
07510 Built-Up Roofing
07500 Membrane Roofing
07530 Elastomeric Roofing
07550 Modified Roofing
07580 Roll Roofing
07590 Roof Removal
07590 Roofing Restoration
07590 Roof Maintenance and Repairs
07600 Flashing and Sheet Metal
07610 Sheet Metal Roofing
07620 Sheet Metal Flashing and Trim
07710 Copings
07710 Counterflashing Systems
07710 Gravel Stops and Facias
07710 Gutters and Downspouts
07710 Reglets
07710 Roof Expansion Assemblies
07710 Scuppers
07720 Roof Curbs
07720 Skylights
07720 Ridge Vents
07720 Roof Hatches
07720 Snow Guards
07460 Siding
07460 Siding
07460 Aluminum Siding
07460 Composition Siding
07460 Hardboard Siding
07460 Mineral Fiber Cement Siding
07460 Vinyl Siding
07460 Plywood Siding
07460 Steel Siding
07460 Wood Siding
07460 Cedar Siding
07460 Soffits and Fascia
07460 Soffits
07460 Fascia
07900 Joint Sealers
07920 Caulking
07920 Window and Door Caulking
07920 Masonry Control Joint Caulking
07920 Concrete Control Joint Caulking
07920 Joint Fillers
07920 Sealants

08000 DOORS AND WINDOWS
08050 Doors
08100 Exterior Doors and Frames
08100 Interior Doors and Frames
08100 Interior Metal Door Frames
08120 Steel Doors and Frames
08180 Metal Screen and Storm Doors
08210 Interior Doors
08210 Wood Doors
08210 Flush Wood Doors
08210 Plastic-Faced Wood Doors
08210 Prefinished Wood Doors
08220 Plastic Laminated Doors
08250 Plastic Door Units
08250 Wood Door Units
08260 Sliding Doors
08280 Screen Doors
08280 Storm Doors
08300 Specialty Doors
08310 Access Doors
08310 Access Panels
08330 Overhead Doors
08350 Folding Doors
08350 Accordion Folding Doors
08350 Panel Folding Doors
08360 Overhead Doors

08370 Vertical Lift Doors
08500 Windows
08500 Exterior Windows
08500 Screens
08500 Interior Windows
08580 Storm Windows
08590 Window Restoration
08590 Window Replacement
08600 Skylights
08610 Roof Windows
08700 Hardware
08710 Door Hardware
08720 Weatherstripping and Seals
08720 Thresholds
08750 Window Hardware
08800 Glass and Glazing
08810 Interior Window Glazing
08810 Interior Window Walls

09000 FINISHES

09200 Gypsum Board Systems
09205 Furring and Lathing
09210 Gypsum Plaster
09220 Portland Cement Plaster
09230 Plaster Work
09250 Gypsum Board Systems
09250 Install Sheetrock
09250 Taping
09280 Plaster Restoration
09300 Tile
09310 Ceramic Tile
09330 Quarry Tile
09340 Paver Tile
09360 Plastic Tile
09370 Metal Tile
09390 Tile Restoration
09500 Ceilings
09510 Suspended Ceiling Systems
09510 Install Ceiling Grid
09510 Ceiling Tile
09510 Acoustical Tile Ceilings
09560 Textured Ceilings
09560 Gypsum Panel Ceilings
09560 Metal Panel Ceilings
09570 Linear Wood Ceilings
09590 Ceiling Assembly Restoration
09600 Flooring
09610 Floor Treatment
09620 Specialty Flooring
09620 Plastic-Laminate Flooring
09630 Masonry Flooring
09630 Brick Flooring
09630 Flagstone Flooring
09630 Granite Flooring
09630 Marble Flooring
09630 Slate Flooring
09630 Stone Flooring
09640 Wood Flooring
09640 Wood Strip Flooring
09650 Resilient Flooring
09650 Base
09650 Sheet Flooring
09650 Resilient Tile Flooring
09680 Carpeting
09680 Carpet Tile
09690 Flooring Restoration
09700 Wall Finishes
09720 Wall Covering
09720 Wallpaper
09750 Stone Facing
09760 Plastic Blocks
09770 Special Wall Surfaces
09790 Wall Finish Restoration
09900 Painting
09910 Exterior Painting
09910 Interior Painting
09930 Staining
09940 Decorative Finishes
09960 High-Performance Coatings
09970 Steel Coatings
09980 Concrete and Masonry Coatings
09980 Epoxy Coatings
09980 Epoxy Floor Coatings
09990 Paint Restoration
09990 Maintenance Coatings
09990 Paint Cleaning

10000 SPECIALTIES
10200 Louvers and Vents
10230 Vents
10230 Soffit Vents
10230 Wall Vents
10290 Pest Control
10300 Fireplaces and Stoves
10350 Flagpoles
10520 Fire Extinguisher Accessories
10530 Awnings
10530 Canopies
10530 Car Shelters
10530 Walkway Coverings
10550 Postal Specialties
10670 Storage Shelving
10710 Exterior Shutters
10715 Storm Panels
10720 Exterior Louvers
10750 Telephone Specialties
10810 Toilet Accessories
10820 Bath Accessories
10820 Shower and Tub Doors
10830 Laundry Accessories
10880 Scales
10900 Wardrobe and Closet Specialties

11000 EQUIPMENT
11050 Library Equipment
11060 Theater Equipment
11450 Residential Equipment
11450 Residential Appliances
11450 Residential Kitchen Equipment
11450 Retractable Stairs
11460 Unit Kitchens
11470 Darkroom Equipment

12000 FURNISHINGS
10250 Fabrics
12110 Art
12120 Wall Decorations
12300 Manufactured Casework
12350 Specialty Casework
12400 Furnishings and Accessories
12480 Rugs and Mats
12480 Floor Mats
12480 Rugs
12480 Runners
12490 Window Treatments
12490 Blinds
12490 Curtains and Drapes
12490 Interior Shutters
12490 Shades
12500 Furniture

13000 SPECIAL CONSTRUCTION
13030 Special Purpose Rooms
13100 Lightning Protection
13150 Swimming Pools
13160 Aquariums
13170 Tubs and Pools
13200 Storage Tanks
13600 Solar and Wind Energy Equipment
13700 Security Access and Surveillance
13850 Fire Alarm Systems
13850 Smoke Alarms
13900 Fire Suppression Systems
13920 Fire Pumps
13930 Wet-Pipe Fire Suppression Sprinklers
13935 Dry-Pipe Fire Suppression Sprinklers
14000 CONVEYING SYSTEMS
14100 Dumbwaiters
14200 Elevators
14410 People Lifts
14420 Wheelchair Lifts
14560 Chutes

15000 MECHANICAL
15100 Plumbing
15140 Rough Plumbing At Walls
15150 Rough Plumbing Under Slab
15150 Floor Drains
15150 Oil Separators
15150 Grease Traps
15150 Sanitary Waste and Vent Piping
15150 Interceptors
15150 Sanitary Piping
15150 Sanitary Piping Separators
15160 Storm Drainage Piping
15160 Roof Drain Piping
15170 Swimming Pool Piping
15170 Fountain Piping
15190 Fuel Piping
15190 Gas Piping

15400 Plumbing Fixtures and Equipment
15410 Plumbing Fixtures
15440 Plumbing Pumps
15440 Pumping Station
15440 Sewage Ejectors
15440 Sump Pumps
15450 Water Storage Tanks
15460 Water Conditioners
15460 Water Softeners
15470 Water Filtrating Equipment
15480 Water Heaters
15480 Water Heaters and Equipment
15480 Water Heat Exchangers
15490 Pool and Fountain Equipment
15490 Fountain Equipment
15490 Pool Equipment
15300 Sprinkler
15300 Rough In Sprinkler Piping
15300 Fire Protection Piping
15300 Cut and Drop Heads
15300 Sprinkler Head Trims
15300 Test and Charge Sprinkler System
15500 Heating Ventilating and Air Conditioning
15090 Mechanical Restoration and Retrofit
15100 Gas Piping
15130 Pumps
15180 Heating and Cooling Piping
15180 Condensate Drain Piping
15180 Water Treatment Equipment
15510 Hot Water Boiler Systems
15530 Furnaces
15540 Fuel-Fired Heaters
15540 Radiant Heaters
15540 Unit Heaters
15550 Gas Vents
15550 Insulated Sectional Chimneys
15700 Heating, Ventilating and Air Conditioning Equipment
15720 Roof Curbs
15720 Rooftop Units
15730 Air Conditioning Equipment
15730 Condensing Units
15730 Room Air Conditioners
15740 Heat Pumps
15740 Rooftop Heat Pumps
15750 Dehumidifiers
15750 Humidifiers
15760 Heating and Cooling Units
15810 Duct Work
15830 Fans
15830 Ceiling Fans
15850 Diffusers, Registers, and Grilles
15850 Louvers and Vents
15915 Thermostats and Controls
15915 Control Wiring
15950 HVAC Testing and Balancing

16000 ELECTRICAL
16090 Electrical Restoration and Repair
16130 Temporary Power
16130 Temporary Lighting
16130 Service Entrance Panels
16130 Energize Power
16140 Wiring Devices (Switches and Outlets)
16210 Electrical Service Entrance (Exterior)
16220 Motors and Generators
16270 Transformers
16400 General Distribution Wiring
16400 Wiring Other Trade Equipment
16400 Equipment Wiring
16400 Specialty Wiring
16430 Switchgear
16500 Exterior Lighting
16510 Interior Lighting
16520 Site Lighting
16520 Walkway Lighting
16530 Emergency Lighting
16550 Special-Purpose Lighting
16550 Display Lighting
16550 Security Lighting
16710 Telephone and Data Wiring
16720 Telephone and Communication Equipment
16800 Sound and Video System
16900 Fire Alarm Systems
16900 Security Alarm Systems
16900 Electrical Specialty Systems

17000 SUBSTANTIAL COMPLETION / C.O.
17005 Punch-list Work
17010 Final Cleaning
17015 Owner Move-In
17020 Turnover Project Information

Appendix F

Residential General

01000	**GENERAL REQUIREMENTS**
01110	Work by Owner
01290	Monthly Requisition for Payment
01400	Quality Requirements
01450	Architectural Inspection
01450	Building Department Inspection
01450	Utility Department Inspection
01700	Execution Requirements
01720	Surveying
01720	As Built Foundation Location Survey
01730	Winter Protection
01740	Final Cleaning
02000	**SITE CONSTRUCTION**
02001	Mobilization
02200	Site Preparation
02220	Site Demolition
02220	Building Demolition
02220	Demolition Waste Removal
02230	Site Clearing
02230	Clearing and Grubbing
02230	Selective Clearing
02300	Earthwork
02310	Temporary Roads and Access
02310	Rough Grading
02310	Finish Grading at Pavement Areas
02310	Pavement Sub Base
02310	Slab Sub Base
02310	Sidewalk Sub Base
02310	Respread Topsoil
02310	Finish Grading Topsoil
02315	Cuts and Fills
02315	Site Bulk Excavation
02315	Site Bulk Fill
02315	Site Excavation
02315	Site Backfill
02315	Foundation Excavation
02315	Foundation Backfill
02315	Foundation Excavation / Backfill
02315	Rough Grade Backfilled Material
02315	Under-slab Trenching and Backfill
02315	Borrow Excavation
02315	Off Site Hauling
02315	Site Compaction
02315	Site Trenching
02315	Site Lighting Excavation and Backfill
02370	Erosion Control
02410	Rock Excavation
02500	Utility Services
02510	Fire Protection Supply
02510	Water Supply Line
02520	Drilled Well
02520	Well Points
02530	Sanitary Sewage Systems
02530	Sewage Force Mains
02540	Septic Tank Systems
02540	Septic Drainage Field
02540	Septic Tank
02585	Electrical / Telephone Entrance Excavation / Backfill
02585	Cable Service Entrance Excavation / Backfill
02585	Gas Service Entrance Excavation / Backfill
02600	Drainage and Containment
02610	Pipe Culverts
02620	Foundation Drainage Piping
02620	Pipe Underdrain and Pavement Base Drain
02620	Retaining Wall Drainage Piping
02620	Subgrade Drains
02630	Storm Drainage Systems
02670	Excavated Wetlands
02700	Paving and Surfacing
02740	Asphalt Pavement
02750	Concrete Pavement
02760	Stamped Concrete
02760	Tack Coating
02770	Gutters
02775	Concrete Sidewalks
02775	Concrete Pads
02775	Concrete Stairs
02780	Pavers
02785	Sealcoating

02800	Site Improvements
02810	Lawn Sprinkler System
02815	Fountains
02820	Fences and Gates
02830	Retaining Walls
02870	Site Furnishings
02900	Landscaping
02910	Mulching
02910	Soil Preparation
02910	Topsoil
02910	Fine Grade Topsoil
02915	Shrub and Tree Transplanting
02920	Lawns and Grasses
02920	Seeding
02920	Hydroseeding
02920	Sodding
02930	Exterior Plants
02930	Ground Covers
02930	Shrubs and Trees
02935	Plant Maintenance
02935	Landscape Maintenance
02935	Fertilizing
02935	Mowing
02935	Watering
03000	**CONCRETE**
03300	Concrete Foundations
03310	Concrete Footings
03310	Concrete Piers / Columns
03310	Concrete Foundation Walls
03310	Transformer Pad
03300	Concrete Slabs
03310	Concrete Slab Preparation
03310	Concrete Slab Placement
03300	Exterior Concrete (If Not Covered under Site)
03310	Concrete Paving
03310	Concrete Sidewalks
03310	Concrete Pads
03310	Concrete Stairs
3310	Stamped Concrete
04000	**MASONRY**
04005	Lintels (Delivery)
04050	Scaffolding
04210	Brick (Delivery)
04210	Brick Masonry
04220	Concrete Block Foundation Wall
04220	Concrete Block (Delivery)
04220	Concrete Block Masonry
04220	Concrete Sills
04880	Masonry Fireplaces
04880	Chimneys
4900	Masonry Restoration and Cleaning
4930	Masonry Cleaning
5 **05000**	**METALS**
05100	Structural Steel
05100	Steel Delivery
05100	Steel Lintels / Angles (Delivery)
05100	Structural Steel Erection
05400	Light Gauge Metal Framing
05500	Miscellaneous Metals
05700	Ornamental Metal
06000	**CARPENTRY**
06100	Rough Carpentry
06110	Wood Framing
06110	Wood Truss Delivery
06110	Furring and Blocking
06110	Ceiling Hatches
06130	Heavy Timber Construction
06150	Wood Decking
06160	Sheathing
06160	Wall Sheathing
06160	Roof Sheathing
06160	Insulating Sheathing
06160	Subflooring
06160	Underlayment
06170	Wood Joists
06170	Wood Truss Erection
06180	Glued-Laminated Construction
06200	Finish Carpentry
06220	Millwork
06250	Pre-finished Paneling
06260	Board Paneling
06270	Closet and Utility Shelving
06400	Architectural Woodwork
06410	Custom Cabinets
06410	Cabinetry
06415	Countertops
06520	Paneling
06430	Wood Stairs and Railings

06430 Ornamental Wood Stairs
06450 Wood Trim
06450 Wood Facias and Soffits
06450 Wood Handrails and Guard Rails
06450 Wood Frames
06450 Exterior Door Wood Frames and Jambs
06450 Fire-Rated Door Wood Frames
06450 Interior Door Wood Frames and Jambs
06470 Screens, Blinds, and Shutters
06600 Plastic Laminate
06910 Wood Restoration and Cleaning
06910 Architectural Woodwork Cleaning
06910 Architectural Woodwork Refinishing
06910 Architectural Woodwork Restoration

07000 THERMAL AND MOISTURE PROTECTION

07100 Damp-proofing and Waterproofing
07110 Foundation Coating
07130 Membrane Waterproofing
07190 Water Repellents
07200 Thermal Protection
07210 Foundation Insulation
07210 Blown-in Insulation
07210 Fiberglass Insulation
07210 Ceiling Insulation
07210 Sound Insulation
07210 Rigid Insulation
07210 Sprayed Insulation
07240 Exterior Insulation Finish Systems (EIFS)
07300 Roofing
07310 Roofing Shingles
07310 Wood Shakes
07400 Roofing Panels
07500 Membrane Roofing
07580 Roll Roofing
07590 Roof Removal
07590 Roofing Restoration
07590 Roof Maintenance and Repairs
07600 Flashing and Sheet Metal
07610 Sheet Metal Roofing
07620 Sheet Metal Flashing and Trim
07710 Copings
07710 Gutters and Downspouts
07720 Roof Curbs
07720 Skylights
07720 Ridge Vents
07720 Roof Hatches
07720 Snow Guards
07460 Siding
07460 Siding
07460 Soffits and Fascia
07900 Joint Sealers
7920 Caulking

08000 DOORS AND WINDOWS

08050 Doors
08100 Exterior Doors and Frames
08100 Interior Doors and Frames
08180 Metal Screen and Storm Doors
08210 Interior Doors
08330 Overhead Doors
08350 Folding Doors
08500 Windows
08500 Exterior Windows
08500 Screens
08500 Interior Windows
08580 Storm Windows
08590 Window Replacement
08600 Skylights
08610 Roof Windows
08700 Hardware
08710 Door Hardware
08720 Weatherstripping and Seals
08750 Window Hardware
08800 Glass and Glazing
08810 Interior Window Glazing
08830 Mirrors

09000 FINISHES

09200 Gypsum Board Systems
09250 Gypsum Board Systems
09250 Install Sheetrock
09250 Taping
09300 Tile
09310 Ceramic Tile
09330 Quarry Tile
09340 Paver Tile
09500 Ceilings
09510 Suspended Ceiling Systems
09510 Install Ceiling Grid
09510 Ceiling Tile
09560 Textured Ceilings
09570 Linear Wood Ceilings

09600 Flooring
09620 Plastic-Laminate Flooring
09630 Masonry Flooring
09640 Wood Flooring
09650 Resilient Flooring
09650 Base
09650 Sheet Flooring
09650 Resilient Tile Flooring
09680 Carpeting
09680 Carpet Tile
09690 Flooring Restoration
09700 Wall Finishes
09720 Wall Covering
09770 Special Wall Surfaces
09790 Wall Finish Restoration
09900 Painting
09910 Exterior Painting
09910 Interior Painting
09930 Staining
09940 Decorative Finishes
09960 High-Performance Coatings
09980 Concrete and Masonry Coatings
09990 Paint Restoration
09990 Paint Cleaning

10000 SPECIALTIES
10200 Louvers and Vents
10230 Soffit Vents
10230 Wall Vents
10290 Pest Control
10300 Fireplaces and Stoves
10520 Fire Extinguisher Accessories
10530 Awnings
10530 Canopies
10670 Storage Shelving
10710 Exterior Shutters
10715 Storm Panels
10720 Exterior Louvers
10810 Toilet Accessories
10820 Bath Accessories
10820 Shower and Tub Doors
10830 Laundry Accessories
10880 Scales
10900 Wardrobe and Closet Specialties

11000 EQUIPMENT
11450 Residential Equipment
11450 Residential Appliances
11450 Residential Kitchen Equipment
11450 Retractable Stairs
11460 Unit Kitchens

12000 FURNISHINGS
10250 Fabrics
12110 Art
12120 Wall Decorations
12300 Manufactured Casework
12350 Specialty Casework
12400 Furnishings and Accessories
12480 Rugs and Mats
12490 Blinds
12490 Curtains and Drapes
12490 Interior Shutters
12490 Shades
12500 Furniture

13000 SPECIAL CONSTRUCTION
03100 Lightning Protection
13150 Swimming Pools
13170 Tubs and Pools
13200 Storage Tanks
13850 Fire Alarm Systems
13850 Smoke Alarms
13900 Fire Suppression Systems
13920 Fire Pumps

14000 CONVEYING SYSTEMS
14200 Elevators
14420 Wheelchair Lifts
14560 Chutes

15000 MECHANICAL
15100 Plumbing
15140 Rough Plumbing at Walls
15150 Rough Plumbing under Slab
15150 Floor Drains
15150 Oil Separators
15150 Grease Traps
15150 Sanitary Waste and Vent Piping
15150 Interceptors
15150 Sanitary Piping

15150 Sanitary Piping Separators
15160 Storm Drainage Piping
15160 Roof Drain Piping
15170 Swimming Pool Piping
15190 Fuel Piping
15190 Gas Piping
15400 Plumbing Fixtures and Equipment
15440 Pumping Station
15440 Sewage Ejectors
15440 Sump Pumps
15450 Water Storage Tanks
15480 Water Heaters and Equipment
15300 Sprinkler
15300 Rough in Sprinkler Piping
15300 Fire Protection Piping
15300 Cut and Drop Heads
15300 Sprinkler Head Trims
15300 Test and Charge Sprinkler System
15500 Heating Ventilating and Air Conditioning
15090 Mechanical Restoration and Retrofit
15100 Gas Piping
15130 Pumps
15180 Heating and Cooling Piping
15180 Condensate Drain Piping
15510 Hot Water Boiler Systems
15530 Furnaces
15550 Gas Vents
15550 Insulated Sectional Chimneys
15700 Heating, Ventilating and Air Conditioning Equipment
15720 Roof Curbs
15720 Rooftop Units
15730 Condensing Units
15730 Room Air Conditioners
15740 Heat Pumps
15750 Dehumidifiers
15750 Humidifiers
15760 Heating and Cooling Units
15810 Duct Work
15830 Ceiling Fans
15850 Diffusers, Registers, and Grilles
15850 Louvers and Vents
15915 Thermostats and Controls
15915 Control Wiring
15950 HVAC Testing and Balancing

16000 ELECTRICAL
16090 Electrical Restoration and Repair
16130 Temporary Power
16130 Temporary Lighting
16130 Service Entrance Panels
16130 Energize Power
16140 Wiring Devices (Switches and Outlets)
16210 Electrical Service Entrance (Exterior)
16400 General Distribution Wiring
16400 Wiring Other Trade Equipment
16400 Equipment Wiring
16400 Specialty Wiring
16500 Exterior Lighting
16510 Interior Lighting
16520 Site Lighting
16520 Walkway Lighting
16530 Emergency Lighting
16550 Special-Purpose Lighting
16550 Display Lighting
16550 Security Lighting
16710 Telephone and Data Wiring
16720 Telephone and Communication Equipment
16800 Sound and Video System
16900 Fire Alarm Systems
16900 Security Alarm Systems
16900 Electrical Specialty Systems

17000 SUBSTANTIAL COMPLETION / C.O.
17005 Punch-list Work
17010 Final Cleaning
17015 Owner Move-In
17020 Turnover Project Information

Appendix G

Residential Summary

01000	General requirements
02000	Site construction
02220	Demolition
02315	Excavation and backfill
02500	Site utilities
02700	Paving and surfacing
02800	Site improvements
02900	Landscaping
03300	Concrete foundation
03300	Concrete slabs
03300	Exterior concrete
03400	Precast concrete
04000	Masonry
04220	Block foundations
05100	Structural steel
06100	Rough carpentry
06200	Finish carpentry
06400	Architectural woodwork
07000	Thermal and moisture protection
07210	Insulation
07240	Exterior Insulation Finish
07300	Roofing
07460	Siding
07900	Caulking
08000	Exterior doors and windows
08210	Interior doors
08500	Exterior windows
08700	Hardware
09000	Interior finishes
09200	Gypsum board systems
09600	Flooring
09720	Wall covering
09910	Exterior painting
09910	Interior Painting
10000	Specialties
11000	Equipment
12000	Furnishings
13000	Special construction
14000	Conveying systems
15000	Mechanical
15100	Plumbing
15300	Sprinkler
15500	HVAC
16000	Electrical
17000	Substantial completion/C.O.
17005	Punch-list work
17010	Final cleaning
17015	Owner move-in
17020	Turnover project information

Appendix H

Reprintable Construction Scheduling Forms

[Production: reprint the following form files]

Job Estimate Summary

Owner ____________________ Contractor ____________________

Property address ____________________ Address ____________________

Estimate by ____________________ Telephone number ____________________

Legal description ____________________ Net bldg. fund __________ Loan number __________

No. Item	Qty	Unit	Material	Labor	Subcontract	Total	Actual cost
1 Excavation							
2 Trenching							
3 Concrete rough							
4 Concrete finish							
5 Asphalt paving							
6 Rough lumber							
7 Finish lumber							
8 Door frames							
9 Windows and glass							
10 Fireplace							
11 Masonry							
12 Roof							
13 Plumbing							
14 Heating							
15 Electric wiring							
16 Electric fixtures							
17 Rough carpentry							
18 Finish carpentry							
19 Lath and plaster							
20 Drywall							
21 Garage door							
22 Doors							
23 Painting							
24 Cabinets							
25 Hardwood floors							
26 Ceramic tile							
27 Formica							
28 Hardware							
29 Linoleum							
30 Asphalt tile							
31 Range and oven							
32 Insulation							
33 Shower door							
34 Temperature facilities							
35 Miscellaneous							
36 Cleanup							
37 Carpet							
38 **Total Cost**							

Estimate Take Off

Company: ______________________ Estimator: ______________________ Date: __________

Project: ______________________ Checked by: ______________________ Date: __________

Address: ______________________ Estimate due: ______________________

Job: ______________________ Estimate #: ______________________

CSI Division/Account: ______________________ Drawing reference: ______________________

Description	Q = Qty	L = Length	W = Width	D = Depth	T = Thickness	H = Height	Calculation	Total / Unit

(Carry description and totals forward to Estimate Detail Sheet)

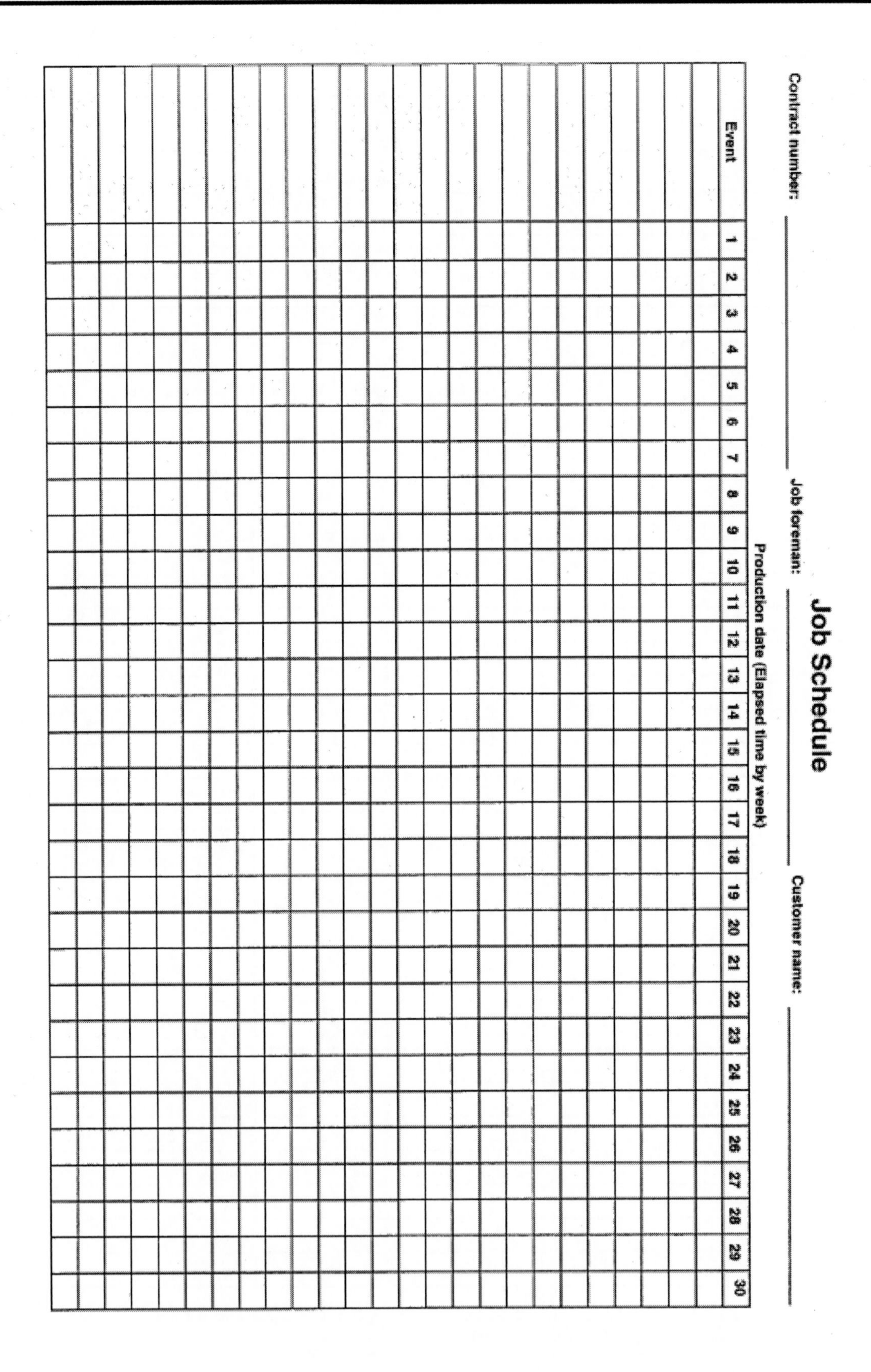

Job Schedule

Contract number: ____________ Job foreman: ____________ Customer name: ____________

Production date (Elapsed time by week)

Event	1	2	3	4	5	6	7	8	9	10	11	12	13	14	15	16	17	18	19	20	21	22	23	24	25	26	27	28	29	30

Job Costs

Job name ______________________ Estimate # ______________

Start date ______________________ Finish date ______________

Quantity & description	Unit	Man-hours	Code #	Material Cost	Labor Cost	Equip-ment	Total Cost

Jobs Worksheet

Job #	Work days		Labor costs		Material costs		Equipment costs		Sub costs		Total costs		Change orders	For:
	Actual	Est	Actual	Est	Actual	Est	Actual	Est	Actual	Est	Actual	Est		
Grand Total														
Over/Under														

Job Summary

Job number ________________ Estimator ________________

Project ________________ Project Manager ________________

Owner ________________ Location ________________

Location ________________

	Labor	Material	Equipment	Subcontract	Total
Actual cost					
Estimate					
Overrun					

Bid amount ________________

Estimated cost ________________

Projected profit ________________

Cost overrun ________________

Bid profit ________________

Change orders ________________

Cost of changes ________________

Total profit ________________

Change Order

Number ____________________

Date:	Subcontract number:
To:	Our job number:
	Our proposal number:
	Architect's C.O. number:
	Effective date of change:

Subject to all the provisions of this Change Order, you are hereby directed to make the following change(s)

The following change(s) will alter the price provided in your subcontract by the ___________ of $ ___________

*Surety Consent date: ____________	Adjusted Subcontract Price Through C.O. Number ____________	$ ____________
This change is approved Name: ____________ By: ____________	Amount this C. O. #: ____________	$ ____________
Title: ____________	Current Adjusted Subcontract Price:	$ ____________

When this Change Order is signed by both parties (and by Subcontractor's surety if subcontract is bonded), it constitutes their agreement: (A) That the subcontract price is adjusted as shown above and that no further adjustment in that price by reason of the change(s) provided herein shall be made; and (B) That all the terms and conditions of the subcontract, except as modified by this and any previous changes, shall remain in full force and effect and apply to the work as so changed.

Accepted and Agreed: Date ____________	____________
By: ____________ Subcontractor	By ____________ Authorized Signature - Contractor
Title: ____________	Title: ____________

*Subcontractor: Sign and return immediately. If subcontract is bonded, obtain consent of surety endorsed thereon. No payment on account of this Change Order will be made until you have complied with the foregoing.

Job Estimate

Owner ______________________ Date ______________ 20 ______

Owner's address ______________________ Telephone ______________________

Job address ______________________ Lot ______ Blk. ______ Tract ______

General Conditions	Labor	Other	Total
Supervision			
Superintendent			
General foreman			
Master mechanic			
Engineer			
Timekeeper			
Assistant timekeepers			
Payroll clerk			
Material checkers			
Watchman			
Waterboy			
Others			
Permits			
Blasting			
Building			
Sidewalk bridge			
Street obstruction			
Sunday work			
Temporary			
Wrecking			
Other			
Bonds			
Completion bond			
Maintenance bond			
Street encroachment bond			
Street repair bond			
Insurance			
Workers' Compensation			
Builder's risk fire insurance			
Completed operations public liability insurance			
Equipment floater insurance			
Public liability insurance			
Truck and automobile license			
Licenses			
Local business license			
State contractor's license			
Taxes			
Excise taxes			
Payroll taxes			
Sales taxes			
Field Office			
Owner's			
Job			
Maintenance of office			
Telephone			
Owner's			
Job			
Job office supplies			
Shanties			
Storage			
Tool			
Watchman's			
Transportation of equipment			
Delivery charges			
Travel expenses			
Temporary utilities			
Power			
Light			

General Conditions	Labor	Other	Total
Subtotal forward			
Temporary utilities (cont.)			
Water			
Heat			
Fuel			
Temporary toilets			
Surveys			
Photographs-damage & progress			
Lost time, weather, etc.			
Removing utilities			
Cutting & adjusting for subcontract			
Repairing damage			
Patching after subcontracting			
Cleanup, general			
Clean windows			
Clean floors			
Cleanup after subcontracting			
Removing debris from job			
Signs			
Pumping			
Protection of construction			
Protection of adjoining land and buildings			
Barricades			
Temporary fences			

Special Conditions	Labor	Material	Total
Plant and equipment (See checklist)			
Acoustical			
Air conditioning			
Architectural concrete			
Architectural terra cotta			
Bins			
Cabinetwork			
Bookcases			
Cases			
Coat closets			
Counters			
Displays			
Linen closets			
Metal cabinets			
Special cabinets			
Stationery cabinets			
Store fixtures			
Telephone booths			
Wardrobes			
Wood cabinets			
Other			
Cellular-steel floors			
Cofferdams			
Concrete			
Admixtures			
Columns			
Curbs and gutters			
Flatwork			
Floors			
Footing			
Foundations			
Frost protection			

Page 1 of 4

Special Conditions	Labor	Material	Total
Subtotal forward			
Concrete (cont.)			
Pumpcrete			
Sidewalks			
Slabs			
Vacuum concrete			
Walls			
Walks			
Other			
Concrete curing compounds			
Contingencies			
Conveyors			
Corrugated steel			
Roofing			
Siding			
Culverts			
Concrete box			
Concrete pipe			
Corrugated metal			
Metal arch			
Doors			
Exterior wood			
Panel			
Flush			
Dutch			
French			
Exterior metal			
Revolving			
Kalamein			
Overhead			
Garage			
Screen			
Tin clad			
Industrial			
Other			
Glass			
Plastic			
Interior metal			
Shop			
Office			
Interior wood			
Panel			
Flush			
Sliding			
Frames			
Metal			
Wood			
Electric fixtures			
Electric wiring			
Building wiring			
Service			
Power wiring			
Motors			
Electric signs			
Other			
Elevators			
Excavation			
Clearing and grubbing			
Removing obstructions			
General excavation			
Blasting			
Shoring			
Structural excavation			
Trench excavation			
Backfill			
Rough grading			
Fine grading			

Special Conditions	Labor	Material	Total
Subtotal forward			
Rock			
Fences and railing			
Fire alarm			
Fireproofing			
Flooring			
Asphalt tile and linoleum			
Composition			
Cork			
Flagstone			
Hardwood			
Marble			
Rubber			
Tile			
Slate			
Terrazzo			
Formwork			
Arches			
Beams			
Beam and slab floors			
Bridge piers			
Caps			
Columns			
Fiber tubes			
Flat slabs			
Floor pans			
Footings			
Foundation walls			
Girders			
Metal pans			
Movable forms			
Other			
Foundations			
Wall footings			
Piers			
Spread footings			
Piles, bearing			
Wood			
Steel			
Concrete			
Piles, sheets			
Wood			
Steel			
Concrete			
Caissons			
Frost protection			
Hardboards			
Hardware			
Rough			
Finish			
Heating			
Incinerator			
Insulation			
House			
Cold storage			
Piping			
Rigid			
Flexible			
Foil			
Other			
Lath			
Metal			
Ceilings			
Exterior-walls			
Interior-walls			
Soffits			
Partitions			

Page 2 of 4

Special Conditions	Labor	Material	Total
Subtotal forward			
Lath (cont.)			
Beads			
Plaster boards			
Walls			
Ceilings			
Arches			
Lift slab			
Loading dock			
Lumber construction			
Heavy			
Beams			
Blocks			
Braces			
Built-up beams			
Caps			
Centering			
Columns			
Falsework			
Girders			
Girts			
Joists			
Lagging			
Laminated members			
Plank and laminated floors			
Planking			
Plates			
Posts			
Sheeting			
Stringers			
Timber purlins			
Trusses			
Trussed beams			
Wales			
Wedges			
Light			
Beams			
Blocking			
Bracing			
Bridging			
Columns			
Cripples			
Dormers			
Fascia			
Furring			
Girders			
Half timber work			
Headers			
Hips			
Jacks			
Joists, floor, roof & ceiling			
Outlooks			
Pier pads			
Plates			
Plywood, flooring, sheathing and roofing			
Posts			
Rafters			
Ribbons			
Ridges			
Roof trusses			
Sheathing-roof, wall			
Sills			
Steps			
Studs			
Subfloor			
Trimmers			

Special Conditions	Labor	Material	Total
Subtotal forward			
Marble			
Manholes			
Masonry			
Ashlar			
Common brick			
Face brick			
Concrete blocks			
Precast concrete panels			
Clay tile			
Dimension stone			
Rubble stone			
Flue			
Metal work			
Art metal work			
Base			
Casings			
Chair rail			
Column guards			
Cornices			
Elevator entrances			
Fire escapes			
Freight doors			
Grillwork			
Information boards			
Linen chutes			
Lintels			
Mail chutes			
Metal doors and frames			
Platforms			
Railings			
Shop front			
Shutter			
Stairs			
Treads			
Trap doors			
Transoms			
Wheel guards			
Wainscoting			
Other			
Millwork and finish			
Base			
Built-ins			
Casings			
Caulking			
Corner boards			
Cornice			
Doors			
Frames			
Jambs			
Mantels			
Molding			
Paneling			
Sash			
Screens			
Shelving			
Siding			
Sills			
Stairs			
Stops			
Storm doors			
Trim			
Windows			
Wood carving			
Other			
Painting and decorating			
Aluminum paint			

Page 3 of 4

Special Conditions	Labor	Material	Total
Subtotal forward			
Painting and decorating (cont.)			
Doors and windows			
Finishing			
Floors			
Lettering			
Masonry and concrete			
Metal			
Paperhanging			
Plaster			
Roofs			
Shingle stain			
Stucco			
Wood			
Pavements			
Asphalt			
Base			
Block			
Brick			
Concrete			
Wood			
Other			
Pipelines			
Cast iron pipelines			
Concrete pipelines			
Corrugated pipelines			
Steel pipelines			
Trenches			
Vitrified tile pipelines			
Plaster			
Bases			
Coves			
Cement			
Exterior			
Interior			
Interior			
Keene's cement			
Models			
Ornamental			
Perlite			
Special finish			
Plumbing			
Interior			
Exterior			
Prestressed concrete			
Railroad work			
Reinforcing steel			
Bars			
Mesh			
Spirals			
Stirrups			
Retaining walls			
Roofing			
Asbestos			
Asphalt shingles			
Built-up			
Concrete			
Copper			
Corrugated			
Aluminum			
Asbestos			
Steel			
Gravel			
Gypsum-poured and plank			

Special Conditions	Labor	Material	Total
Subtotal forward			
Roofing (cont.)			
Shingles			
Slate			
Steel			
Tile			
Tin			
Sandblasting			
Service lines			
Sheet metal			
Shutters			
Skylights			
Sound deadening			
Stacks			
Stalls			
Structural steel			
Anchors for structural steel			
Bases of steel or iron			
Beams, purlins and girts			
Bearing plates and shoes			
Brackets			
Columns of steel, iron or pipe			
Crane rails and stops			
Door frames			
Expansion joists			
Floor plates			
Girders			
Grillage beams			
Hangars of structural steel			
Lintels			
Monorail beams			
Painting steel and remove rust			
Rivets			
Steel stacks			
Welding			
Engineering and shop details			
Inspection			
Freight			
Unloading			
Tanks			
Test holes			
Thresholds			
Tile			
Tilt-up concrete			
Trench shoring			
Vaults and vault doors			
Wallboards			
Waterproofing			
Well drilling			
Windows			
Frames			
Wood			
Steel			
Other			
Sash			
Wood			
Steel			
Aluminum			

Subtotal:			
Contingency			
Profit			

Total:			

Page 4 of 4

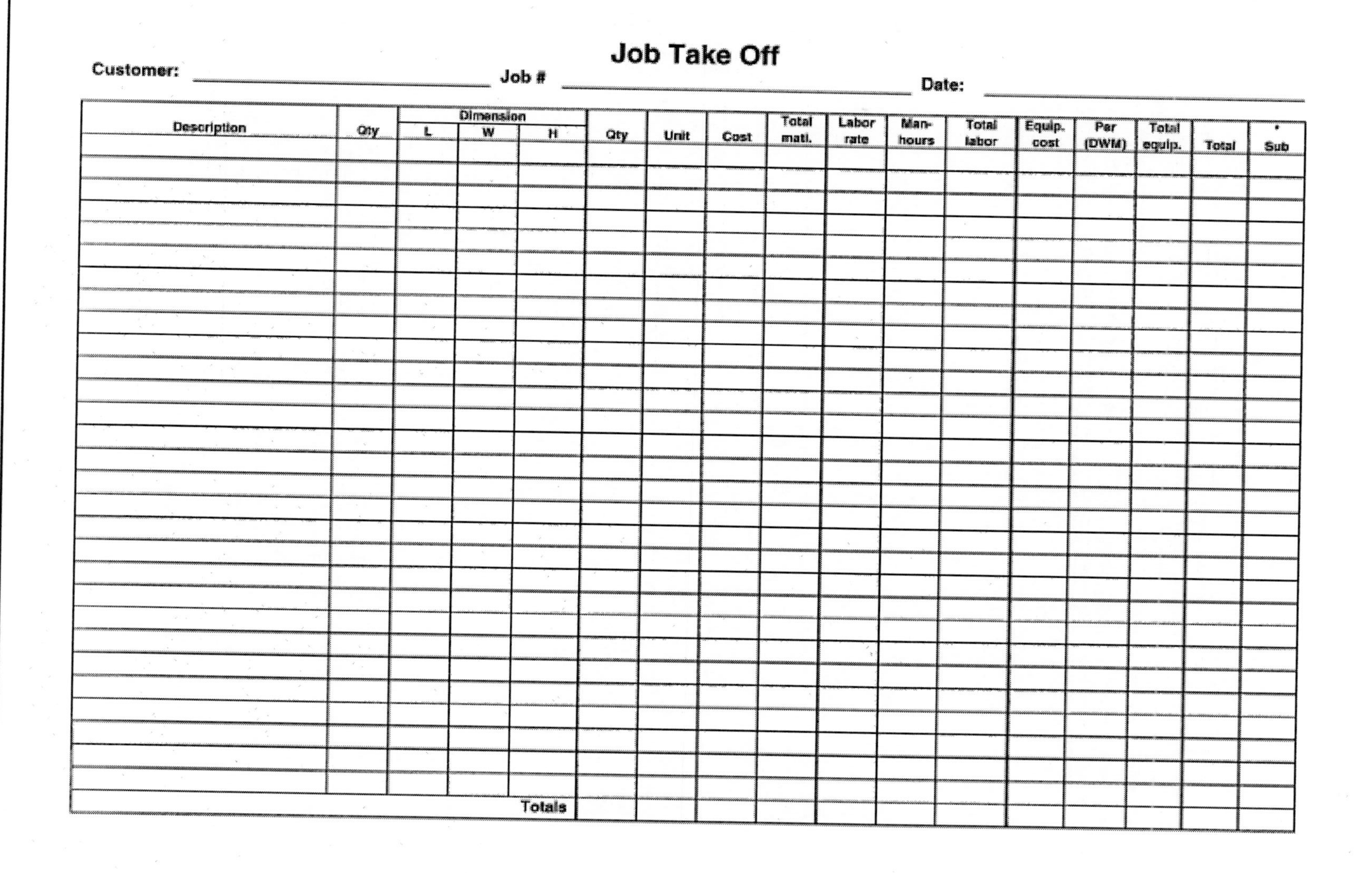

Job Take Off

Customer: ________ Job # ________ Date: ________

Description	Qty	Dimension L	W	H	Qty	Unit	Cost	Total matl.	Labor rate	Man-hours	Total labor	Equip. cost	Per (DWM)	Total equip.	Total	* Sub
Totals																

Substantial Completion Certificate

Project: ______________________ **Date of submittal:** ______________________

Location: ______________________

Contractor: ______________________ **Address:** ______________________

Owner: ______________________ **Address:** ______________________

The Contractor considers the work on the Project to be completed in that it is suitable for the Owner to use for its intended purpose. The date of this certificate shall be the date of commencement of all warranties and shall be considered as approval for application for final payment as set forth in the contract. This certificate does not relieve the Contractor from any obligation of the Contract.

Work yet to be completed:

Any other deficiencies or requests for service should be made in writing to the address above. Please be as specific as possible about the defect. Any claim concerning an appliance or fixture under warranty should be made directly to the manufacturer.

I (we) have inspected the work performed and find the job completed to our satisfaction with the exception of the items listed above as yet to be completed.

Date of completion: ______________________ Owner's signature: ______________________

Subcontractor Agreement

Subcontract No. ______________________________

This Agreement, made and entered into at ______________ this ________ day of ____________ 19 _____ by and between ______________________________ hereinafter called Contractor, with principal office at ______________________, ______________ and ____ hereinafter called a Subcontractor.

Recitals

On or about the ____________________ day of ______________, 19 _____, Contractor entered into a prime contract with ____ hereinafter called Owner, whose address is ____ to perform the following construction work: ____ ____. Said work to be performed in accordance with the prime contract and the plans and specifications. Said plans and specifications have been prepared by or on behalf of ____, Architect.

Section 1 - Entire Contract

Subcontractor certifies and agrees that he is fully familiar with all of the terms, conditions and obligations of the Contract Documents, as hereinafter defined, the location of the job site, and the conditions under which the work is to be performed, and that he enters into this Agreement based upon his investigation of all of such matters and is in no way relying upon any opinions or representations of Contractor. It is agreed that this Agreement represents the entire agreement. It is further agreed that the Contract Documents are incorporated in this Agreement by this reference, with the same force and effect as if the same were set forth at length therein, and that Subcontractor and his subcontractors will be and are bound by any and all of said Contract Documents insofar as they relate in any part or in any way, directly or indirectly to the work covered by this Agreement. Subcontractor agrees to be bound to Contractor in the same manner and to the same extent as Contractor is bound to Owner under the Contract Documents, to the extent of the work provided for in this Agreement, and that where, in the Contract Documents, reference is made to Contractor and the work or specification therein pertains to Subcontractor's trade, craft, or type of work, then such work or specification shall be interpreted to apply to Subcontractor instead of Contractor. The phrase "Contract Documents" is defined to mean and include:

____

Section 2 - Scope

Subcontractor agrees to furnish all labor, services, materials, installation, cartage, hoisting, supplies, insurance, equipment, scaffolding, tools and other facilities of every kind and description required for the prompt and efficient execution of the work described herein and to perform the work necessary or incidental to complete

____

for the project in strict accordance with the Contract Documents and as more particularly, though not exclusively, specified in:

____

____

Section 3 - Contract Price

Contractor agrees to pay Subcontractor for the strict performance of his work, the sum of:

____ (____________________),

subject to additions and deductions for changes in the work as may be agreed upon, and to make payment in accordance with the Payment Schedule, Section 4.

Section 4 - Payment Schedule

Contractor agrees to pay Subcontractor in monthly payments of ____________ % of labor and materials which have been placed in position and for which payment has been made by Owner to Contractor. The remaining ________ % shall be retained by Contractor until he receives final payment from Owner, but not less than thirty-five days after the entire work required by the prime contract has been fully completed in conformity with the Contract Documents and has been delivered and accepted by Owner, Architect and Contractor.

Page 1 of 2

Subject to the provisions of the next sentence, the retained percentage shall be paid Subcontractor promptly after Contractor receives his final payment from Owner. Subcontractor agrees to furnish, if and when required by Contractor, payroll affidavits, receipts, vouchers, release of claims for labor, material and subcontractors performing work or furnishing materials under this Agreement, all in form satisfactory to Contractor, and it is agreed that no payment hereunder shall be made, except at Contractor's option, until and unless such payroll affidavits, receipts, vouchers or releases; or any or all of them, have been furnished. And payment made hereunder prior to completion and acceptance of the work, as referred to above, shall not be construed as evidence of acceptance of any part of Subcontractor's work.

Section 5 - General Provisions

1. Subcontractor agrees to begin work as soon as instructed by the Contractor, and shall carry on said work promptly, efficiently and at a speed that will not cause delay in the progress of the Contractor's work or work of other subcontractors. If, in the opinion of the Contractor, the Subcontractor falls behind in the progress of the work, the Contractor may direct the Subcontractor to take such steps as the Contractor deems necessary to improve the rate of progress, including, without limitation, requiring the Subcontractor to increase the number of shifts, personnel, overtime operations, days of work, equipment, amount of plant, or other remedies and to submit to Contractor for Contractor's approval an outline schedule demonstrating the manner in which the required rate of progress will be regained, without additional cost to the Contractor. Contractor may require Subcontractor to prosecute, in preference to other parts of the work, such part or parts of the work as Contractor may specify.

The Subcontractor shall complete the work as required by the progress schedule prepared by the Contractor, which may be amended from time to time. The progress schedule may be reviewed in the office of the Contractor and sequence of construction will be as directed by the Contractor.

The Subcontractor agrees to have an acceptable representative (an officer of Subcontractor if requested by the Contractor) present at all job meetings and to submit weekly progress reports in writing if requested by the Contractor. Any job progress schedules are hereby made a part of and incorporated herein by reference.

2. Reserved Gate Usage

Subcontractor shall notify in writing, and assign its employees, material men and suppliers, to such gates or entrances as may be established for their use by Contractor and in accordance with such conditions and at such times as may be imposed by Contractor. Strict compliance with Contractor's gate usage procedures shall be required by the Subcontractor who shall be responsible for such gate usage by its employees, material men, suppliers, subcontractors, and their material men and suppliers.

3. Staggered Days and Hours of Work and for Deliveries

Subcontractor shall schedule the work and the presence of its employees at the jobsite and any deliveries of supplies or materials by its material men and suppliers to the jobsite on such days, and at such times and during such hours, as may be directed by Contractor. Subcontractor shall assume responsibility for such schedule compliance not only for its employees but for all its material men, suppliers and subcontractors, and their material men and suppliers.

Section 6 - Special Provisions

____

____

Contractors are required by law to be licensed and regulated by the Contractors' State License Board. Any questions concerning a contractor may be referred to the registrar of the board.

In Witness Whereof: The parties hereto have executed this Agreement for themselves, their heirs, executors, successors, administrators, and assignees on the day and year first above written.

Subcontractor	Contractor
____________________	____________________
____________________	____________________
By ____________________	By ____________________
Name Title	Name Title
☐ Corporation ☐ Partnership ☐ Proprietorship	
Contractor's State License No. __________	Contractor's State License No. __________

Page 2 of 2

Job Change Estimate

Quotation to ______________________ Date ______________________
Address ______________________ Change Order Number ______________________
Address ______________________ Job Number ______________________
City, ST, ZIP ______________________ ______________________
Attention ______________________ ______________________
Job Name ______________________ ______________________
Reference ______________________

A. Material and equipment: $ ____________
B. Sales tax: $ ____________
C. Direct labor: $ ____________
D. Indirect costs: $ ____________
E. Equipment and tools: $ ____________
F. Subtotal: .... $ ____________
G. Overhead at _____ % of line F: $ ____________
H. Subcontracts: $ ____________
I. Overhead at _____ % of line H: $ ____________
J. Subtotal .... $ ____________
K. Profit at _____ % of line J: $____________
L. Subtotal: .... $ ____________
M. Bond premium at _____ % of line L: $ ____________
N. Service reserve at _____ % of line L: $ ____________
O. Total cost estimate, lines L thru N: ¨ Add ¨ Deduct $____________
P. Exclusions from this estimate: ______________________

Q. ¨ This quotation is valid for _____ days.
R. ¨ We require _____ days extension of the contract time.
S. ¨ We are proceeding with this work per your authorization.
T. ¨ Please forward your confirming change order.

Signed by ______________________

Project Manager

Index

A

B

C

T

U